JAGUAR Mk1 & Mk2
Gold Portfolio
1955-1969

Compiled by
R.M.Clarke

ISBN 1 85520 2433

BROOKLANDS BOOKS LTD.
P.O. BOX 146, COBHAM,
SURREY, KT11 1LG. UK

BROOKLANDS BOOKS

BROOKLANDS ROAD TEST SERIES

Abarth Gold Portfolio 1950-1971
AC Ace & Aceca 1953-1983
Alfa Romeo Giulietta Gold Portfolio 1954-1965
Alfa Romeo Giulia Berlinas 1962-1976
Alfa Romeo Giulia Coupés 1963-1976
Alfa Romeo Giulia Coupés Gold P. 1963-1976
Alfa Romeo Spider 1966-1990
Alfa Romeo Spider Gold Portfolio 1966-1991
Alfa Romeo Alfasud 1972-1984
Alfa Romeo Alfetta Gold Portfolio 1972-1987
Alfa Romeo Alfetta GTV6 1980-1987
Allard Gold Portfolio 1937-1959
Alvis Gold Portfolio 1919-1967
AMX & Javelin Muscle Portfolio1968-1974
Armstrong Siddeley Gold Portfolio 1945-1960
Austin A30 & A35 1951-1962
Austin Healey 100 & 100/6 Gold P. 1952-1959
Austin Healey 3000 Gold Portfolio 1959-1967
Austin Healey Sprite 1958-1971
BMW Six Cyl. Coupés 1969-1975
BMW 1600 Collection No.1 1966-1981
BMW 2002 Gold Portfolio1968-1976
BMW 316, 318, 320 (4 cyl.) Gold P. 1975-1990
BMW 320, 323, 325 (6 cyl.) Gold P .1977-1990
BMW M Series Performance Portfolio1976-1993
BMW 5 Series Gold Portfolio1981-1987
Bristol Cars Gold Portfolio 1946-1992
Buick Automobiles 1947-1960
Buick Muscle Cars 1965-1970
Cadillac Automobiles 1949-1959
Cadillac Automobiles 1960-1969
Chevrolet 1955-1957
Chevrolet Impala & SS 1958-1971
Chevrolet Corvair 1959-1969
Chevy El Camino & SS 1959-1987
Chevelle & SS Muscle Portfolio 1964-1972
Chevrolet Muscle Cars 1966-1971
Chevy Blazer 1969-1981
Chevrolet Corvette Gold Portfolio 1953-1962
Chevrolet Corvette Sting Ray Gold P. 1963-1967
Chevrolet Corvette Gold Portfolio 1968-1977
High Performance Corvettes 1983-1989
Camaro Muscle Portfolio 1967-1973
Chevrolet Camaro Z28 & SS 1966-1973
Chevrolet Camaro & Z28 1973-1981
High Performance Camaros 1982-1988
Chrysler 300 Gold Portfolio 1955-1970
Chrysler Valiant 1960-1962
Citroen Traction Avant Gold Portfolio 1934-1957
Citroen 2CV Gold Portfolio 1948-1989
Citroen DS & ID 1955-1975
Citroen DS & ID Gold Portfolio 1955-1975
Citroen SM 1970-1975
Cobras & Replicas 1962-1983
Shelby Cobra Gold Portfolio 1962-1969
Cobras & Cobra Replicas Gold P. 1962-1989
Cunningham Automobiles 1951-1955
Daimler SP250 Sports & V-8 250 Saloon Gold Portfolio 1959-1969
Datsun Roadsters 1962-1971
Datsun 240Z 1970-1973
Datsun 280Z & ZX 1975-1983
The De Lorean 1977-1993
De Tomaso Collection No. 1 1962-1981
Dodge Charger 1966-1974
Dodge Muscle Cars 1967-1970
Dodge Viper on the Road
Excalibur Collection No. 1 1952-1981
Facel Vega 1954-1964
Ferrari Dino 1965-1974
Ferrari Dino 308 1974-1979
Ferrari 308 & Mondial 1980-1984
Fiat 500 Gold Portfolio 1936-1972
Fiat 600 & 850 Gold Portfolio 1955-1972
Fiat Pininfarina 124 & 2000 Spider 1968-1985
Fiat-Bertone X1/9 1973-1988
Ford Consul, Zephyr, Zodiac Mk.I & II 1950-1962
Ford Zephyr, Zodiac, Executive, Mk.III & Mk.IV 1962-1971
Ford Cortina 1600E & GT 1967-1970
High Performance Capris Gold P. 1969-1987
Capri Muscle Portfolio 1974-1987
High Performance Fiestas 1979-1991
High Performance Escorts Mk.I 1968-1974
High Performance Escorts Mk.II 1975-1980
High Performance Escorts 1980-1985
High Performance Escorts 1985-1990
High Performance Sierras & Merkurs Gold Portfolio 1983-1990
Ford Automobiles 1949-1959
Ford Fairlane 1955-1970
Ford Ranchero 1957-1959
Ford Thunderbird 1955-1957
Ford Thunderbird 1958-1963
Ford Thunderbird 1964-1976
Ford GT40 Gold Portfolio 1964-1987
Ford Bronco 1966-1977
Ford Bronco 1978-1988
Holden 1948-1962
Honda CRX 1983-1987
Isetta 1953-1964
ISO & Bizzarrini Gold Portfolio 1962-1974
Jaguar and SS Gold Portfolio 1931-1951
Jaguar XK120, 140, 150 Gold P. 1948-1960
Jaguar Mk.VII, VIII, IX, X, 420 Gold P.1950-1970
Jaguar Mk.1 & Mk.2 Gold Portfolio 1959-1969
Jaguar E-Type Gold Portfolio 1961-1971
Jaguar E-Type V-12 1971-1975
Jaguar XJ12, XJ5.3, V12 Gold P. 1972-1990
Jaguar XJ6 Series II 1973-1979
Jaguar XJ6 Series III 1979-1986
Jaguar XJS Gold Portfolio 1975-1990
Jeep CJ5 & CJ6 1960-1976
Jeep CJ5 & CJ7 1976-1986
Jensen Cars 1946-1967
Jensen Cars 1967-1979
Jensen Interceptor Gold Portfolio 1966-1986
Jensen Healey 1972-1976
Lagonda Gold Portfolio 1919-1964
Lamborghini Cars 1964-1970
Lamborghini Countach & Urraco 1974-1980
Lamborghini Countach & Jalpa 1980-1985
Lancia Fulvia Gold Portfolio 1963-1976
Lancia Beta Gold Portfolio 1972-1984
Lancia Stratos 1972-1985
Land Rover Series I 1948-1958
Land Rover Series II & IIa 1958-1971
Land Rover Series III 1971-1985
Land Rover Discovery 1989-1994
Lincoln Gold Portfolio 1949-1960
Lincoln Continental 1961-1969
Lincoln Continental 1969-1976
Lotus Sports Racers Gold Portfolio 1953-1965
Lotus & Caterham Seven Gold P. 1957-1989
Lotus Elite 1957-1964
Lotus Elite & Eclat 1974-1982
Lotus Elan Gold Portfolio 1962-1974
Lotus Elan Collection No. 2 1963-1972
Lotus Elan & SE 1989-1992
Lotus Cortina Gold Portfolio 1963-1970
Lotus Europa Gold Portfolio 1966-1975
Lotus Elite & Eclat 1974-1982
Lotus Turbo Esprit 1980-1986
Marcos Cars 1960-1988
Maserati 1965-1970
Maserati 1970-1975
Mercedes 190 & 300 SL 1954-1963
Mercedes 230/250/280SL 1963-1971
Mercedes Benz SLs & SLCs Gold P. 1971-1989
Mercedes S & 600 1965-1972
Mercedes S Class 1972-1979
Mercury Muscle Cars 1966-1971
Messerschmitt Gold Portfolio1954-1964
Metropolitan 1954-1962
MG Gold Portfolio 1929-1939
MG TC 1945-1949
MG TD 1949-1953
MG TF 1953-1955
MGA & Twin Cam Gold Portfolio 1955-1962
MG Midget Gold Portfolio1961-1979
MGB Roadsters 1962-1980
MGB MGC & V8 Gold Portfolio 1962-1980
MGB GT 1965-1980
Mini Cooper Gold Portfolio 1961-1971
Mini Muscle Cars 1961-1979
Mini Moke Gold Portfolio1964-1994
Mopar Muscle Cars 1964-1967
Morgan Three-Wheeler Gold Portfolio 1910-1952
Morgan Plus 4 & Four 4 Gold P. 1936-1967
Morgan Cars 1960-1970
Morgan Cars Gold Portfolio 1968-1989
Morris Minor Collection No. 1 1948-1980
Shelby Mustang Muscle Portfolio 1965-1970
High Performance Mustang IIs 1974-1978
High Performance Mustangs 1982-1988
Oldsmobile Automobiles 1955-1963
Oldsmobile Muscle Cars 1964-1971
Oldsmobile Toronado 1966-1978
Opel GT 1968-1973
Packard Gold Portfolio 1946-1958
Pantera Gold Portfolio 1970-1989
Panther Gold Portfolio 1972-1990
Plymouth Barracuda 1964-1974
Plymouth Muscle Cars 1966-1971
Pontiac Tempest & GTO 1961-1965
Pontiac Muscle Cars 1966-1972
Pontiac Firebird & Trans-Am 1973-1981
High Performance Firebirds 1982-1988
Pontiac Fiero 1984-1988
Porsche 356 Gold Portfolio1953-1965
Porsche 911 1965-1969
Porsche 911 1970-1972
Porsche 911 1973-1977
Porsche 911 Carrera 1973-1977
Porsche 911 Turbo 1975-1984
Porsche 911 SC 1978-1983
Porsche 914 Collection No. 1 1969-1983
Porsche 914 Gold Portfolio 1969-1976
Porsche 924 Gold Portfolio 1975-1988
Porsche 928 1977-1989
Porsche 944 Gold Portfolio1981-1991
Range Rover Gold Portfolio 1970-1992
Reliant Scimitar 1964-1986
Riley Gold Portfolio 1924-1939
Riley 1.5 & 2.5 Litre Gold Portfolio 1945-1955
Rolls Royce Silver Cloud & Bentley 'S' Series Gold Portfolio 1955-1965
Rolls Royce Silver Shadow Gold P. 1965-1980
Rover P4 1949-1959
Rover P4 1955-1964
Rover 3 & 3.5 Litre Gold Portfolio 1958-1973
Rover 2000 & 2200 1963-1977
Rover 3500 1968-1977
Rover 3500 & Vitesse 1976-1986
Saab Sonett Collection No.1 1966-1974
Saab Turbo 1976-1983
Studebaker Gold Portfolio 1947-1966
Studebaker Hawks & Larks 1956-1963
Avanti 1962-1990
Sunbeam Tiger & Alpine Gold P. 1959-1967
Toyota MR2 1984-1988
Toyota Land Cruiser 1956-1984
Triumph TR2 & TR3 1952-1960
Triumph TR4, TR5, TR250 1961-1968
Triumph TR6 Gold Portfolio 1969-1976
Triumph TR7 & TR8 Gold Portfolio 1975-1982
Triumph Herald 1959-1971
Triumph Vitesse 1962-1971
Triumph Spitfire Gold Portfolio 1962-1980
Triumph 2000, 2.5, 2500 1963-1977
Triumph GT6 Gold Portfolio1966-1974
Triumph Stag 1970-1980
TVR Gold Portfolio 1959-1990
VW Beetle Gold Portfolio1935-1967
VW Beetle Gold Portfolio1968-1991
VW Beetle Collection No.1 1970-1982
VW Karmann Ghia 1955-1982
VW Bus, Camper, Van 1954-1967
VW Bus, Camper, Van 1968-1979
VW Bus, Camper, Van 1979-1989
VW Scirocco 1974-1981
VW Golf GTI 1976-1986
Volvo PV444 & PV544 1945-1965
Volvo Amazon-120 Gold Portfolio 1956-1970
Volvo 1800 Gold Portfolio 1960-1973

BROOKLANDS ROAD & TRACK SERIES

Road & Track on Alfa Romeo 1949-1963
Road & Track on Alfa Romeo 1964-1970
Road & Track on Alfa Romeo 1971-1976
Road & Track on Alfa Romeo 1977-1989
Road & Track on Aston Martin 1962-1990
R & T on Auburn Cord and Duesenburg 1952-84
Road & Track on Audi & Auto Union 1952-1980
Road & Track on Audi & Auto Union 1980-1986
Road & Track on Austin Healey 1953-1970
Road & Track on BMW Cars 1966-1974
Road & Track on BMW Cars 1975-1978
Road & Track on BMW Cars 1979-1983
R & T on Cobra, Shelby & Ford GT40 1962-1992
Road & Track on Corvette 1953-1967
Road & Track on Corvette 1968-1982
Road & Track on Corvette 1982-1986
Road & Track on Corvette 1986-1990
Road & Track on Datsun Z 1970-1983
Road & Track on Ferrari 1975-1981
Road & Track on Ferrari 1981-1984
Road & Track on Ferrari 1984-1988
Road & Track on Fiat Sports Cars 1968-1987
Road & Track on Jaguar 1950-1960
Road & Track on Jaguar 1961-1968
Road & Track on Jaguar 1968-1974
Road & Track on Jaguar 1974-1982
Road & Track on Jaguar 1983-1989
Road & Track on Lamborghini 1964-1985
Road & Track on Lotus 1972-1981
Road & Track on Maserati 1952-1974
Road & Track on Maserati 1975-1983
R & T on Mazda RX7 & MX5 Miata 1986-1991
Road & Track on Mercedes 1952-1962
Road & Track on Mercedes 1963-1970
Road & Track on Mercedes 1971-1979
Road & Track on Mercedes 1980-1987
Road & Track on MG Sports Cars 1949-1961
Road & Track on MG Sports Cars 1962-1980
Road & Track on Mustang 1964-1977
R & T on Nissan 300-ZX & Turbo 1984-1989
Road & Track on Peugeot 1955-1986
Road & Track on Pontiac 1960-1983
Road & Track on Porsche 1951-1967
Road & Track on Porsche 1968-1971
Road & Track on Porsche 1972-1975
Road & Track on Porsche 1975-1978
Road & Track on Porsche 1979-1982
Road & Track on Porsche 1982-1985
Road & Track on Porsche 1985-1988
R & T on Rolls Royce & Bentley 1950-1965
R & T on Rolls Royce & Bentley 1966-1984
Road & Track on Saab 1972-1992
R & T on Toyota Sports & GT Cars 1966-1984
R & T on Triumph Sports Cars 1953-1967
R & T on Triumph Sports Cars 1967-1974
R & T on Triumph Sports Cars 1974-1982
Road & Track on Volkswagen 1951-1968
Road & Track on Volkswagen 1968-1978
Road & Track on Volkswagen 1978-1985
Road & Track on Volvo 1957-1974
R&T - Henry Manney at Large & Abroad
R&T - Peter Egan's "Side Glances"

BROOKLANDS CAR AND DRIVER SERIES

Car and Driver on BMW 1955-1977
Car and Driver on BMW 1977-1985
C and D on Cobra, Shelby & Ford GT40 1963-84
Car and Driver on Corvette 1956-1967
Car and Driver on Corvette 1968-1977
Car and Driver on Corvette 1978-1982
Car and Driver on Corvette 1983-1988
C and D on Datsun Z 1600 & 2000 1966-1984
Car and Driver on Ferrari 1955-1962
Car and Driver on Ferrari 1963-1975
Car and Driver on Ferrari 1976-1983
Car and Driver on Mopar 1956-1967
Car and Driver on Mopar 1968-1975
Car and Driver on Mustang 1964-1972
Car and Driver on Pontiac 1961-1975
Car and Driver on Porsche 1955-1962
Car and Driver on Porsche 1963-1970
Car and Driver on Porsche 1970-1976
Car and Driver on Porsche 1977-1981
Car and Driver on Porsche 1982-1986
Car and Driver on Saab 1956-1985
Car and Driver on Volvo 1955-1986

BROOKLANDS PRACTICAL CLASSICS SERIES

PC on Austin A40 Restoration
PC on Land Rover Restoration
PC on Metalworking in Restoration
PC on Midget/Sprite Restoration
PC on Mini Cooper Restoration
PC on MGB Restoration
PC on Morris Minor Restoration
PC on Sunbeam Rapier Restoration
PC on Triumph Herald/Vitesse
PC on Spitfire Restoration
PC on Beetle Restoration
PC on 1930s Car Restoration

BROOKLANDS HOT ROD 'MUSCLECAR & HI-PO ENGINES' SERIES

Chevy 265 & 283
Chevy 302 & 327
Chevy 348 & 409
Chevy 350 & 400
Chevy 396 & 427
Chevy 454 thru 512
Chrysler Hemi
Chrysler 273, 318, 340 & 360
Chrysler 361, 383, 400, 413, 426, 440
Ford 289, 302, Boss 302 & 351W
Ford 351C & Boss 351
Ford Big Block

BROOKLANDS RESTORATION SERIES

Auto Restoration Tips & Techniques
Basic Bodywork Tips & Techniques
Basic Painting Tips & Techniques
Camaro Restoration Tips & Techniques
Chevrolet High Performance Tips & Techniques
Chevy Engine Swapping Tips & Techniques
Chevy-GMC Pickup Repair
Chrysler Engine Swapping Tips & Techniques
Custom Painting Tips & Techniques
Engine Swapping Tips & Techniques
Ford Pickup Repair
How to Build a Street Rod
Land Rover Restoration Tips & Techniques
MG 'T' Series Restoration Guide
Mustang Restoration Tips & Techniques
Performance Tuning - Chevrolets of the '60's
Performance Tuning - Pontiacs of the '60's

BROOKLANDS MILITARY VEHICLES SERIES

Allied Military Vehicles No.1 1942-1945
Allied Military Vehicles No.2 1941-1946
Complete WW2 Military Jeep Manual
Dodge Military Vehicles No.1 1940-1945
Hail To The Jeep
Land Rovers in Military Service
Off Road Jeeps: Civ. & Mil. 1944-1971
US Military Vehicles 1941-1945
US Army Military Vehicles WW2-TM9-2800
VW Kubelwagen Military Portfolio1940-1990
WW2 Jeep Military Portfolio 1941-1945

1434

BROOKLANDS BOOKS

CONTENTS

Page	Title	Publication	Date	Year
5	A 2.4 Litre Jaguar	Autosport	Sep. 30	1955
6	The Jaguar 2.4	Motor	Sep. 28	1955
12	Brief Encounter Road Impressions	Autocar	Nov. 4	1955
14	Two Point Four Jaguar Road Test	Sports Cars Illustrated	Oct.	1956
20	The Jaguar "Two-Point-Four" Road Test	Autosport	Dec. 14	1956
23	3.4 Litre Jaguar for Dollar Markets	Autocar	Mar. 1	1957
24	The Sensational 3.4 Litre Jaguar	Autosport	Apr. 26	1957
26	The Jaguar 3.4 Litre Road Test	Motor	Apr. 10	1957
30	Jaguar 2.4	Motor Trend	July	1956
32	The First 2,000	Motor	July 17	1957
36	Jaguar 3.4 Road Test	Motor Trend	July	1957
39	Mike Hawthorn's Jaguar 3.4 Road Impressions	Motor Racing	Feb.	1959
40	Next Year's Jaguars	Motor	Sep. 4	1957
42	After 7,000 Miles the 2.4 Still Rates Tops!	Wheels	Oct.	1957
44	Jaguar 3.4 Road Test	Modern Motor	June	1958
47	The 3.4 Jaguar	Sporting Motorist	May	1959
51	Jaguar Widens its Range	Autocar	Oct. 2	1959
54	Improved 2.4 and 3.4 Jaguars	Motor	Oct. 7	1959
58	Jaguar 3.8 Mk2 Overdrive Road Test	Autocar	Feb. 26	1960
62	March Hare	Motor	Apr. 13	1960
66	Jaguar's Glossy New Claws	Wheels	May	1960
70	Jaguar 3.8 Sedan Road Test	Road & Track	Aug.	1960
74	Jaguar 3.8 Road Test	Sports Cars Illustrated	Sep.	1960
76	Jaguar 3.8 Mk2 Road Test	Motor Sport	Sep.	1960
78	The Jaguar 3.8 is Strictly Fabulous Road Test	Cars Illustrated	Dec.	1960
82	The Best Value in the World	Autosport	Jan. 13	1961
84	A Pride of Jaguars Road Test	Modern Motor	Apr.	1961
86	The Jaguar 3.4 Mk2 Road Test	Motor	Aug. 16	1961
92	Jaguar 3.8 Road Test	Motor Trend	Sep.	1961
94	The 2.4 Litre Jaguar Mk2 Road Test	Car South Africa	Feb.	1960
98	3.8 - Sir William's Big-Bore Cats Road Test	Wheels	Sep.	1961
102	The Daddy of All Jaguars	Wheels	Apr.	1962
105	Used Cars on the Road - 1958 Jaguar 3.4	Autocar	Feb. 10	1961
106	Jaguar 3.8 Sedan Mk2 Road Test	Sports Car Graphic	Aug.	1962
110	Charming Jaguar Cub Road Test	Wheels	Mar.	1962
114	106 m.p.h. for 10,000 Miles	Motor	Mar. 20	1963
116	Jaguar 3.8 Mk2 Automatic Road Test	Autocar	Apr.	1963
121	Jaguar 3.8 Mk2 Automatic Road Test	Cars Illustrated	Aug.	1963
124	Mk2 Saloon 3.4 Litre Jaguar Road Test	Car South Africa	Aug.	1964
128	The Jag Belt	Motor	Sep.	1964
132	Jaguar 3.8 Mk2 Saloon	The World Car Catalogue		1965
133	Best Luxury Compact - Jaguar 3.8 Mk2	Car and Driver	May	1964
134	The Jaguar 3.4 and 3.8S	Motor Sport	Oct.	1964
136	Used Cars on the Road - 1961 Jaguar Mk2 3.8	Autocar	Dec. 27	1963
137	Used Cars on the Road - 1960 Jaguar 2.4 Mk2	Autocar	May 15	1964
138	After the New Wears Off - 3.8 Jaguar Sedan	Road & Track	Apr.	1966
139	Jaguar Range Revised	Autocar	Sep.	1967
141	Jaguar 240 2,483cc Road Test	Autocar	Jan. 4	1968
147	Jaguar 240 - Value for Money Road Test	Motor	Jan. 27	1968
154	Jaguar 340 - High Performance at a Modest Price	Autosport	Feb. 16	1968
156	50,000 In a 3.8 Mk2	Foreign Car Guide	Dec.	1968
158	Jaguar 240 Long Term Test	Autocar	Apr. 17	1969
162	Jaguar 240 Long Term Test	Autocar	Feb. 12	1970
164	Jaguar 340 Used Car Test	Autocar	July 15	1971
166	The Old One-Two Profile	Classic and Sportscar	May	1985
172	Jaguar - A Stylish Performer Buyer's Guide	Practical Classics	Oct.	1992
178	Jaguar Mk2 Owners View	Classic and Sportscar	May	1985
179	1961 Jaguar Mk2 3.4	Classic and Sportscar	July	1992

BROOKLANDS BOOKS

ACKNOWLEDGEMENTS

Sales of our earlier Brooklands Road Test book covering the Mk.2 Jaguars proved to us just how popular these cars are among enthusiasts, and so when it came time to consider a reprint we decided instead to upgrade the book into a Gold Portfolio. In this new volume, we have therefore been able to include not only all the material which was in the earlier editions, but also a great deal of extra material about Mk.2s which was not. Better still, we thought enthusiasts would also appreciate some documentation on the Mk.1s, and so here it all is in one volume. Regular readers of Brooklands Books will know that we depend on the generous co-operation of those who hold the copyright of the material included. For this book, we are pleased to record our thanks to *Autocar, Autosport, Car and Driver, Car South Africa, Cars Illustrated, Classic and Sportscar, Foreign Car Guide, Modern Motor, Motor, Motor Racing, Motor Sport, Motor Trend, Practical Classics, Road & Track, Sporting Motorist, Sports Car Graphic, Sports Cars Illustrated, Wheels,* and *The World Car Catalogue.* Our thanks also go to motoring writer James Taylor for the few words of introduction which follow.

R. M. Clarke

For proof that the compact Jaguar saloons of the 1960s were an enduring design, you have only to look at the pages of the classic car press today and to count the number of specialists willing to work on them. And not only will these specialists restore an ailing example: many of them will actually rebuild a car to better than new specification, adding in modern components to improve the steering, transmission and brakes and endow the car with the dynamic qualities of a modern luxury saloon.

Most of the enthusiasm for the compact Jaguars centres on the Mk.2 models, and for several reasons. They had better rear suspension than the Mk.1s of the 1950s, more graceful styling, better brakes (shared with some of the last of their predecessors) and the option of even more power. In addition, they did not suffer from the cheapening touches which affected the final versions of the compact Jaguars, the 240 and 340 models introduced in 1967.

Nevertheless, the Mk.1 compact Jaguars have their own special charm, and the run-out models of the late 1960s are still first-rate driver's cars. So, too, is the so-called Daimler-Jaguar, a compact Mk.2 fitted with Daimler's splendid little V8 engine and marketed more as a small luxury saloon than as a sporting model. This model is fully dealt with in our Daimler SP 250 Sports and V8 250 Saloon Gold Portfolio. After 1969, when Jaguar withdrew their remaining compact models and settled for a single saloon in the shape of the XJ6 family, the small Jaguar was much missed. So powerful has its appeal remained that today, some 25 years on, Jaguar are said to be planning at long last a new compact saloon to take its place.

James Taylor

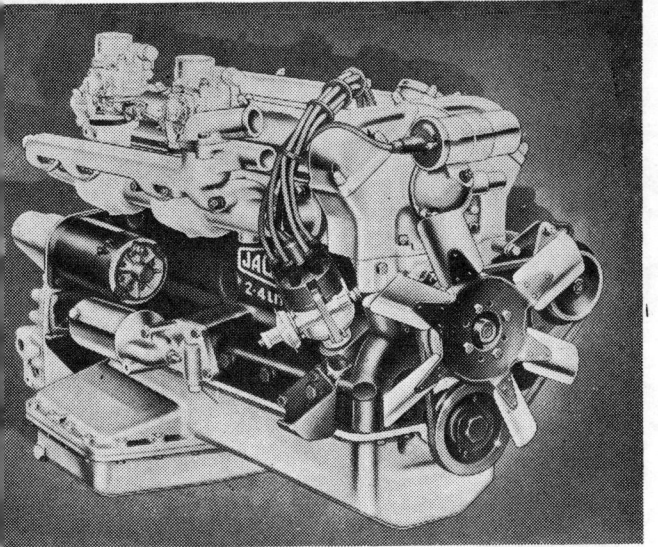

CAR AND ENGINE: The new Jaguar five-seater saloon certainly follows that marque's distinguished tradition for good looks. (Left) The 2,483 c.c. six-cylinder, twin o.h.c. power unit is a development of the well-proved XK 140 design, with shorter stroke, making it "over-square". Twin Solex carburetters are fitted.

A 2·4-LITRE JAGUAR

Exciting New High-Performance Five-seater Saloon with Twin o.h.c. six-cylinder engine

At last the long-awaited "small" Jaguar has been revealed, and a most interesting and promising vehicle it appears to be. As far as external appearance is concerned, the new 2.4-litre saloon differs considerably from the familiar 3½-litre Mark VII saloon, and many may consider it to be a "better looker". The five-seater body has clean lines and the front end uses the neat XK 140 grille. The suspension, too, has little in common with the earlier car, helical springs being used instead of torsion bars for the front unit, and cantilever springs with radius arms are used at the rear. This system is cleverly adapted to modern construction, and extensive use of rubber in the mountings eliminates vibration and noise. Side location of the axle is by simple Panhard rod. Novel, too, in a Jaguar, is the integral body-frame construction, which affords both strength and lightness.

However, the new engine is a development of the well-tried XK 140 3½-litre power unit, and in fact uses the same cylinder head and many other components including the same size bearings, which latter should make for a large margin of safety and durability. The difference in capacity has been obtained by using a shorter stroke with the same bore, making the new unit "over-square". Twin Solex downdraught carburetters are

FRONT VIEW (right) while embodying an XK 140-style grille has a pleasing individuality.

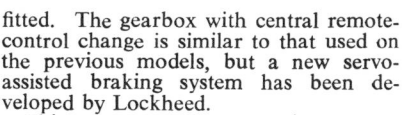

fitted. The gearbox with central remote-control change is similar to that used on the previous models, but a new servo-assisted braking system has been developed by Lockheed.

This new Jaguar, incorporating the first really new engine since the introduction of the XK 120 in 1949, should prove an immediate best-seller at its reasonable price of £1,269 0s. 10d. including P.T.—and one hopes it is more than merely coincidental that the current Grand Prix formula is also 2½ litres.

Specification

Engine: 6 cylinder, twin o.h.c., 83 mm. x 76.5 mm. bore and stroke, 2,483 c.c. (R.A.C. rating 25.6 h.p.), 8:1 compression (7:1 optional), 112 b.h.p. at 5,750 r.p.m., twin Solex d/d. carburetters, Lucas fluid-cooled coil ignition.
Transmission: Borg & Beck single dry-plate clutch, four-speed gearbox (ratios 4.55, 6.21, 9.01 and 15.35 to 1), Hardy Spicer propeller shaft, Salisbury hypoid final drive (4.55 to 1).
Body/chassis: Integral construction.
Suspension: Front, independent by helical springs and wishbones, rear, trailing link-type by cantilever, semi-elliptic leaf springs and radius arms. Girling telescopic dampers.
Brakes: Lockheed hydraulic, self-adjusting, servo-assisted, 11¼ ins. drums.
Wheels: Pressed steel bolt-on, 6.40 x 15 ins. tyres.
Steering: Burman recirculating ball.
Dimensions: Wheelbase, 8 ft. 11⅜ ins.; track, (front) 4 ft. 6⅝ ins., rear 4 ft. 2⅛ ins. Overall length, 15 ft. 0¾ ins.; width, 5 ft. 6¾ ins.; height, 4 ft. 9½ ins.; ground clearance, 7 ins. laden; turning circle, 33 ft. 6 ins. Dry weight, approx. 25 cwt.
Performance Data: Piston area, sq. in./ton, 40.3; brake lining area, sq. in./ton, 125.5; top gear m.p.h. per 1,000 r.p.m., 17.0; top gear m.p.h., at 2,500 ft./min. piston speed, 85; litres/ton mile, dry, 3,510.

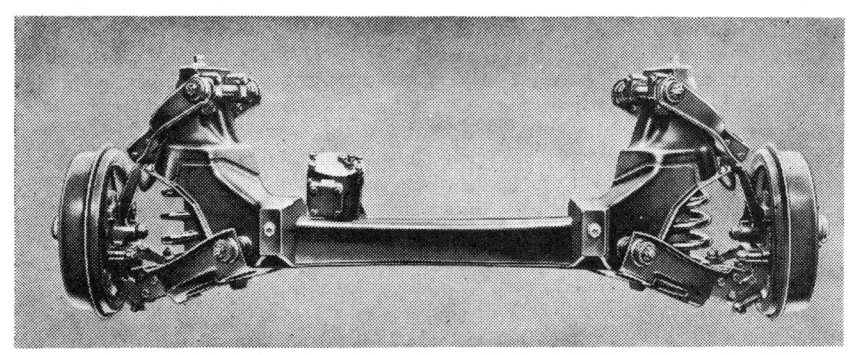

DISQUALIFICATIONS AT MONZA

After finishing first, second and third in the up to 1,300 c.c. class of the Coppa Inter-Europa production car race at Monza on 11th September, the Type 356 Porsches were disqualified by the organizers. Grounds were that the wheelbase and track of Von Hanstein's winning car differed from those of the standard Type 356 Porsche, and that Von Frankenberg's second place car used a fuel other than that permitted by the regulations.

Both Von Hanstein and Von Frankenberg have appealed against their disqualification to the F.I.A. The Porsche concern state that the cars concerned were private entries, and that they can only assume the Italians are in error regarding the dimensions of Von Hanstein's car. Hanstein stated that his car had been checked by a technical expert of the A.C. Suisse, who confirmed that the wheelbase and track are standard for the model. Von Frankenberg has informed Porsche that he used a German brand of fuel containing benzol, which was under 90 oct. rating, and as such was in accordance with regulations.

ALAN BRUCE.

FRONT SPRINGING of the 2.4-litre Jaguar is by helical springs and wishbone links, and not by torsion bars, as on the larger models.

1956 CARS

The JAGUAR 2·4-litre

IN THE past five years the Jaguar company have built a larger number of high-powered cars than any other constructor outside America, and in America they have secured a larger volume of dollar sales than any other European motor manufacturer. We are now privileged to publish the first full description of a supplementary model which is offered to those buyers who seek a more compact car with lower running costs. It has been the aim of Mr. W. M. Heynes, technical director, and his assistants to provide such a car and yet to retain maximum speed and acceleration comparable with the larger model; to retain or even improve upon existing standards of smoothness and quiet running; to offer high standards of roadworthiness; and to make possible a sensible diminution in selling price which has been fixed at £895 in standard form and £916 for a lavishly equipped special model, or, with purchase tax in Britain, £1,269 0s. 10d. and £1,298 15s. 10d.

The technical means which have been adopted to attain these ends give the car as great an interest to the student of design as it will undoubtedly have for many tens of thousands of potential buyers throughout the world. A point of prime interest is the adoption in principle of a one-piece structure for the main carcass to which all the mechanical elements are attached.

It has long been recognized that such a structure offers the greatest opportunity for securing an optimum stiffness/weight ratio but in practice the overall rigidity is diminished by the inevitable "cut-outs" for doors, gearbox and propeller shaft tunnels, and luggage locker lids, and this weakness increases (*caeteris paribus*) at an increasing power as the distance between the stress points is enlarged. In addition to these torsional problems adequate beam stiffness on a relatively large car is dependent upon considerable compression and tension loadings in the sheet metal used for the roof and floor which may thus become stretched sounding boards for noises originating at the road or in the engine transmission.

To avoid these difficulties the hull of the 2.4 Jaguar is based upon two straight longitudinal members, mated to the floor by welding to give a box section, and tied into several transverse box sections. These are placed at the extreme nose of the car, beneath the scuttle: adjacent to the attachment point of the front suspension elements and

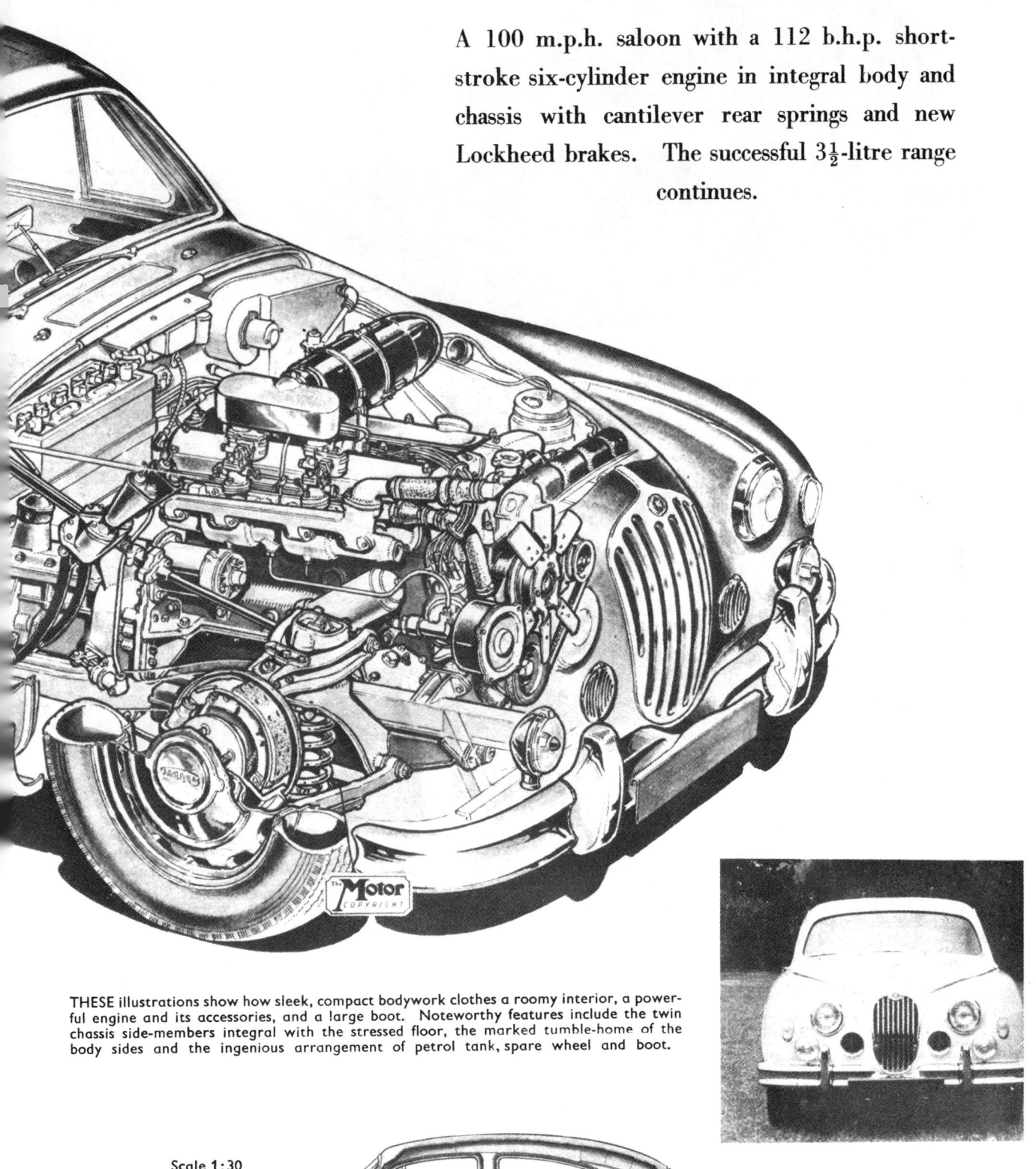

A 100 m.p.h. saloon with a 112 b.h.p. short-stroke six-cylinder engine in integral body and chassis with cantilever rear springs and new Lockheed brakes. The successful 3½-litre range continues.

THESE illustrations show how sleek, compact bodywork clothes a roomy interior, a powerful engine and its accessories, and a large boot. Noteworthy features include the twin chassis side-members integral with the stressed floor, the marked tumble-home of the body sides and the ingenious arrangement of petrol tank, spare wheel and boot.

Scale 1:30

GRILLE design owes more to XK than to Mk. VII. The car has notably smooth lines and has a compactness which makes it entirely suitable for the average private garage.

beneath the front seats with the rear engine mounting pendant therefrom. Behind the rear seats there is powerful structural reinforcement from the conjunction of wheel arches, seat pan, and a transverse pressing.

The body sills provide additional box section members on each side and full use is made of the scuttle, which is, naturally, shaped and positioned to reduce torsional movement in the fore part of the car.

To sum up, this is a one-piece, but not a highly-stressed-skin, structure and the ingenuity evident in the design is reflected in an all-up dry weight of only 25 cwt.

As one might expect, the sheet metal is coated with sound deadening material but coincidentally with making the carcass itself as stiff and as sound-proof as possible great efforts have been made to reduce the bending moments applied, and the sound-exciting vibrations fed into it. The front and rear suspension arrangements are relevant to both these issues.

Each front wheel on the 2.4 Jaguar is positioned by rear-inclined wishbones of unequal length, the lower and longer, forgings, the upper, pressings. The wheels pivot on ball joints and are turned through a Burman steering box by means of a three-piece track rod. Rubber bushes are used for the centre section of the track rod, for the front anti-roll bar, and for the inner end of the wishbones, reducing the number of steering assembly grease points to nine.

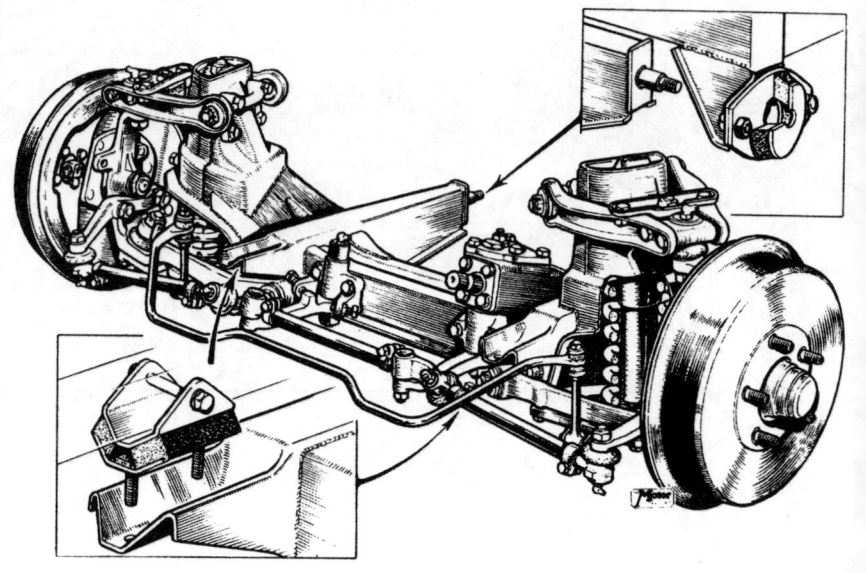

For spatial reasons coil springs embracing Girling telescopic dampers have been preferred to torsion bars, but the novelty of the whole arrangement resides in the controlled flexibility introduced between the cross-member upon which the springs and wishbones are mounted, and the main structure to which it is attached. This flexibility is conferred in two planes.

As shown in the drawing above, vertical and transverse loads are transmitted by bonded rubber blocks fitting indentations which are formed in the supporting cross-member in line with the axes of the wheels.

Longitudinal location and resistance to braking torque are controlled by two additional rubber blocks at the front end of the assembly which are thereby subject to tension, compression, and shear loads in the performance of their double duties.

The whole lay-out has been designed and developed in the light of some years of theoretical research and *ad hoc* road work which demonstrated that the road noises emerging from the front wheels arose not directly but as a resonance built up by a rela-

THE JAGUAR 2·4-LITRE

Engine dimensions		Chassis details	
Cylinders	6	Brakes	Lockheed Brakemaster hydraulic, vacuum assisted
Bore	83 mm.		
Stroke	76.5 mm.		
Cubic capacity	2,483 c.c.	Brake drum diameter	11¼ in.
Piston area	50.4 sq. in.	Friction lining area	157 sq. in.
Valves	Overhead (twin o.h.c.)	Suspension:	
Compression ratio	8/1 (7/1 optional)	Front	Independent (coil and wishbone)
Engine performance			
Max. power	112 b.h.p.	Rear	Cantilever (with radius arms)
at	5,720 r.p.m.		
Max. b.m.e.p.	140 lb./sq.in.	Shock absorbers	Girling telescopic
at	2,000 r.p.m.	Wheel type	Pressed steel, bolt-on
B.H.P. per sq. in. piston area	2.22	Tyre size	6.40-15 in. (tubeless)
Peak piston speed, ft. per min.	2,880	Steering gear	Burman re-circulating ball
Engine details		Steering wheel	17-in. (adjustable)
Carburetter	Two Solex downdraught (type B32. PBI—S/S)	**Dimensions**	
Ignition	Lucas coil (fluid cooled)	Wheelbase	8 ft. 11⅜ in.
Plugs- make and type	Champion N8B (L10S with 7/1 C.R.)	Track:	
		Front	4 ft. 6⅝ in.
Fuel pump	S.U. electric (type AUA 57)	Rear	4 ft. 2⅛ in.
Fuel capacity	13½ gallons	Overall length	15 ft. 0¾ in.
Oil filter	Tecalemit full-flow	Overall width	5 ft. 6¾ in.
Oil capacity	9½ pints	Overall height	4 ft. 9¾ in.
Cooling system	Pump thermostat and fan	Ground clearance	7 in.
Water capacity	20 pints	Turning circle	33 ft. 6 in.
Electrical system	12 volt	Dry weight	25 cwt. (approx.)
Battery capacity	51 amp.hr.	**Performance data**	
Transmission		Piston area, sq. in. per ton	40.3
Clutch	9-in. Borg and Beck s.d.p.	Brake lining area, sq. in. per ton	125
Gear ratios:		Top gear m.p.h. per 1,000 r.p.m.	17.0 (overdrive 21.8)
Top	4.55 (optional o'drive 3.55)		
3rd	6.21		
2nd	9.01	Top gear m.p.h. at 2,500 ft./min. piston speed	85.0 (o'drive 109)
1st	15.35		
Rev.	15.35		
Prop. shaft	Hardy Spicer, open	Litres per ton-mile, dry	3,510 (overdrive 2,730)
Final drive	Salisbury hypoid bevel		

tion between the pitching frequency of the axle unit around its centre of percussion and the frequency of the road noise itself. To thwart this unhappy combination the main mounting rubbers are made from a relatively hard, and the forward blocks of a much softer, compound, but the necessity to avoid any flexibility which would impair precision of control has constantly been borne in mind.

It may at this juncture be sensible to remark that noise is prevented from coming up the steering column by two universal joints and that the top end of the column carries a Bluemel steering wheel with fore and aft adjustment.

As at the front, so at the back, unorthodox design features can be seen; of even greater degree, perhaps, but

downward loads from the rear axle are transmitted to the fore part of the spring. Here they are transferred to a thick rubber button which bears upon an abutment, but is not positively attached thereto; there is no end movement at this point, so the buttons are subject solely to compression loading and are not a source for squeaks.

It will be seen that with this arrangement noise flowing from the rear wheels, or the transmission, is damped by the rubber bushes between axle and spring, by the rubber spring mounting at pivot point, and by the rubber button reaction stops respectively. Despite this massive insulation, road experiments proved the flexibility permitted to the centre pivot to be of cardinal importance in securing the last degrees of refined running.

FRONT SUSPENSION arrangements of the 2.4-litre Jaguar include coil springs which enclose the dampers, rear-inclined wishbones of unequal length, and an anti-roll bar. The mounting of the suspension cross-member has been the subject of special study, the composition and disposition of the rubber mountings having been planned to absorb high-frequency noise-inciting vibrations whilst also giving good transverse and fore-and-aft location.

UNCONVENTIONAL.—Cantilever springs with parallel torque arms and a Panhard rod form the basis of the rear suspension which, besides concentrating stresses in the main structure forward of the tail, also facilitates low build. As on the front, special steps have been taken to damp out transmitted vibration.

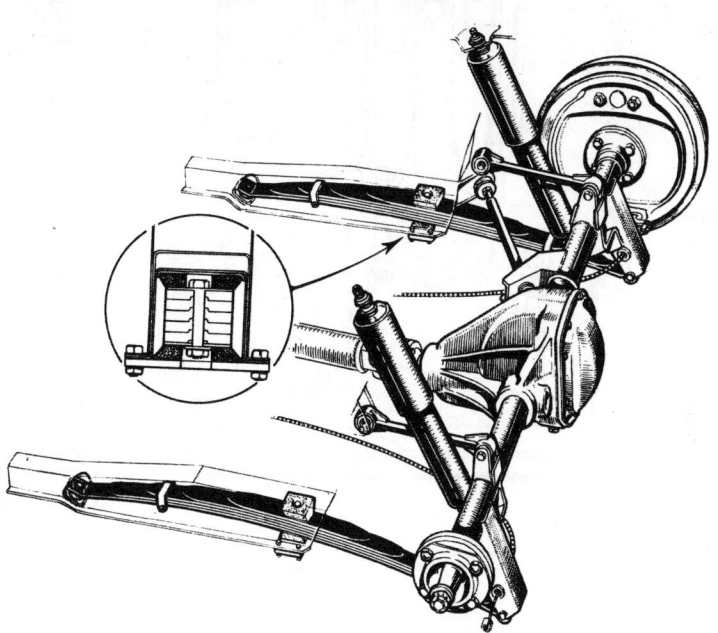

introduced, similarly, to reduce the level of running noise.

In order to diminish the distance between the points at which cross racking loads are applied to the structure (conventionally the wheel hub centre at the front and the rear spring shackles at the back) some modern cars have had coil, torsion, or in rare cases transverse leaf, springs to support the rear axle. The use on the Jaguar of cantilever springs revives a practice fallen into desuetude for the past quarter century and yet fits admirably the designer's purpose.

The springs themselves have five leaves and are identical with conventional semi-elliptics except that they are placed upside down and are differently mounted.

At their back, or trailing end, they are attached to the live rear axle through bonded rubber bushes. They are fixed to the main structure through a centre pivot of ingenious design. The leaves themselves are keyed together by a bolt passing through a dimpled hole in each leaf and all five are firmly clamped between two rubber blocks. This central assembly performs three functions. It locates the spring sideways and longitudinally so that it performs the function of a lower radius arm; it gives a limited pre-calculated fore and aft movement; and it acts as a pivot so that upward or

The control of the rear axle is completed by trailing arms, a Panhard rod, and Girling telescopic dampers, and the rear springs support 46% of the unladen weight of the car. The five-stud pierced disc wheels carry Dunlop tubeless tyres on a wide base and surround $11\frac{1}{8}$-inch diameter brake drums, part of the Lockheed Brakemaster servo system designed to abate fade which is separately described in this issue (pages 287-288).

After consideration had been given to an engine with four cylinders of similar capacity to those used on the $3\frac{1}{2}$-litre Mark VII model, the decision was made to use six cylinders of identical bore to the $3\frac{1}{2}$-litre engine, and thus of equal aggregate piston area. Thus slight disadvantages in bulk, weight and cost were accepted to gain the more important advantages of flexibility, smoothness, higher ultimate power, the benefit of years of experience on road and track, and a substantial measure of interchangeability with many important existing components.

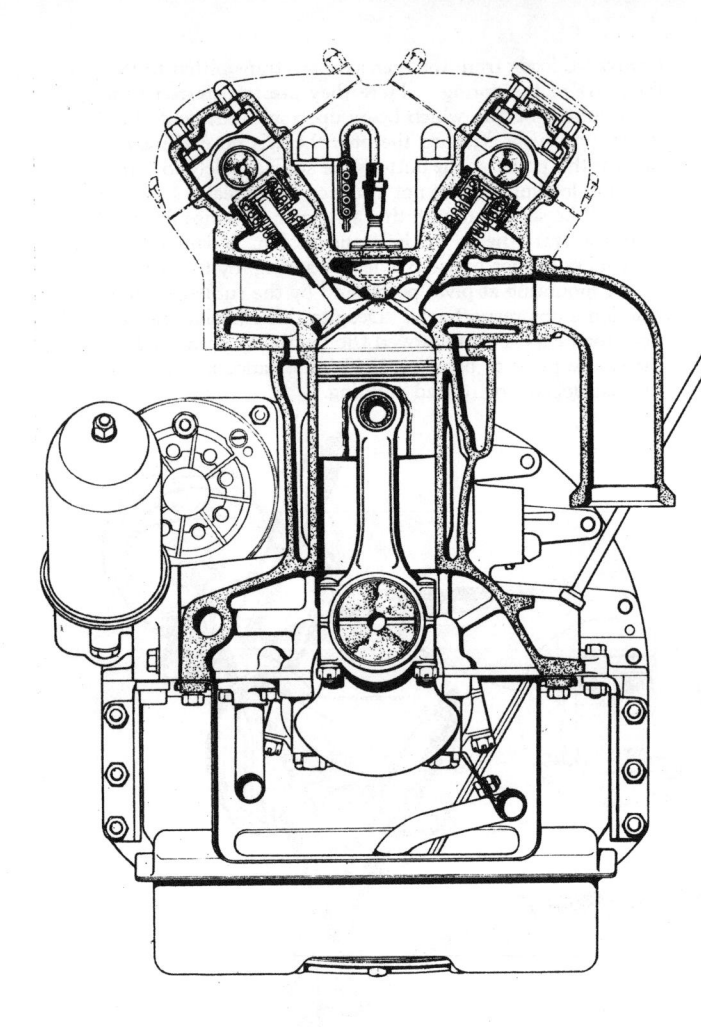

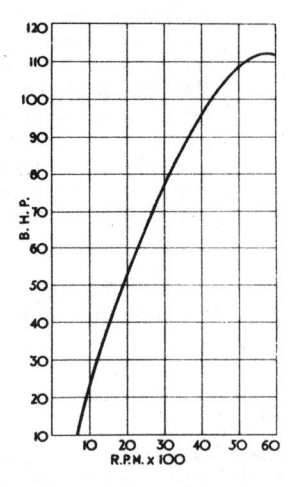

GRAPH which tells its own story of the power curve. Peak output of 112 b.h.p. is reached at 5,750 r.p.m., at which speed the piston speed is 2,880 ft. per min.

CROSS-SECTION of the new 2.4-litre six-cylinder engine which emphasizes its ancestry. Actually, the unit is a short-stroke edition of the "XK" engine and many parts, including the light-alloy cylinder head with its hemispherical combustion chambers, are interchangeable with those of the larger unit.

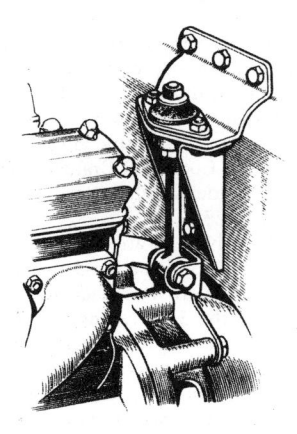

VERTICAL VIBRATIONS due to the up-and-down movements (as opposed to torque reaction) of the mass of the engine-gearbox unit are resisted by this specially designed Metalastik damper in the 2.4-litre Jaguar, thus eliminating a form of front-end shake found on some cars.

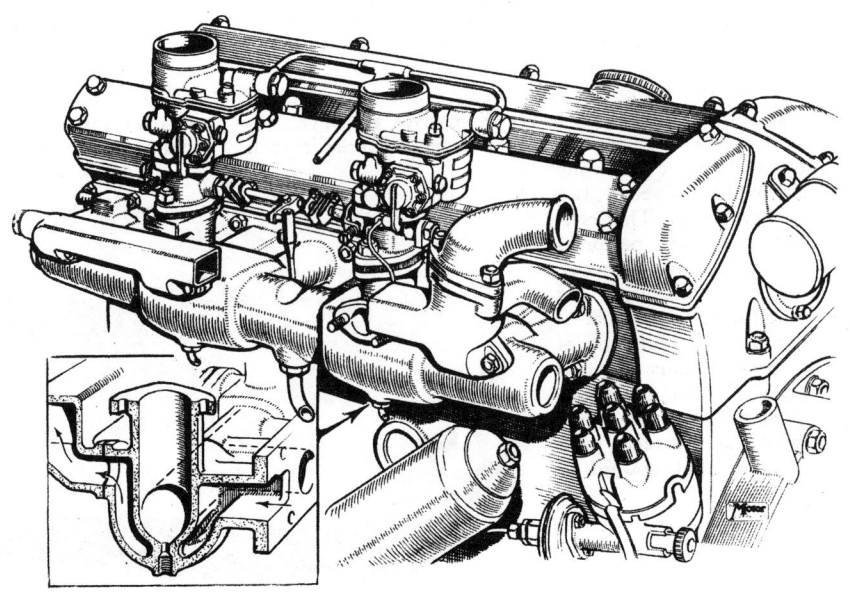

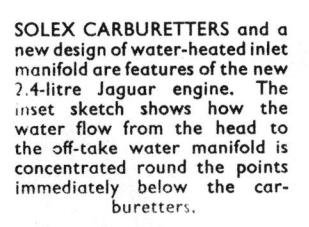

SOLEX CARBURETTERS and a new design of water-heated inlet manifold are features of the new 2.4-litre Jaguar engine. The inset sketch shows how the water flow from the head to the off-take water manifold is concentrated round the points immediately below the carburetters.

The reduction in crank throw which has brought the stroke of the engine down from 106 mm. to 76.5 mm. has diminished the capacity to 2,483 c.c. and the output to 112 b.h.p. at 5,750 r.p.m., the equal of a b.m.e.p. of 102 lb./ sq. in. at 2,880 ft./min.: both conservative figures, which should aid exceptionally long engine life.

The already unusually stiff engine structure has gained rigidity by a reduction in block height from 11.5 inches to 8.85 inches, and, as on the XK140, the 2¾-inch diameter undamped crankshaft turns in seven steel-back bearings. The whole of the cylinder-head is common with the larger size of engine, but the inlet camshaft is identical with an earlier type used on the XK120. A new system of manifolding with two downdraught Solex carburetters is employed and cool air is ducted from the front of the car through an orifice placed opposite to the intake to the air cleaner. The manifold is warmed in the area immediately below the carburetter intakes by short jackets receiving water from two of the four off-take orifices.

Power is transmitted through the hydraulically operated Borg and Beck mono-plate clutch to a single-helical-pinion gearbox in which the four synchronized speeds are engaged by a short central lever. This may be supplemented by a Laycock-de Normanville unit reducing engine speed by 25%, and if this is not fitted a simple light-alloy housing carries an extension shaft to the front universal joint.

The engine, clutch and four-speed gearbox unit weighs, with electrical components, 51 lb. less than the 3½-litre version, and all 630 lb. are mounted in the frame by a single rear mounting with rubber in shear and two vee-mounted rubbers at the front.

To avoid vibrations which are often imputed to structural weakness, movement of the engine in a vertical plane is constrained by a damper placed between the dash structure and the clutch housing. This refinement, developed by Metalastik, Ltd., with Jaguar engineers, carries no load, but uses a rubber compound of high creep value which adjusts to the static position of the engine and simultaneously provides effective damping.

Turning now from the many technical features of interest in the car to the body interior, a point immediately noticeable is that the full-width body styling is reflected by corresponding interior roominess, the front passengers having 56 in. and the back 57 in. available at elbow level. The front seats are separately mounted on inclined planes so as to rise as they go forward and have cut-away bases to provide additional toe room at the back. Head room is 37 in. at the front and 35 in. at the back, and all doors have fixed armrests.

The seats have foam rubber cushions covered in high-grade leather and on the floor a pile carpet has a thick felt underlay.

Behind the rear seats is a rear locker giving 13½ cu. ft. of luggage space, the spare wheel being recessed into the 13½-gallon fuel tank and the S.U. electric fuel pump situated in a left-hand rear wing recess.

Both the rear-hinged bonnet top and the boot lid are spring counterbalanced and need no struts, and the rear locker has an automatic interior light.

Passenger Amenities

In conformity with Jaguar tradition, polished walnut is used for the facia panel, screen pillars and other interior garnishings, and, despite the low build, a large glass area gives good visibility, the front screen being 15 in. deep and the rear window 9 in. deep, with widths of 48 in. and 44 in. respectively. All four doors have stay-open hinges and ventilating panels, and the rear wheels, which are set 4½ in. closer together than the front, are covered by panels removable by half a turn on two Dzus fasteners concealed by the rear door.

Standard equipment embraces stop/tail lamps-cum-reflectors, reversing lamps, self-cancelling flashing indicators, twin blended horns and four-point jacking. Map pockets are in all four doors and cubbyholes on each side of the instrument panel, that on the passenger side having a lockable lid. The standard instrument panel carries a 5-in. diameter speedometer, ammeter, and oil, water temperature, and fuel gauges. The Special Equipment model has, in addition, a 5-in. tachometer, an electric clock, and a cigar lighter. It is identifiable externally by a Jaguar radiator mascot and is equipped with fresh-air interior heating, folding centre armrest at the back, screen washers, adjustable twin fog lamps, courtesy switches for the interior rear quarter lamps, and vitreous enamel finish for the exhaust manifolds.

With the small changes thus enumerated the two models are identical, and the photographs reproduced show that their harmonious and attractive form matches the technical interest of the design for the engineer and the excellent value for money offered to the buyer.

WELL-FOUND.—Polished woodwork and well-shaped leather seats follow the Jaguar tradition. Shown above and below is the Special Equipment model which includes extra instruments and accessories and a rear folding armrest in its specification.

SPACE AND STRUCTURE.—(*Below*) Wheel arches, seat pan and transverse member form a rigid reinforcement at the rear, stiffness being imparted by extensive indentations. The spare wheel is recessed into a well in the flat petrol tank and is covered by a hinged flap.

ROAD IMPRESSIONS

BRIEF ENCOUNTER

Two Fleeting Hours in the new 2.4 Jaguar

With the aid of an adjustable steering column and a seat that can be varied in both height and distance from the controls, the driver of the 2.4 can make himself really comfortable. In top, the gear lever lies back at an acute angle

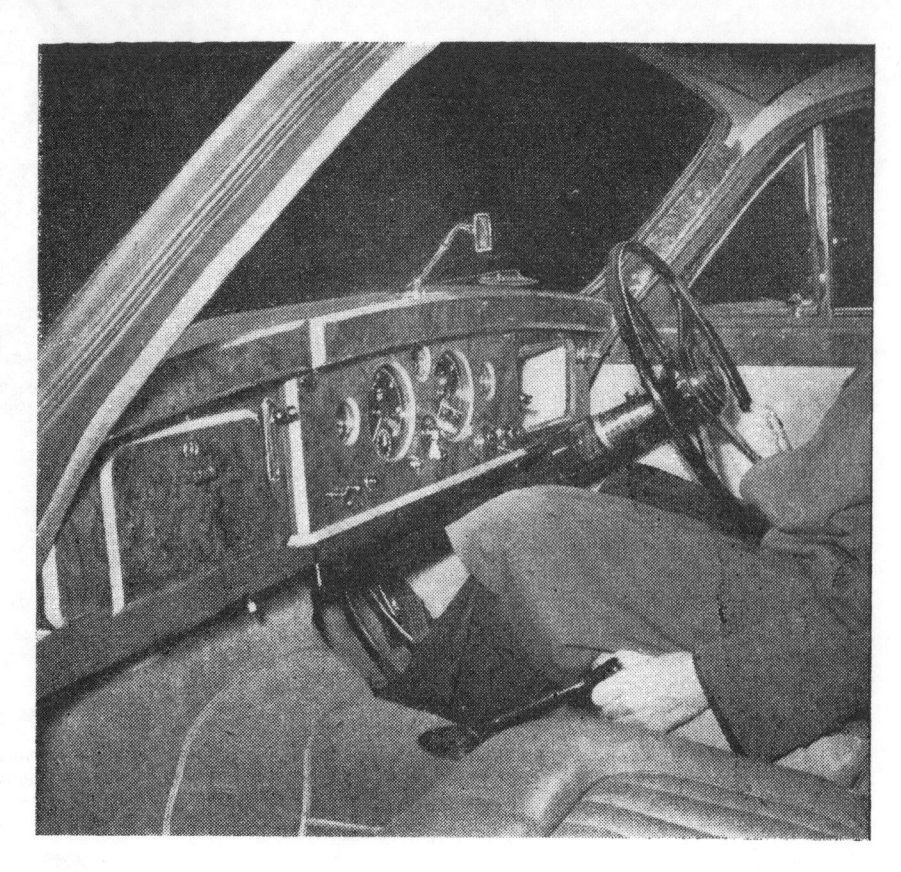

AMONG the new models introduced at each Motor Show, there always seems to be at least one which makes members of *The Autocar* staff say "Ah, I must drive that soon"; it was with very much satisfaction that I took the opportunity to put in a couple of hours of exciting motoring with a colleague on the new 2.4-litre Jaguar.

On the journey to our meeting place to take over the car, it was interesting to contemplate whether the main characteristics of the larger Mark VII would be retained. Points which sprang to mind were the comfortable ride achieved by the use of a relatively soft front suspension, coupled with the ability to corner really fast without undue roll. The light steering of the Mark VII has always attracted me, particularly the high degree of self-centring which reduces the amount of work required from the driver. Self-centring is comparatively easy to provide, but it is often accompanied by a heaviness completely foreign to the Jaguar.

Yes, these characteristics were there with, I thought, improvements. The walnut facia and the slight yet not unpleasant whine from the gear box when running in the intermediate ratios confirmed that the family tradition is maintained, as also is the four-spoked, leaf-spring steering wheel—a defiance to modern stylists.

The temperature had been near freezing at sunrise, and it was still early when I took the driving seat. The car had been in the open all night, and the starting choke was placed in the uppermost of its three positions; the intermediate one is the half-choke position, marked "normal," and the lower one is the running position when the engine has reached its operating temperature. The engine started at the first touch, without assistance from the throttle pedal.

The seats, which are mounted on an inclined ramp, were set to suit, as was the position of the adjustable steering wheel. As one who likes the full arm length driving style I set the steering wheel at its maximum forward position initially. The first quarter of a mile on the road demonstrated that this arrangement caused the right knee to touch the steering wheel rim when transferring the foot from the throttle to the brake pedal. The throttle pedal is set well forward from the level of the clutch and brake pedals, and this necessitates a considerable lift of the foot

A good lock and ready visibility of the road in front enables the 2.4 to be manœuvred with accuracy and with confidence

12

during the change from driving to braking conditions. Otherwise, there is no criticism of the pendant pedals for clutch and brake operation.

Pendant pedals have many advantages for the designer who is seeking to save room, but there are conflicting arcs of travel between the motion of the pedals and the foot which can cause discomfort. This, fortunately, is avoided on the new Jaguar, and the operating loads for each are light. Although there is quite a bulge around the engine and gear box resulting from the forward seating position and dropped floor level, there is plenty of room to rest the left foot clear of the pedals.

The central gear change lever required rather long travel to engage first gear, but this can be achieved without the driver displacing himself or reaching his hand beyond comfortable stretch. In second and fourth gears the lever attains a near horizontal position between the seats, with the knob below cushion level and just to the rear of the seat edge. Visibility from the driver's seat is good in spite of the fairly thick transmission front pillars, because these are well to the side and the degree of rake of the screen is wisely chosen. The top of the steering wheel is only an inch or so above the bottom screen rail, which itself is low in relation to seat height. As a result, forward vision from both front and rear seats is unusually good.

During the journey traffic conditions provided ideal opportunities for road impressions. For the first three or four miles the traffic was fairly dense, and it was a joy to use the gear box to weave a pattern through the sometimes hesitant throng. The engine had taken a little time to warm up and perhaps I had been a little too eager to place the choke control lever in the running position, which resulted in a slight hesitancy on two occasions when maximum acceleration was demanded in second gear.

As traffic thinned out towards more open country the car really came into its own. The outstanding impression of the power unit was its smoothness. It was necessary to look at the rev counter to believe that it was hovering in the red segment between 5,500 and 6,000 r.p.m. when maximum use was made of the intermediate gears. Furthermore, when the foot was taken off the throttle at these speeds, the engine retained its turbine-like smoothness and there was not the slightest trace of drum inside the body.

In top gear 5,000 r.p.m. was recorded very quickly during one break in the traffic stream. This corresponded to a reading of 90 m.p.h. on the speedometer, and up to the moment when braking became necessary the car was still accelerating. The braking, left deliberately late when approaching a roundabout, resulted in a good straight line stop without any sign of deviation.

But for prior knowledge that servo-assistance is used it would have been impossible to detect its effect on the brake pedal. It was light and progressive under all conditions, and there was never any feeling of the braking effort being taken over completely by the servo system, as sometimes happens when the degree of assistance is high. Furthermore, there was no lag. The suspension appears very well balanced for all conditions. Bumpy stretches, taken fast, produced a seemingly cushioned movement of the passengers, with a well controlled recovery and quick levelling out. On corners taken deliberately fast there was a degree of roll to give the driver the necessary feel, but this seemed to be quickly arrested—in fact it provides a firmness which gives confidence that the car is always well under control. The light, yet not too low geared steering provided, in conjunction with a high degree of self-centring, demands little effort on the part of the driver. With three up the impression of two separate drivers was that the steering was neutral, without bias towards either under- or over-steering effects.

Getaway in second gear was tried with three up, and although there was an occasional hesitancy low down until the engine was really warm, there was no trace of transmission judder. On the return journey through the incoming early morning traffic to London the car proved very flexible. In top gear it pulled evenly at a speed as low as 12 m.p.h. and with good pick up, devoid of snatch, when accelerating away.

Although it had not been possible to try for maximum speed, the 90 m.p.h. came up very quickly, so that one feels that the speed of 104 m.p.h. recorded by the Jaguar experimental department would be no exaggeration. Up towards three figures the car really comes into its own. It was Hilary who described the unique world of the fighter pilot in his great war book, and in a lesser way the good fast car—usually the race-bred fast car—transports its driver to another, more exhilarating, sphere than that occupied by the rest of the road users. The 2.4 certainly does so. The engine takes hold, a throaty note comes from the direction of the bonnet, and the car is alive in the driver's hands, rolling the road back under the wheels and straightening the bends, with the compactness to weave through the potterers until the time comes for the brakes to take hold and bring you back once more to the common tempo.

In a description of the new car (September 30) the steps taken to eliminate road noise were stressed, and I can say without hesitation that the Jaguar engineers have attained their object. Over rough stretches of road and cobbles, taken at various speeds, there was no trace of the annoying rumbles so often experienced on this type of road. The driver was aware of the surface through a slight reaction at the wheel, but it was obviously well damped out by the rubber coupling incorporated in the steering column.

My encounter with this car was all too brief, and I look forward to a much longer association in the not too distant future.

H. M.

The newcomer is as English as its background—the Round Tower of Windsor Castle. Yet it is as smart as anything from overseas

13

SCI ROAD TEST:

Two point four Jaguar

Two-point-four rolls to the side going around test curve. Jaguar's latest export is more a sports family car than a sports or sports touring car.

IT'S NOT often that the sports car clan takes a serious interest in the latest four-door sedans, but the newest Jaguar "saloon" has caused a lot of excitement in those quarters. Part of it is the result of Jaguar's reputation as a builder of sports cars, which promises good things in the performance and handling departments. Also, as the first major new model from Coventry since the Mark VII, the 2.4 reveals a lot of advanced thinking and points strongly toward Jag's future sports car plans. It intrigued us, anyway, so in search of a dual personality we took over a pastel blue machine from Jaguar Cars North American for a few days' testing.

Following a reliable model change plan, the 2.4 emerges as a tastefully balanced blend of tried principles with some of the unusual techniques learned in six years of Le Mans competition. The biggest switch from English convention is the use of integral chassis-body construction on a car of this size, which brought advantages of its own and forced complete reconsideration of a lot of suspension and silencing problems.

As in most early ventures in integration, a vestigial frame still angles beneath the stressed floorboards, though the boxed sills and cowl add a great deal to the overall strength. Tightly stretched panels act like drum heads in picking up and broadcasting small vibrations, and for this reason the 2.4 makes only limited use of fully stressed skin sections. To cut road noises at their source, the whole front suspension is mounted on a pressed sub-frame, which in turn is joined to the chassis by four bonded rubber blocks. Wholly new to Jaguar, this system made torsion bars very inconvenient, so a conventional coil spring setup was adopted. This has the happy result, though, that this front assembly could handily be lightened and welded into the chassis of a special.

The rear axle location smacks strongly of that used on the D-type, since the housing is guided by parallel trailing arms and a short lateral arm. In this case, the bottom arms are the rear halves of cantilever leaf springs, which are heavily insulated from the body on their total of four mounting points. There is, in other words, a lot of flexibility provided at both ends, and little resemblance to older production Jag methods.

All this attention both on the drawing board and the test track has given the 2.4 a remarkable feeling of solidity and almost complete freedom from rattles. It rides silently and smoothly over a variety of surfaces, with a minimum of pitching in both front and rear seats. Freedom from joggling is obtained without losing a degree of firmness and stability on the road.

With an anti-roll bar and 57 per cent of the weight in front, the 2.4 can be expected to understeer, which it does

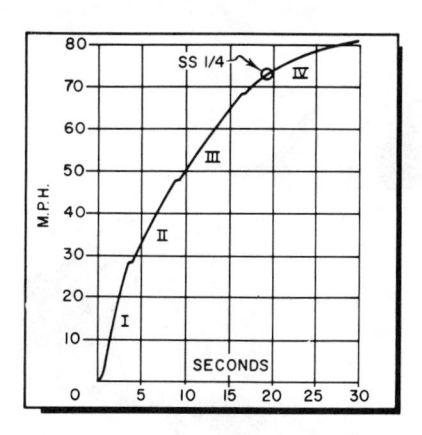

From the front, the 2.4 gives several hints of Jaguar styling, but in a diminutive aspect. Flame throwers on bumper are optional equipment.

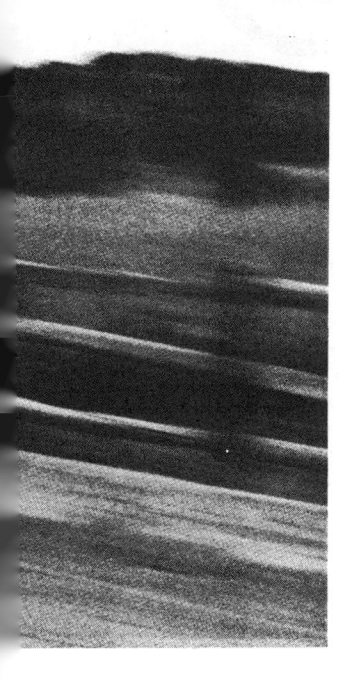

Interior is rich in comfort and appointments. Tach sits at left almost before driver with other instruments well placed. Shift lever is far forward.

with a vigor that's exaggerated by the somewhat slow steering ratio. You need a lot of helm to hold it into a bend, but only moderate effort. The car as a whole sticks nicely in fast highway maneuvers, and it can be very satisfying on fast corners. If it's thrown around more vigorously, however, the steering wheel response becomes delayed and erratic, due in part to the roll angle assumed. Unfortunately, there is always some tire noise, even at the front/rear pressures of 30/28 psi recommended in the manual for fast driving.

Taken to extremes, though, the car tends to drift on all fours, and rear wheel slides can be provoked but don't pop up without warning. Cornering on bumpy surfaces is good except at very high lateral G's, when some hopping occurs at both ends of the car. Briefly, the solidity of the car itself isn't always matched by the connection between car and road.

Thanks to this Jag's inherent stability, tracking on straight roads is easy and true, and is aided by a steering gear that is pleasantly sensitive for this class of car. Feel is direct at most speeds without excessive road reaction at the wheel. Characteristic of Jaguar is a strong caster action, and the steering is not light at low and parking speeds, though complete manageability is retained.

Competition know-how makes a welcome appearance

Typically Jaguar, the 150 cubic incher follows design of previous engines with double overhead cams.

15

In profile, this first major new model from Coventry since the Mark VII combines the distinctive form of its predecessors, and the utility of sleek, advanced lines.

inside the 2.4, which has its adjustable steering wheel placed at the authentic Grand Prix angle and distance, and along with the big, four-spoke design this may have induced us to push a little harder than usual. Heel-and-toe downshifting is also made easy by the pedal placement, the suspended pedals generally being fine and allowing acres of left foot room. The gearbox housing pushes up a big lump in the 2.4 floorboards, but not quite big enough to make a comfortable rest for the accelerator foot. One of the neatest things we've seen in a long time is the placement of the handbrake between the seat and the door at the driver's left hand. It's completely out of the way, yet right at hand and powerful when needed.

The service brakes are Lockheed's new Brakemaster system, which features power boost, automatic adjustment, and freedom from fade. On first acquaintance they didn't impress us, having an initially soft feel which stiffened up after application, as well as a tendency to be touchy at very low speeds. In our ten-stop-test, though, they were always on the job with a reasonably straight stop, and

PERFORMANCE

TOP SPEED:
Two-way average .... 99 mph
Fastest one-way run .... 99 mph

ACCELERATION:
From zero to   Seconds
30 mph 4.6
40 mph 6.5
50 mph 10.2
60 mph 13.4
70 mph 17.2
80 mph 27.9

Standing ¼ mile 19.2
Speed at end of quarter .... 73 mph

SPEED RANGES IN GEARS:
I .... 0—28
II .... 1—48
III .... 3—68
IV .... 10—99

SPEEDOMETER CORRECTION:
Indicated   Actual
30 31
40 40
50 50
60 59
70 68
80 78

FUEL CONSUMPTION:
Hard driving .... 16.5 mpg.
Average driving (under 60 mph).. 22 mpg.

BRAKING EFFICIENCY:
(10 successive emergency stops from 60 mph, just short of locking wheels):
1st stop 62
2nd stop 60
3rd stop 61
4th stop 61
5th stop 57
6th stop 58
7th stop 55
8th stop 57
9th stop 58
10th stop 62

SPECIFICATIONS

POWER UNIT:
Type .... six cylinder, in-line
Valve Arrangement .... V-inclined, twin overhead cams
Bore & Stroke (Engl. & Met.).... 3.27 x 3.01 ins. (83 x 76.5 mm.)
Stroke/Bore Ratio .... 0.923 to 1
Displacement (Engl. & Met.) .... 151 cu. ins. (2483 cc.)
Compression Ratio .... 8 to 1
Carburetion by .... Z Solex downdraft, type B32.PB1-5/5
Max. bhp @ rpm .... 112 bhp @ 5720 rpm.
Max. Torque @ rpm .... 140 lb-ft. @ 2000 rpm.
Idle speed .... 600 rpm.

DRIVE TRAIN:
Transmission ratios   Rev.
Rev. 3.375
I 3.375
II 1.962
III 1.367
IV 1.0

Final drive ratio (test car) 4.55
Other available final drive ratio. Optional overdrive gives 3.55
Axle torque taken by .... Radius rods and leaf springs

CHASSIS:
Wheelbase .... 107¾ ins.
Front Tread .... 54⅝ ins.
Rear Tread .... 50⅛ ins.
Suspension, front .... Independent, coil spring and wishbones
Suspension, rear .... Cantilever leaf springs and radius rods
Shock absorbers .... Girling telescopic
Steering type .... Burman recirculating ball
Steering wheel turns L to L ... 4¼
Turning diameter .... 33.5 ft.
Brake type .... Lockheed Brakemaster hydraulic, vacuum servo
Brake lining area .... 157 sq. ins.
Tire size .... 6.40 x 15

GENERAL:
Length .... 180¾ ins.
Width .... 66¾ ins.
Height .... 57½ ins.
Weight, test car .... 2960 lbs.
Weight distribution, F/R 57/43
Weight distribution, F/F, with driver 57/43
Fuel capacity — U. S. gallons ... 14.5

RATING FACTORS:
```
Bhp per cu in. ................... 0.743
Bhp per sq. in. piston area ...... 2.21
Torque (lb-ft) per cu. in. ....... 0.924
Pounds per bhp—test car .......... 26.4
Piston speed @ 60 mph ............ 1740 ft./min.
Piston speed @ max bhp ........... 2880 ft./min.
Brake lining area per ton
   (test car) .................... 106. sq. ins.
```

The front suspension consists of rear-inclined wishbones of unequal length. Steering is through a Burman box linked to a three-piece tie-rod supported by an idler arm. Bonded rubber blocks in the cross member transmit vertical and transverse loads from body to springs and shocks.

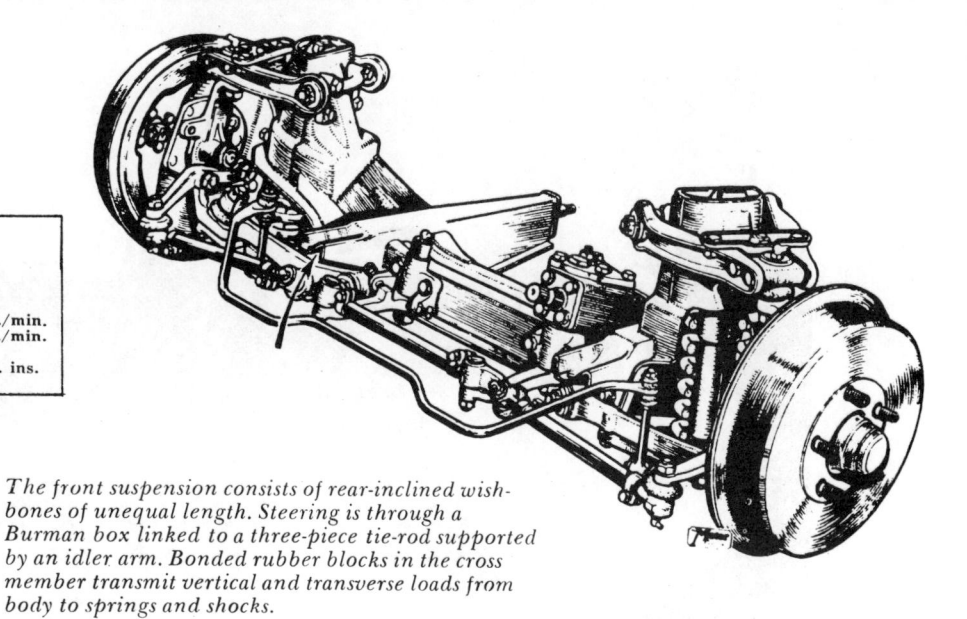

held their moderate power well.

As usual with Jaguar, the instrumentation is comprehensive on the special equipment model we drove. There's a simpler, no-tach version available in England, but the 2.4 would be lonely without that instrument, which is placed on the driver's side. Although the dials themselves are well graduated and readable, visibility continues to be sacrificed to symmetry, and items like the water temperature are remote. Gauges are well lit at night, but there's no front map light, and the two interior lights are in the rear quarters but controlled from the dashboard.

The small dash switches are now pushbutton-type; which are not entirely reliable, and the carburetor starting control works well, but is needlessly distant. The starter button is conveniently close to the ignition switch, from which the keys inevitably hang into the open ashtray. Interior storage space is exceptionally generous, with pockets in all doors and a roomy lockable glove compartment. Change for tolls can be kept handy in the open shelf on the driver's side.

From the driver's standpoint, then, the 2.4 controls are sporting but the car's reactions to them are more sedate and sedanlike. It narrowly misses being the perfect machine for the enthusiastic driver who needs family room, but taken strictly as a roomy five-seater, it's exceptionally safe and stable. The rear seat, of course, is fully usable with plenty of footroom, and thanks to both the level ride and adequate headroom, there's no tendency to rap skull against headliner as in many smaller imports. A hinge-down center armrest and two ashtrays complete the amenities.

All doors open from the rear by pushbutton, and have strong hold-open stays. When you walk up to the 2.4 you're pleasantly surprised by its compact lowness, which imposes some sport-car-like motions on getting in. The seats are raked well back, which contributes to the G.P. stance, and they have a just-right feel of firm cushioning. They don't form-fit too closely or support the shoulders well, though, and some extra fatigue on long hauls results. The front-seat passenger in particular finds himself clutching for support. Adjustment is made easy by a long spring-loaded track that fairly flings you into the dash when released.

Though the hood is impressively long to the onlooker, it falls away rapidly and allows very good forward vision with both fender peaks defining the limits of the car. The

Rear axle setup shows cantilever spring, angled shock absorber, upper locator arm, and part of the gas tank.

At 55 mph around bend, two point four seems to be sliding toward outside. Actually car stayed glued to road.

CONTINUED ON PAGE 38

TOURING TRIAL No. 4   JAGUAR 2.4

(Special Equipment Model)

(ABOVE) *Since tested the Jaguar 2.4 has been equipped with the wider grille of the 3.4* (BELOW) *The comfortable and well finished interior is restful on long journeys*

SALOON cars which exceed the once magic figure of 100 m.p.h. are now commonplace, but a distinction can be made between cars in this class. This is demonstrated by two factors; one, the speed with which the "ton" can be reached and, secondly, the comfort offered to the passengers and ease of control for the driver when the top speed has been achieved. The "Two-Point-Four" Jaguar can undoubtedly be placed near to the head of the list of cars in this class for it meets both the above conditions admirably.

Considering the speed aspect first, we found that the 2.4 litre Jaguar has a maximum speed in overdrive top gear of 102 m.p.h. whilst in direct top it will reach 92 m.p.h. with ease. The second gear maximum speed is 48 m.p.h. and third gives a very useful 65 m.p.h. It should be made quite clear that all of these maximum speeds were achieved at the recommended engine speed limit of 5,500 r.p.m.

Power comes from the twin overhead camshaft engine which almost fills the space below the bonnet. This 2,483 c.c. unit is very potent, developing its 112 b.h.p. without the least fuss and with a remarkable degree of silence. It is moreover, exceptionally well finished, as is the case with all Jaguars.

Accessibility to the main auxiliaries can be said to be reasonable. The long dip stick can be reached without difficulty but the external, full flow oil filter is rather hemmed in by other equipment. The line of six spark plugs between the two camshaft covers can be removed with ease except for numbers three and four which are rather close to the mounting brackets for the large transversely placed air cleaner which ducts air to the twin Solex downdraught carburetters on the off-side of the compartment.

Inside the neat and graceful steel bodyshell the furnishings are of luxury standard. Upholstery is carried out in top-grade hide and pile carpets cover the floor. The facia

18

Leaning well over the Jaguar 2.4 turns into a side road at high speed; the degree of stability is high

There is excellent stability at speed and the full force of acceleration does not appear to be apparent to the passenger. The extent of this acceleration maybe judged from the following figures—

From standing start to 30 m.p.h. took 5.2 sec.; to 40 m.p.h., 7.9 sec.; to reach 50 m.p.h., 11.3 sec. and 60 m.p.h., 15.0 sec.

Acceleration at full throttle showed the following—

	2nd Gear	3rd Gear	Top Gear	O/Drive Top
20–40 m.p.h.	4.7 sec.	5.1 sec.	7.1 sec.	10.5 sec.
30–50 m.p.h.	—	6.0 sec.	7.3 sec.	10.9 sec.
40–60 m.p.h.	—	7.5 sec.	9.4 sec.	11.4 sec.
50–70 m.p.h.	—	—	11.4 sec.	13.1 sec.

Recognizing the need for good braking on high performance cars, Jaguar now offer disc brakes as an optional extra on this model but the Lockheed hydraulic brakes with vacuum servo assistance that were fitted to the test car proved well up to their work. Operating on 11⅛ in. diameter drums the 157 sq. in. of lining halted this heavy car from a steady 30 m.p.h. in a distance of 31 ft. At no time did fade or any other suggestion of reduced efficiency appear despite hard braking for some miles.

Roadholding and handling is of the highest order and praise is due to the suspension system comprising coil springs and wishbones at the front and cantilever springs with radius arms at the rear. The unawareness of passengers of the speed at which the car is travelling can be illustrated by the fact that two carefully timed journeys were completed without causing the slightest concern to the four occupants. One of the journeys of 52 miles, was undertaken in daylight and the average speed obtained was 46 m.p.h. The other journey, at night, covered a distance of 60 miles and the average speed was 42.4 m.p.h. In both cases the journeys were cross country and included some main roads with fairly heavy traffic and less populated by-roads.

Luggage accommodation is generous for the size of the car but unfortunately the contents of the boot have to be removed if the spare wheel or tool kit are required. Both these items are housed in a well beneath the floor and access to the compartment is gained by lifting a portion of the locker floor.

The lighting equipment of this model is very complete and includes twin foglamps and illumination for the boot. The interior lighting is restricted to two lamps recessed in the rear quarters of the body and these are controlled by courtesy switches on the front doors or a pushbutton on the facia. It is a pity some form of illumination has not been provided in the front compartment as it is difficult to read a map in the diffused light from the rear compartment lamps.

So far, fuel consumption has not been mentioned. We found this varied considerably with the type of driving but the test car did give 31.6 m.p.g. when in direct top at a constant 30 m.p.h. and 36.4 m.p.g. when in overdrive at the same speed. At a constant 60 m.p.h. the "Two-Point-Four" travelled 18 miles on each gallon of fuel when using direct top and 26.4 m.p.g. when the overdrive was in use. At the end of the test after which the car had been driven hard for some 500 miles, we found that the overall consumption was 21.1 m.p.g.

Visibility to the front is first class but rather thick screen pillars spoil the driver's side vision. The high-set rear window is too shallow to give a perfect view of the road behind when parking in city streets. However, for city use, the 2.4 is an ideal car with surprising top-gear flexibility and a light steering.

The Two-Point-Four Jaguar is a very compact and sleek looking saloon, tailored for the owner who likes to indulge in high speed motoring or take part in some sporting activities but who, at the same time, requires the vehicle to serve as a family car. It is a car which will give the enthusiastic driver plenty of fun. The Special Equipment model retails at £1529 17s. 0d. but a standard model is available at £1495 7s. 0d., both prices including purchase tax. ★

panel and door cappings are finished in polished walnut and cloth is used for the roof lining. All the deep cushioned seats are very comfortable but the front bucket seats are outstanding in this respect. Well shaped squabs minimize fatigue and prevent lateral movement of the body when cornering at speed.

The test car was the "Special Equipment" model and therefore had a full range of instruments in the centre of the facia. The most prominent is the speedometer with its decimal trip and total distance registers and the tachometer with an inset electric clock. The recorder was accurate, as was the speedometer at 30 m.p.h. but at 60 m.p.h. this read 2 per cent. fast. Flanking these are the fuel contents gauge which incorporates a low-level warning lamp, and the combined oil pressure and water temperature gauge, whilst between the large dials lies the ammeter and main lighting switch.

In a line near the base of the facia are the minor control switches and these, with the exception of the knob for the two-speed electric wipers, are all of the push-button type. Just by the driver's left hand is the cigar lighter and a concealed pull-out type ashtray.

The clarity of the white on black instruments is very good and they are well illuminated at night. On either side of the instrument panel are cubbyholes of reasonable capacity and on the one in front of the passenger there is a lockable lid. A parcel shelf is fitted beneath the wide but shallow rear window and map pockets are provided on all doors.

In the driving seat we found the main controls well placed with the 17-in., four-spoke steering wheel on a telescopically adjustable column. The short, direct acting gear lever is conveniently located in the centre of the floor of the front compartment and it was a joy to use. Initially the gear change was inclined to stiffness. In conjunction with the hydraulically operated single dry plate clutch, however, very speedy changes of ratio in the four-speed gearbox proved possible. The fifth ratio, that of the Laycock de-Normanville overdrive top gear, is controlled by a fingertip switch on the facia panel. This switch is illuminated when the overdrive is in operation.

The layout of the pendant pedals is good and adequate space is given for the left foot, the toe of which rests happily on the dipper switch.

A treadle type throttle pedal is used but we noted that it was too upright for complete comfort, and if long distances were covered at moderate speed the muscles at the top of the foot began to ache a little. As with all high performance cars, too heavy a right foot will cause wheelspin and on wet surfaces the tail of the "Two-Point-Four" will slide. Should this happen, however, full control can be maintained and instant correction effected through the pleasantly geared Burman recirculating ball type steering box. The Jaguar's steering has a smooth action and little lost motion, but road shocks are transmitted to the driver's hands. With a turning circle of approximately 33 ft., it requires 3½ turns of the steering wheel between locks. Throttle control must be accurate as in any fast car, for high speed motoring in the rain.

19

HOUNDS, GENTLEMEN, PLEASE! The Jaguar Two-Point-Four stands outside the Vigo Inn, near Wrotham, Kent, while the West Kent Hunt prepare to move off.

JOHN BOLSTER TESTS

THE JAGUAR "TWO-POINT-FOUR"

And Declares it the Best "All-Round" Car Yet Tested by AUTOSPORT

In the last year or two, it has become apparent that two sizes of car are likely to supersede all other categories, except for specialized purposes. The first of these, of course, is the economy car, which is now expected to carry four people at 70 m.p.h. on a 40 m.p.g. budget. The second type is the roomy saloon with an engine of rather over 2-litre capacity. This is assumed to be a car of considerable luxury, and the better makes will encompass a full 100 m.p.h.

One of the latest examples of this latter category is the Jaguar 2.4-litre. This is a brilliant design, breaking new ground technically in several important respects. All its teething troubles have been systematically overcome, and the "Two-Point-Four" is now in full production.

To dissect the whole design in detail would occupy more space than I am allowed. May I, therefore, state the results which have been achieved in practice, and then comment on a few features that have made these results possible? First and foremost comes the absence of road noise from the interior of the body, which can only be described as sensational. In this respect, the new Jaguar excels the finest luxury cars from America, Germany and our own country.

The second milestone is the combination of a very soft ride with exceptional roadholding, particularly on wet and greasy roads. Finally, comes the performance angle, and in this respect the Jaguar is a little faster than any of its rivals while being at least their equal in the important matter of fuel consumption.

As would be expected, the absence of road noise and the excellence of the suspension are both bound up together. Most important in this respect is the body-chassis structure. This is extremely rigid, but the body panels themselves carry a lower stress than is usual in unitary construction, it being held that a highly loaded panel becomes a sounding board. There are, in fact, a pair of longitudinal frame members built into the floor pressings, which are shaped to incorporate cross members. The highly stressed structure terminates in a rugged cross member formed of the seat pan and wheel arches, behind which the body becomes a lightly stressed envelope.

Having produced a light but rigid body-cum-chassis with little tendency to pick up and amplify noise, the next step was to evolve a suspension system which would not "telephone" noise and vibration into the main carcase. In front, the complete suspension and steering assembly is mounted on its own cross member. It is a wishbone and helical spring layout, with a slight rearward inclination of the wishbones, and ball joints for the wheel pivots. There is an anti-roll torsion bar, and the recirculating ball steering box operates through a three-piece track rod with slave arm.

This whole assembly is mounted on rubber, with one pair of blocks to take vertical and lateral loads, and another pair, at the end of forward facing arms, to look after fore and aft positioning and to take the braking torque. Of course, these bonded rubber blocks only absorb high-frequency vibrations of low amplitude, and do not allow any appreciable movement. A pair of rubber universal joints effectively insulate the steering column.

At the rear, it was necessary to design a suspension system which would carry the weight forward to the main rear seat pan-cum-bulkhead, no rearward anchor point existing for conventional semi-elliptic springs. The solution chosen was effectively the same as the pairs of twin radius arms that most racing cars have, except that in this case the lower radius arms are in the form of cantilever leaf springs. The cantilever spring has many virtues, including reduced unsprung weight, but it was superseded in the past because of its poor lateral rigidity. In this case, lateral location is by a Panhard rod, so the springs are not asked to perform this duty. Once again, a system of rubber bushes and blocks assures complete sound insulation.

It is splendid that the makers have resisted the modern tendency towards using tiny wheels. The large 6.40-15 ins. tyres assure good road adhesion and avoid that rather ridiculous appearance which 13 ins. wheels impart to a fairly large car. The brakes are on the new Lockheed Brakemaster system, with leading and trailing shoes and a vacuum-operated hydraulic servo.

Turning from the chassis to the motive power, we at once see the reason for the speed bonus. The engine is a beautiful over-square unit with twin overhead camshafts, incorporating all the know-how learned in Jaguar's racing successes. Economy is assured by the high compression ratio and also by the comparatively small Solex carburetters. As a genuine 100 m.p.h. can easily be exceeded, it would be pointless to fit larger carburetters and obtain still more speed at the expense of a higher petrol consumption. The conventional four-speed gearbox has synchromesh on the upper three ratios and has a Laycock-de Normanville

SMOOTH, SLEEK LINES of the Jaguar's steel bodywork belie a car which can carry five people and their luggage in comfort, yet be easily parked in crowded city streets.

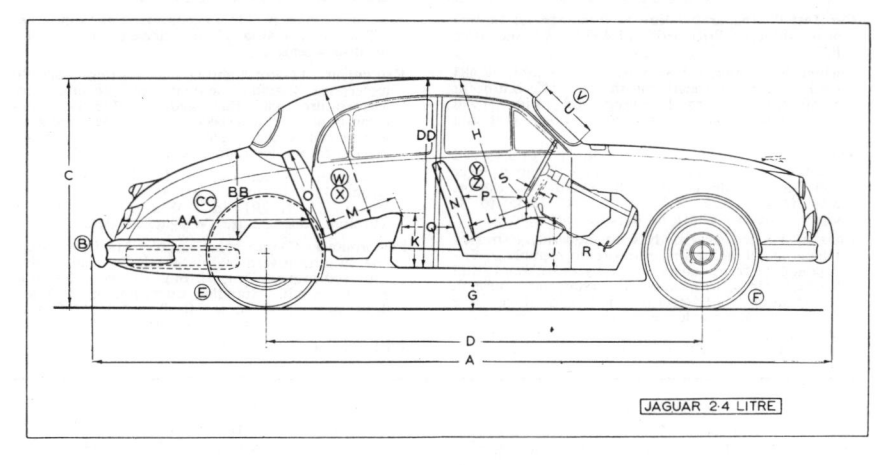

DIMENSIONS OF THE JAGUAR 2·4

A Overall length, 15 ft. 1 in.
B Overall width, 5 ft. 6¾ ins.
C Overall height, 4 ft. 9¼ ins.
D Wheelbase, 8 ft. 11½ ins.
E Rear track, 4 ft. 2¼ ins.
F Front track, 4 ft. 6⅝ ins.
G Ground clearance, 7 ins.
H Front head room, 3 ft. 7 ins.
I Rear head room, 3 ft. 0½ in.
J Height of front seat cushion, 1 ft. 0 in.
K Height of rear seat cushion, 11½ ins.
L Depth of front seat, 1 ft. 6 ins.
M Depth of rear seat, 1 ft. 6 ins.
N Height of front seat squab, 1 ft. 10 ins.
O Height of rear seat squab, 1 ft. 11½ ins.
P Steering wheel to seat squab, min. 1 ft. 0 in.
    max. 1 ft. 7 ins.
Q Rear seat to front seat distance, min. 6 ins.
    max. 1 ft. 1 in
R Pedals to seat cushion, min. 1 ft. 3½ ins.
    max. 1 ft. 9½ ins.
S Steering wheel to seat cushion, 5 ins.
T Steering wheel adjustment, 3 ins.
U Windscreen depth, 1 ft. 3 ins.
V Windscreen overall width, 3 ft. 11 ins.
W Width between arm rests (rear seat), 4 ft. 1 in.
X Overall width (rear seat), 4 ft. 10½ ins.
Y Width between arm rests (front seat), 4 ft. 4 ins.
Z Overall width (front seat), 4 ft. 10 ins.
AA Luggage compartment max. length, 3 ft. 10 ins.
BB Luggage compartment max. height, 1 ft. 9 ins.
CC Luggage compartment max. width, 4 ft. 2 ins.
DD Max. interior height, 3 ft. 8½ ins.

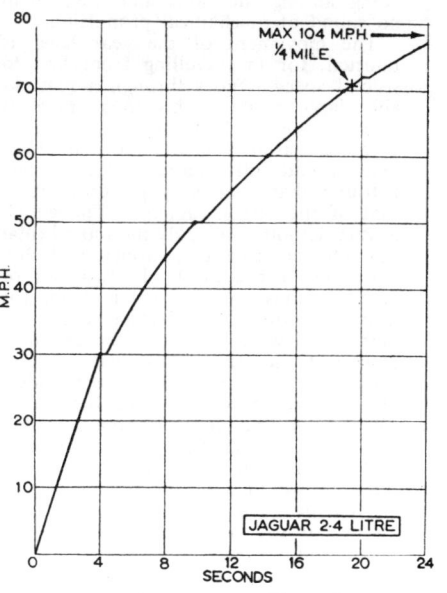

Acceleration Graph

overdrive mounted as an extension at the rear.

So much for a very brief survey of an enthralling piece of engineering; now let us get down to some actual motoring.

On taking my seat, I was delighted to find that it had a large enough range of adjustment to accommodate the tallest or the shortest driver. The steering column is also adjustable, and one can quickly assume a driving position that is perfectly comfortable. I admired the short central gear lever, and also the proper and very effective brake lever; a car like this would be ruined by one of those pathetic little umbrella handle devices.

Of course, the interior is finished in typical Jaguar fashion, and the walnut dashboard carries a rev. counter beside the large speedometer. The windscreen pillars are perhaps a little thicker than one would expect, and the rear window is shallow by modern standards. Nevertheless, the all-round visibility is adequate.

It is easy to misjudge the potency of the engine. As I drove off, I at first felt that I was going to miss the "punch" of the bigger Jaguars. It took time to realize that this short-stroke unit gives its power in a different way, and in fact the performance is very fine indeed. This is assisted by the close-ratio gearbox, and the high speeds that one can attain on the indirects are a joy. The engine is silky at low speeds, and never becomes rough. It does produce a refined but purposeful hum at the higher revolutions, but all sound virtually disappears when the overdrive is engaged at cruising speeds.

Yet, the engine has another side to its nature, for it is far more flexible than any other unit of comparable size. For instance, when driving in London I often started in second gear, and then lazily eased the gear lever into top in a couple of car's lengths. At about 15 m.p.h. I engaged the overdrive, and trickled

RACE-BRED (above). The 2.4-litre twin overhead camshaft engine fills the bonnet—a power unit which is far more flexible than others of similar size, yet which can propel the car at over 100 m.p.h.

SEATING (left). Looking across from the driver's seat, both the centre-mounted gear lever and "proper" handbrake are visible.

along among the taxis and buses with no sound of mechanical propulsion.

The movement of the gear lever is unusual, for in travelling from first to second speed it goes through a remarkably large angle. This long travel is disconcerting at first, but is completely forgotten after the first few minutes. Once I had fully mastered the change, I found that I actually preferred it to that of the larger Jaguars. The clutch is very smooth, and only the most brutal gear changes caused a suspicion of slip.

In the instruction book, two sets of tyre pressures are given—a low one for normal touring and a harder setting for high speed work. I found the car most responsive to the adjustment of tyre pressures, even 2 lbs. making a considerable difference to the road behaviour. The low setting, in my opinion, is too soft, the handling then lacking crispness and the steering being a little heavier than one would wish, though the cornering remains good. I finally chose a figure about halfway between the two in the book, and the steering immediately became light, sensitive and accurate.

The ride is very good indeed, particularly in the back seat. Thanks to the accurate location of the axle and the low unsprung weight, the rear wheels hold the road exceptionally well; one might almost imagine that there was "independence" behind. The car corners very fast for a biggish saloon, and the behaviour on wet roads is superb. Brands Hatch was in that treacherous half-wet state, but even when driven almost ridiculously hard the Two-Point-Four clung tenaciously to the road. High average speeds in complete safety are certainly the forte of this car, and the powerful brakes seem well up to their work.

There is a feeling of sheer quality and luxury, and I enjoyed all the thoughtful touches that make for driver comfort. There are trivial points to criticize, of course. The clutch and brake pedals are too close together, and the control for the flashing direction indicators is too far away from the wheel. If one has an adjustable steering column the lever for the indicators should be adjustable too, so that it can be moved with a fingertip without taking a hand from the wheel.

This type of Jaguar has many virtues, and one of these is its size. It is large enough to carry five people and all their luggage, but small enough to park easily. It feels quite a narrow car to drive in traffic and can be weaved in and out between other vehicles—however, one must sit up very straight to catch a glimpse of the nearside mudguard. The moderate fuel consumption is another good point, and about 25 m.p.g. should

SPECIFICATION AND PERFORMANCE DATA

Car Tested: Jaguar 2.4-litre Saloon (Special Equipment Model). Price £976 (£1,465 7s. 0d. including P.T.).

Engine: Six cylinders, 83 mm. x 76.5 mm. (2,483 c.c.). Twin overhead camshafts. 112 b.h.p. at 5,750 r.p.m. 8 to 1 compression ratio. Twin Solex downdraught carburetters. Lucas coil and distributor.

Transmission: Single dry plate clutch. Four-speed gearbox with short central remote control lever and Laycock de Normanville overdrive: Ratios, 3.54 (o/d.), 4.55, 6.22, 9.01 and 15.35 to 1. Open propeller shaft to hypoid rear axle.

Chassis: Pressed steel body with moderately stressed panels and longitudinal reinforcing members. Independent front suspension by wishbones and helical springs, with anti-roll torsion bar. Burman recirculating ball steering box and three-piece track rod. Rear axle on cantilever springs, radius arms, and Panhard rod. Girling telescopic dampers all round. Lockheed hydraulic brakes with vacuum servo in 11¼ ins. drums, total lining area 157 sq. ins. 6.40 x 15 ins. tubeless tyres on bolt-on disc wheels.

Equipment: 12-volt lighting and starting. Speedometer, rev. counter, ammeter, oil pressure, water temperature and fuel gauges. Electric clock, cigar lighter. Two-speed self-parking wipers and windscreen washer. Heater and demister. Flashing indicators, spotlamps, etc.

Dimensions: Wheelbase, 8 ft. 11¼ ins. Track (front) 4 ft. 6⅝ ins., (rear) 4 ft. 2⅛ ins. Overall length, 15 ft. 0¾ in. Width, 5 ft. 6¾ in. Turning circle, 33 ft. Weight, 27 cwt.

Performance: (Damp road surface) Maximum speed, 104 m.p.h. (overdrive). Speeds in gears: direct top, 100 m.p.h.; 3rd, 72 m.p.h.; 2nd, 50 m.p.h.; 1st, 30 m.p.h. Standing quarter-mile, 19.5 secs. 0-30 m.p.h., 4 secs.; 0-50 m.p.h., 9.8 secs.; 0-60 m.p.h., 14.2 secs.; 0-70 m.p.h., 18.8 secs.

Fuel Consumption: (Driven hard) 23 m.p.g.

be expected on the average run. As I have always remarked about Jaguars, I just do not see how they make it for the money.

The Jaguar 2.4-litre saloon is an economical fast touring car of exceptional refinement. It also combines a most comfortable ride with the cornering ability of a more sporting type of car. Above all, it is a real engineering job and beautifully made and finished.

Car design is a compromise, but outside the competition sphere I think that this machine has more solid everyday virtues than any of its competitors. Yet, it has beauty too, and the glamour that goes with a famous race-proved name. There is a postscript to this article: After a tough and exhaustive road test I sent the following telegram: "SIR WILLIAM LYONS, JAGUAR, COVENTRY. CONGRATULATIONS YOUR NEW TWO-POINT-FOUR IS BEST ALL-ROUND CAR EVER TESTED BY AUTOSPORT".

INSTRUMENT PANEL has a full range of dials, including tachometer, speedometer, and fuel, water temperature and oil pressure gauges, and is, traditionally, of wood.

SPARE WHEEL is recessed into the floor of the large boot, while the tool-kit is, in turn, recessed into the centre of the wheel, the kit being housed in a circular container.

3.4 Litre JAGUAR for Dollar Markets

The wider, larger-area radiator grille is noticeable at once—if you are looking for it—and the absence of spotlamps tidies the front. From behind, a pair of exhaust pipes on the right-hand side of the car gives a clue to the model

DESPITE the recent fire which destroyed the assembly tracks and many near-complete cars, the Jaguar Company produced a total of 93 cars in the first full week of production following the disaster. The 100 mark was passed early last Monday morning, and at lunch-time on Tuesday the total stood at 124. The company is confident that by tomorrow 200 cars will have been built, among them some of the new 3.4-litre saloons which are destined, for the time being, for the export market only. The new model is basically similar to the 2.4-litre Jaguar but it is fitted with a B type 3,442 c.c. engine with twin S.U. carburettors which has a power output of 210 b.h.p. at 5,500 r.p.m.

Externally the only means of distinguishing a 3.4- from a 2.4-litre model is by the slightly larger radiator grille, the cutaway spats round the rear wheels and the twin exhaust outlets at the rear right corner.

Judging by a short road experience of this very potent *gran turismo* saloon, steering and cornering characteristics appear to be very similar to those of the smaller-engined saloon. The most outstanding feature of the car is its acceleration, and the manufacturers' claim of 17.9sec for a standing quarter-mile and the ability to reach 60 m.p.h. from rest in 11.7sec would appear to be justified. A maximum speed of over 120 m.p.h. is claimed, and rather surprisingly, an average fuel consumption of 24-26 m.p.g.

Mechanical alterations include modified engine bearers and a larger radiator block, and the front coil springs have been stiffened to deal with the small increase in weight. Overdrive on top gear is an optional extra, and the Borg Warner automatic transmission is offered as an alternative. In this case the control is placed in the lower centre of the facia panel instead of on the steering column. The very short lever works in a horizontal arc between the usual positions of P, N, D, L and R. The Jaguar-designed transmission feature for holding intermediate ratio is fitted and is operated by a facia-mounted switch.

The body interior is familiar to those who know their Jaguars. The leather-covered foam rubber of the seats, the highly polished wood facia and window rails, and the carpeted floor are well-known features of these fine cars. Separate front seats are standard, but a split bench front seat is used on the automatic transmission models. Two-speed wipers, a screen washer, cigar lighter, clock, and heating and demisting equipment are amongst the standard fittings.

Production of the 3.4-litre model commenced some weeks ago and over 200 cars had been shipped to the U.S.A. to fill up the dealer pipe line before the official announcement date.

The 210 b.h.p. (gross) 3,442 c.c. engine certainly occupies the available space, yet all the fillers oil, water, and screen-washer, and the dip-stick are accessible

Opinions will differ as to the improvement, or otherwise, of appearance by replacing the rear wheel covers with partially enclosing spats. We prefer to see the wheels, and feel that wire wheels, as fitted by most Americans to their European cars, would look well. This is also the type of vehicle which, in due course, will lend itself well to the fitting of disc brakes

BRIEF SPECIFICATION

ENGINE. Six cylinder, 3,422 c.c., double overhead camshafts, compression ratio 8.1 to 1. 210 b.h.p. (gross) at 5,500 r.p.m.

TRANSMISSION. Four speed single helical synchromesh gear box. Overall ratios: top, 3.77; third, 4.55; second, 6.56; and first, 11.2 to 1. Overdrive 2.93 to 1. Borg Warner automatic, direct 3.54; intermediate 5.3-10.95; low 8.16-17.6.

BRAKES. Lockheed Brakemaster servo-assisted hydraulic. 11⅛in dia. x 2¼in wide.

STEERING. Burman re-circulating ball, 3½ turns lock to lock, turning circle 33½ft, left or right hand drive.

TYRE SIZE. 6.40 x 15 on steel disc wheels.
TANK CAPACITY. 12 Imp. gallons.
DIMENSIONS. Wheelbase 8ft 11⅜in; track front 4ft 6⅝in; rear 4ft 2¼in; overall length 15ft 0¾in; width 5ft 6¾in; height 4ft 9½in; ground clearance 7in; dry weight 27 cwt approx.
PRICE. British basic £1,114, purchase tax £558 7s, total £1,672 7s. Automatic transmission model £1,864 7s. Overdrive model £1,739 17s.

WOODED setting at Glengarven, Long Island, brings out the elegance and compactness of the 3.4-litre Jaguar, which is proving highly popular in America for its speed and comfort.

THE SENSATIONAL 3.4-LITRE JAGUAR

by GREGOR GRANT

Photography by Manny Greenhaus, N.Y.

Once in a while there arrives a motor car which takes a country by storm. In the case of the 3.4-litre Jaguar, here we have a closed car, supremely comfortable, compact, beautifully finished, and with a performance which not only matches, but surpasses the large-capacity vehicles of the majority of the Detroit factories engaged in the horsepower race. It is, of course, based on the 2.4-litre car, but whereas the former is acknowledged as being a high-speed, medium-capacity touring machine, the "3.4", with its 210 b.h.p. motor, enters the grand touring type of production vehicle category, with the added attraction of a four-door body, not to mention generous luggage accommodation.

The ideal fast-tourer should have the performance and handling characteristics of a sports car, allied to silence and flexibility. This is not so easy of achievement as it would appear, but Sir William Lyons and his men have done it. I cannot remember having driven a car which combines so many admirable features. It is as ideally suited to heavy town traffic conditions as it is to highway cruising, and its remarkable acceleration makes it one of the safest cars to drive on the freeways and turnpikes of the American continent. U.S.A. is, of course, entirely speed-cop conscious, and maximum speed limits are strictly enforced, not only by radar checks, but by grim-faced characters in sedans and on motor-cycles. Very high speeds are possible only in comparatively few states, the accent being on getaway and super-efficient braking. I can tell you it is quite disconcerting to drive along the Hollywood Freeway in a four-lane traffic stream, almost bumper-to-bumper, at 70 m.p.h., which is 10 m.p.h. above the permitted limit. When someone stops, everyone stops, often with alarming results, and I can quite understand why those outsize stop lamps are considered as necessities on American-built cars.

Racing experience has gone into the make-up of the "3.4", the road-holding of which is beyond criticism. There is no disconcerting pitching, no wavering nor shimmying, and a lightness of steering which makes for completely effortless motoring. This is extremely important in U.S.A. and Canada, where the vast distances involved make 600-700 miles in a day fairly commonplace. So you see that a car which is non-fatiguing to drive must have a considerable advantage over machines which, although admirable in many respects, do not quite match up to "3.4" standards.

TWO MEN and a car. AUTOSPORT'S *Editor Gregor Grant, John Gordon Benett, Vice-President of the U.S. Jaguar Corporation, who once raced a Talbot in the Jersey road race, and the 3.4-litre Jaguar.*

Yes, indeed, the new Jaguar is fun to drive. It gives one that pride of ownership feeling which is more important in the overseas market than is generally realized. Folk do like something different, and this Coventry product must be regarded as a most desirable piece of automobile engineering—a connoisseur's car in fact!

As a matter of general interest, several acceleration figures were taken. These are mean times, and the speedometer was checked for accuracy, so they should give a pretty fair idea of the Jaguar's performance. It should also be stressed that the machine had covered only about 3,000 miles, and that no attempt was made to do "straight-through" gear changes.

```
0-30 m.p.h.   3.6 secs.
0-40 m.p.h.   6.0 secs.
0-50 m.p.h.   8.0 secs.
0-60 m.p.h.  10.0 secs.
0-70 m.p.h.  14.3 secs.
0-80 m.p.h.  18.0 secs.
0-90 m.p.h.  22.1 secs.
```

It will be interesting to compare these times with those obtained when John Bolster does his more scientific road test of the car in Europe. Although the magic "100" comes up with ease, owing to a variety of conditions, it was not possible to clock zero to 100 m.p.h. figures. Something always happened, such as other cars pulling on to the road, and having to take one's foot off for road irregularities and so on.

As regards maximum speed, the

Road Impressions in USA of a Most Important Addition to Britians's High-performance Export Market

Jaguar is slightly faster in direct top gear (3.54 to 1), than overdrive (2.76 to 1). A shade over 120 m.p.h. was obtained, and in overdrive, 114-115 m.p.h. Maximum speeds obtained in the gears were first, 40 m.p.h.; second, 66 m.p.h.; third, 96 m.p.h. At no time was 5,400 r.p.m. exceeded, and there was never the slightest suggestion of valve-crash. In point of fact, the 3.4-litre, twin-o.h.c. engine is as smooth as silk right throughout its range. On a trip up to Connecticut, John Gordon Benett, of Jaguar's American company, demonstrated the complete silence of the engine and transmission, by cutting into neutral at 55 m.p.h. cruising. There was no discernible difference, and it was extremely difficult to tell when he snicked it into gear again.

Economy and high-performance do not always go together, but the 3.4-litre Jaguar is extremely light on fuel. Normal highway cruising results in around 25 miles per U.S. gallon, and even in heavy traffic consumption does not appear to drop below 20 m.p.g.

The car is very similar in appearance to the existing 2.4-litre machine, but points of difference include a wider radiator grille, a cleaner frontal aspect owing to the absence of fog lamps, and a twin-exhaust system. The standard 3,442 c.c. six-cylinder engine is equipped with 8 to 1 pistons, but 7 to 1, and 9 to 1 compression ratios are available. Output is 210 b.h.p. (standard) at 5,500 r.p.m., with a B.M.E.P. of 155 at 3,000 r.p.m., resulting in excellent torque characteristics. Twin HD6 S.U. carburetters are fitted, drawing fuel from a 13-gallon (Imperial) tank via S.U. electric pump. Transmission is taken through a 10 ins. single-plate Borg and Beck clutch to a four-speed synchromesh gearbox. Actually three types of transmission are available, namely, normal, Laycock de Normanville overdrive, and Borg-Warner automatic. Optional gear ratios are supplied, and a 3.77 to 1 final drive can be fitted if so desired. A car with the Borg-Warner two-pedal system was tried, this transmission being very smooth in operation, with an obvious appeal to transatlantic buyers. In the case of the overdrive, the switch is mounted on the facia panel.

A monocoque body is used, being of all-steel construction and having immense rigidity. Front suspension is independent by means of wishbones and helical springs; at the rear, radius arms are used in conjunction with semi-elliptic springs. The powerful Lockheed hydraulic brakes work in 11⅛ ins. drums and are servo-assisted. Steering is by means of the well-tried Burman, recirculating ball system.

Now how about your garage? With an overall length of 15 ft. 0¾ in., width of 5 ft. 6¾ ins., front track of 4 ft. 6⅝ ins., height of 4 ft. 9½ ins., and wheelbase of 8 ft. 11⅜ ins., the Jaguar takes up far less room than a 3.4-litre car has a right to do. Its compactness makes for easy parking, which is a very strong selling point in these United States of America. Some people might want power-assisted steering, but personally I found the car so light to handle that the fitment seems to me to be unnecessary. However, that is a purely personal opinion, and the addition of power-operation might make even more difference than I can realize.

The cars supplied with normal and/or overdrive transmission are fitted with comfortable bucket-type seats, and the automatic transmission job is obtainable with larger, and more close together bench-type seats owing to the absence of the central gear lever. Interior finish is in the highest quality hide, and the facia panel is carried out in polished walnut. General equipment is lavish, and includes speedometer, tachometer, ammeter, water temperature, oil pressure and fuel gauges, electric clock and cigar lighter. The standard heater-cum-ventilator is most efficient, and the central fitting of the radio blends in with the general air of luxury to be found within this magnificent vehicle.

It is small wonder that people like Briggs Cunningham, John ("Road and Track") Bond, famous radio man Art Peck, to name a few, enthuse over the "3.4", and, after trying it, immediately rushed to acquire one. Nevertheless, a sobering thought is that Jaguar Cars North American Corporation will absorb so many of them, that they will be few and far between on the home market. Its potentialities as a rally car are tremendous, and I fervently hope that the U.S.A. distributors will not demand every available car, so that they can be seen in the important European events.

TALKING IT OVER: (Right) The new 3.4-litre model at Jaguar's New York showrooms, with (l. to r.) John Gordon Benett, Briggs Cunningham, Gregor Grant and Everett Martin, Jaguar publicity executive in the United States.

Technical Data

Engine: Special six-cylinder 3.4-litre "blue-top" engine with "B" type head. High lift cams. Twin S.U. type HD6 carburetters and double exhaust system. Develops 210 b.h.p. at 5,500 r.p.m. Seventy deg. twin-overhead camshafts driven by two-stage roller chain. 83 mm. bore x 106 mm. stroke. Cubic capacity 3,442 c.c. (210 cu. ins.). Compression ratio 9 to 1 (8 to 1 optional). Cooling by pump and fan with by-pass thermostat control. Forced lubrication by submerged pump system incorporating full-flow filter. Chrome iron cylinder block. Cylinder head of high-tensile aluminium alloy with hemi-spherical combustion chambers. Aluminium alloy pistons. Steel connecting rods. 2¼ ins. diameter counterweighted crankshaft carried in seven large steel-backed bearings.

Transmission (manual gearbox): Four-speed single helical synchromesh gearbox. Centrally positioned gear-shift lever. Gear ratios: first and reverse, 11.2; second, 6.56; third, 4.55; fourth, 3.77. Borg and Beck 10 ins. single dry-plate clutch with hydraulic operation. Hardy-Spicer propeller shaft. Hypoid rear axle.

Transmission (automatic transmission model): Borg-Warner automatic transmission system giving following ratios: direct, 3.54; intermediate, 5.3 to 10.95; low, 8.16 to 17.6. Hardy-Spicer propeller shaft. Hypoid rear axle.

Body/chassis: Integral construction.

Suspension: Front: independent by helical springs and wishbones; rear: trailing link-type by cantilever leaf springs and radius arms, Girling telescopic dampers.

Brakes: Lockheed hydraulic, self-adjusting, servo-assisted.

Wheels: Pressed steel bolt-on, 6.40 x 15 ins. tyres.

Steering: Burman recirculating ball.

Dimensions: Wheelbase, 8 ft. 11⅜ ins.; track, front 4 ft. 6⅝ ins., rear 4 ft. 2⅛ ins. Overall length, 15 ft. 0¾ ins.; width, 5 ft. 6¾ ins.; height, 4 ft. 9½ ins.; ground clearance, 7 ins. laden; turning circle, 33 ft. 6 ins.

Performance (figures as given by factory): 10-30 m.p.h. in top gear, 6 secs.; 10-30 m.p.h. in 3rd gear, 4.2 secs.; 0-60 through the gears, 11.7 secs.; standing quarter-mile, 17.9 secs.; fuel consumption, 24-26 m.p.g.

Car available in standard form with four-speed synchromesh gearbox, with Laycock de Normanville overdrive, or with Borg-Warner automatic transmission. Prices in Britain, when car becomes available, are: Standard, £1,672 7*s*. (£1,114 basic); Overdrive, £1,739 17*s*. (£1,159); Automatic Transmission, £1,864 7*s*. (£1,242).

The Motor Road Test No. 16/57 (Continental

Make: Jaguar

Type: 3.4-litre (with Automatic transmission)

Makers: Jaguar Cars Ltd., Coventry

Test Data

CONDITIONS: Weather: Mild and dry with strong diagonal breeze. (Temperature 50°—57°F., Barometer 29.7–29.8 in. Hg). Surface: Smooth concrete (Ostend—Ghent motor road). Fuel: Belgian premium-grade pump petrol, approx. 92 Research Method Octane Rating.

INSTRUMENTS
Speedometer at 30 m.p.h. .. .. 1% slow
Speedometer at 60 m.p.h. .. .. accurate
Speedometer at 90 m.p.h. .. .. accurate
Distance recorder .. .. .. accurate

WEIGHT
Kerb weight (unladen, but with oil, coolant and fuel for approx. 50 miles) .. 29 cwt.
Front/rear distribution of kerb weight 58/42
Weight laden as tested .. .. 33 cwt.

MAXIMUM SPEEDS
Flying Half Mile
Mean of four opposite runs .. 119.8 m.p.h.
Best one-way time equals .. 121.6 m.p.h.

"Maximile" Speed. (Timed quarter mile after one mile accelerating from rest.)
Mean of four opposite runs .. 113.6 m.p.h.
Best one-way time equals .. 115.4 m.p.h.

Speeds in Gears. (Automatic change-up speeds at full throttle).
Max. speed in 2nd gear .. .. 81 m.p.h.
Max. speed in 1st gear .. .. 43 m.p.h.

FUEL CONSUMPTION
29.5 m.p.g. at constant 30 m.p.h. on level
28.5 m.p.g. at constant 40 m.p.h. on level
28.0 m.p.g. at constant 50 m.p.h. on level
27.0 m.p.g. at constant 60 m.p.h. on level
23.5 m.p.g. at constant 70 m.p.h. on level
21.0 m.p.g. at constant 80 m.p.h. on level
18.5 m.p.g. at constant 90 m.p.h. on level
15.5 m.p.g. at constant 100 m.p.h. on level

Overall Fuel Consumption for 1,564 miles, 81.2 gallons, equals 19.25 m.p.g. (14.7 litres/100 km.)

Touring Fuel Consumption (m.p.g. at steady speed midway between 30 m.p.h. and maximum, less 5% allowance for acceleration), 21.1 m.p.g.
Fuel tank capacity (maker's figure) 12 gallons.

STEERING
Turning circle between kerbs:
Left .. .. 34¾ ft.
Right .. .. 33¾ ft.
Turns of steering wheel from lock to lock 4⅓

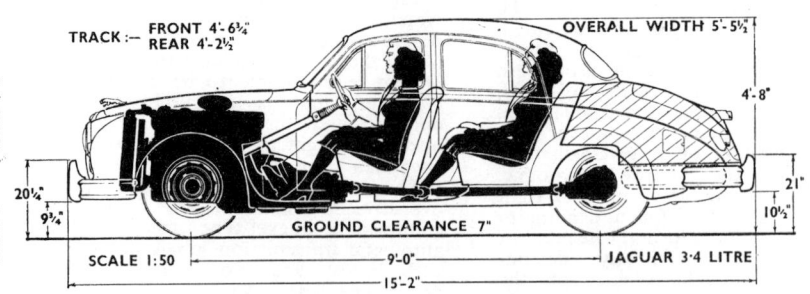

BRAKES from 30 m.p.h. (tested in neutral with engine idling)
0.93g retardation (equivalent to 32½ ft. stopping distance) with 125 lb. pedal pressure
0.90g retardation (equivalent to 33½ ft. stopping distance) with 100 lb. pedal pressure
0.80g retardation (equivalent to 37½ ft. stopping distance) with 75 lb. pedal pressure
0.63g retardation (equivalent to 48 ft. stopping distance) with 50 lb. pedal pressure
0.39g retardation (equivalent to 77 ft. stopping distance) with 25 lb. pedal pressure

ACCELERATION TIMES from standstill

0-30 m.p.h.	4.5 sec.
0-40 m.p.h.	6.5 sec.
0-50 m.p.h.	8.7 sec.
0-60 m.p.h.	11.2 sec.
0-70 m.p.h.	14.2 sec.
0-80 m.p.h.	17.9 sec.
0-90 m.p.h.	23.0 sec.
0-100 m.p.h.	30.3 sec.
Standing quarter mile	18.0 sec.

ACCELERATION TIMES from rolling start (in Drive range, using kick-down pressure on accelerator).

0-20 m.p.h.	3.0 sec.
10-30 m.p.h.	3.1 sec.
20-40 m.p.h.	3.5 sec.
30-50 m.p.h.	4.2 sec.
40-60 m.p.h.	4.7 sec.
50-70 m.p.h.	5.5 sec.
60-80 m.p.h.	6.7 sec.
70-90 m.p.h.	8.8 sec.
80-100 m.p.h.	12.4 sec.

HILL CLIMBING at sustained steady speeds

Max. gradient on top gear, approx. 1 in 8.2 (Tapley 270 lb./ton)

Max. gradient on 2nd gear, approx. 1 in 5.1 (Tapley 430 lb./ton)

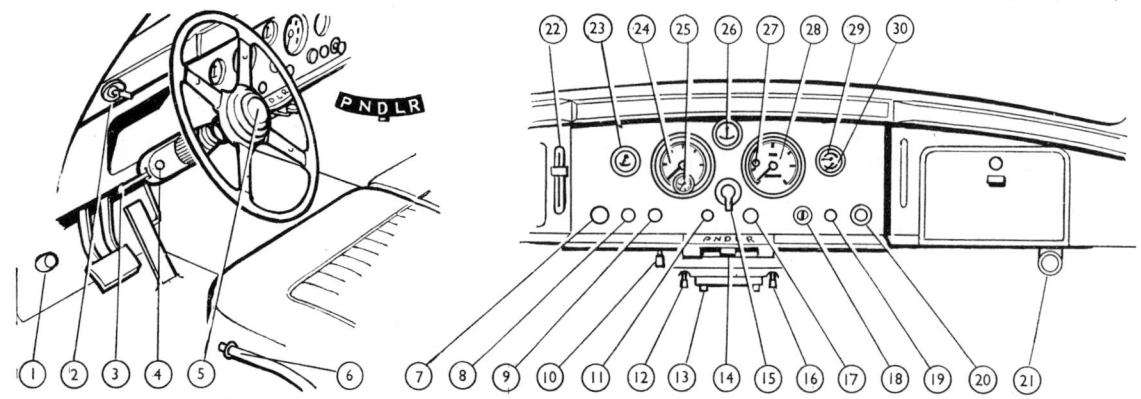

1, Headlamp dip switch. 2, 2nd gear hold-in switch. 3, Direction indicator switch and 4, warning light. 5, Horn button. 6, Hand brake. 7, Windscreen wipers switch. 8, Heater fan switch. 9, Panel light switch. 10, Scuttle vent control. 11, Windscreen washer button. 12, Clock regulator. 13, Heater flaps control. 14, Transmission selector quadrant. 15, Lights switch. 16, Mileage trip resetting control. 17, Interior light switch. 18, Ignition switch. 19, Starter button. 20, Cigar lighter. 21, Bonnet catch release. 22, Heater. 23, Fuel contents gauge. 24, Tachometer. 25, Clock. 26, Ammeter. 27, Dynamo charge warning light. 28, Speedometer and distance recorder. 29, Oil Pressure gauge. 30, Water thermometer.

The JAGUAR 3.4-litre (With Automatic Transmission)

SLEEK body lines which minimize air drag are well set off by the 3.4-litre Jaguar's large radiator intake grille.

An Extremely Comfortable and Quiet Touring Car, with Vivid Acceleration and 120 m.p.h. Maximum Speed

WHEN the factory which had built four winners of the Le Mans 24-hour race harnessed the latest version of its 3.4-litre engine to a 2.4-litre saloon body shell, thereby obtaining a top speed of 120 m.p.h. with acceleration to match, it was easy to assume that the car produced would be a rip-snorting sports model with a lid on it. Such predictions were in fact very wide of the mark.

More relevant to a correct assessment of the 3.4-litre Jaguar is the knowledge that the factory from which it emerges has been uniquely successful in selling European-style large cars to Americans, and that the native factories of America have recently all begun to offer specially tuned V-8 engines in their cars for customers wanting the last ounce of performance. With its optional fully automatic transmission, the Jaguar 3.4-litre saloon is comparable in roominess, refinement and ease of driving with the latest and lowest American cars, adds a very genuine 15 m.p.h. to the top speed hitherto offered by the Mark VII Jaguar saloon together with extra acceleration to match, yet is compact in overall size and retains all the qualities of a high-grade British touring car which have given the make its unique appeal to so many thousands of Americans. Undoubtedly the racing experience of Jaguar engineers has helped them to design the engine, suspension and low-drag body of this model, but its virtues are primarily those of an ultra-fast touring saloon rather than those of a sports car.

Various choices are offered to the buyer of this model, and our test was of a car having the Borg-Warner fully automatic transmission (with exclusive Jaguar refinements) and 3.54/1 axle ratio; options are a four-speed synchromesh gearbox and the same axle ratio, or the synchromesh gearbox with overdrive and 3.77 axle gearing. Our test model had left-hand drive as supplied to America, and was run almost exclusively with the Dunlop "Road-Speed" tyres inflated to the pressures recommended for very fast touring although not to the even higher figure advised for sustained maximum speed.

Of the seats in this four-door saloon body, the driving seat is the one which most buyers are likely to occupy. Quite high above the floor, it is a comfortable individually adjustable unit which gives just enough lateral support yet can be set level with the passenger seat to form a wide bench. Some passengers would prefer a little more lateral support than their seat gives them when being driven fast along a winding road. Over a polished wood facia panel the driver looks along a broad bonnet, high enough just to obscure vision of the farther front wing for most people. A broad and high hump over the automatic gearbox nevertheless leaves adequate width available around the pedals, which are offset slightly to the side of the car. The screen has rather thick pillars which impede vision during town driving, but the rear window is wide enough to make reversing fairly easy even if its height is limited by squeezing between a comfortably high rear seat backrest and a sharply falling roofline. Easy to enter, the rear compartment seats three passengers in comfort or can be divided by a folding central armrest; kneeroom and headroom are fully adequate and the central hump over the propeller shaft very modest in dimensions, but heel-wells or greater cutaways under the front seat could with advantage improve footroom even further. Without being gigantic, the luggage locker has very useful capacity and is easy to load through a lift-up lid. Only two rather small cubby holes are provided for minor parcels, but there is quite a capacious pocket for maps or guide books inside each of the four doors.

In mild spring weather the Jaguar engine started instantly from cold, the automatic choke, after brief initial hesitation, giving a slow and uneven but quite reliable tick-over. The automatic transmission selector lever, which moves sideways across a neat and convenient quadrant below the centre of the facia panel, could

---

In Brief
Price (including automatic transmission as tested): £1,242 plus purchase tax £622 7s. 0d. equals £1,864 7s. 0d.
Price with synchromesh gearbox (including purchase tax): £1,672 7s. 0d.
Capacity ... ... 3,442 c.c.
Unladen kerb weight ... 29 cwt.
Acceleration:
  20-40 m.p.h. in drive range 3.5 sec.
  0-50 m.p.h. through gears 8.7 sec.
Maximum direct top gear gradient, approx. ... ... 1 in 8.2
Maximum speed ... 119.8 m.p.h.
"Maximile" speed ... 113.6 m.p.h.
Touring fuel consumption .. 21.1 m.p.g.
Gearing: 21.9 m.p.h. in top gear at 1,000 r.p.m.; 31.5 m.p.h. at 1,000 ft./min. piston speed.

CLASSIC in style with circular instruments on a polished wood panel, the facia carries a neat gear selector quadrant on two-pedal cars such as the example tested.

FUNCTIONAL good looks distinguish the twin overhead camshaft engine, with twin S.U. carburetters, which fills the bonnet. Cool air is taken in through the very effective silencer seen above the engine.

The JAGUAR 3.4-litre

promptly be moved to "Drive" or "Reverse" and the car driven briskly away.

To drive, the Jaguar 3.4-litre is very much of a dual personality car. A light touch on the accelerator pedal will take it quietly and smoothly away from rest with very little fuss indeed, the engine speed not rising above 2,000 r.p.m., and whilst firmer pressure on the accelerator gives faster acceleration there is nothing fierce about the response. Press really hard on the accelerator pedal, however, and the vast power of the twin-camshaft Jaguar engine is really unleashed, the automatic gearbox then letting the rev. counter run round to rather more than 5,000 r.p.m. before making the upward changes from first to second and second to top gears, and the car really leaping forwards. From a standstill on a dry road, 60 m.p.h. is attained in 11.2 seconds and a matter of 200 yards, 80 m.p.h. in a total of 17.9 seconds and about 440 yards from rest, 110 m.p.h. within just under a mile of starting from rest. The initial getaway from rest is impressive, the manner in which the acceleration is then sustained right up to well over 100 m.p.h. fantastic for such a roomy and refined car, whilst it requires emphasis that obtaining maximum acceleration merely involves pressing the accelerator hard without any need to use a clutch pedal or gear lever skilfully.

Easy Overtaking

The level-road maximum speed of this car, 119.8 m.p.h. as the timed mean of opposite runs made with the handicap of a strong cross wind, is something which many buyers may never attain, although some will in fact discover the possibility of attaining even higher speeds without over-revving the engine when conditions are favourable. One valuable advantage which accompanies the high top speed of this car will be appreciated by all buyers, however, this being the fact that 80 m.p.h. can be exceeded in the middle ratio of the automatic three-speed gearbox when overtaking other traffic, so that a slower car can safely be passed on surprisingly short stretches of clear road.

Our test mileage included Continental motor roads, and such surroundings emphasize that the Jaguar 3.4-litre is an exceptionally quiet car in which to sustain high speeds, especially if the windows are closed but by no means only so. Both wind and engine noise are surprisingly subdued, and rough roads do not induce resonances in the integral steel body. This is a comfortably sprung car for the individual sharp bumps encountered on modern concrete roads, and also rides very well over really bad surfaces such as Belgian pavé, but at speed on some sorts of wavy going the suspension bump stops come into audible action rather more easily than might be desired by hard-driving owners.

The automatic gearbox on this model has been tuned to change out of direct drive more readily on part-throttle at medium speeds than do some units of the same basic design, to suit an engine which revs freely and to give quick response to sudden pressure on the accelerator: this means that double changes of gear, down and then very quickly up again, happen not infrequently, but this slight extra fussiness at times is not a heavy price to pay for really vivid performance. Slightly jerky at times is the automatic engagement of first gear just before the car stops.

On the normal control quadrant, the only alternative to "Drive" range is the low gear which, giving rather over 40 m.p.h. at an engine speed of 5,000 r.p.m., is needed only to help the brakes down exceptionally long and steep hills. An extra switch is provided near to the steering wheel rim of the Jaguar, to prevent top gear engaging, so that middle ratio can be held in use for hard driving along a winding or hilly road: this will be appreciated by keen drivers, but on the over-run a free-wheel inside the gearbox lets the engine speed drop to that equivalent to top gear, so that unfortunately no extra engine braking effect can be secured by use of this switch. An interesting item of equipment is a link-up between transmission and brakes which, when the car is stopped with the foot-brake, holds the pressure in the hydraulic brake system until the accelerator is again pressed (or the ignition switched off) so that the handbrake need not be used at traffic checks even on appreciable gradients; this device only comes into action when the car has actually stopped, and does not operate when the brakes are used only for slowing down.

Regrettably, the brakes of the model submitted for test fell considerably below desirable standards of performance. Self-adjusting and with vacuum-servo assisted application, they would give a good emergency stop from 100 m.p.h., but even a single hard application from high speeds produced some judder accompanied by slight directional hesitancy on the part of the car, and frequent hard brake applications would accentuate these symptoms. The more gentle driver can enjoy light pedal pressures and smooth braking, but a braking system able to withstand harder usage without protest would greatly widen the appeal of this car.

For fast touring, the Jaguar 3.4-litre has very pleasant steering, heavy only at the very lowest car-park speeds, cushioned effectively against road shocks at the expense of a very mild loss of precision, and not too low geared at $4\frac{1}{3}$ turns from lock to lock since this figure relates to a very compact turning circle. Cornering at

COMFORT for passengers is provided by the broad and easy-to-enter rear compartment.

28

fast touring speeds is, if not absolutely roll-free, at any rate accompanied by only the slightest change of lateral attitude. The driver who seeks to handle this model as a sports car, however, cornering at the limit of adhesion between tyres and road, will find it behaving in rather less tidy fashion, especially on wet roads when it is exceptionally easy to provoke wheelspin by even quite moderate acceleration.

So much does this Jaguar offer in performance, carrying capacity and refinement that it is inevitably judged against very high standards indeed. A conscious effort must be made by the tester to remember that this is a car costing less than £1,250, exclusive of purchase tax, and that it cannot be expected to match in every detail cars which cost several times as much. Given sound fundamentals, it is

GENEROUS accommodation for luggage is contrived within the smoothly tapering tail of the body. The spare wheel, tools and 12-gallon petrol tank are located below the luggage floor.

possible for an owner to modify details of any car to his own particular tastes, and in this case scope for improvement seems to lie in such details as the lack of any body interior light near enough to the driver for map-reading purposes, and lack of rheostat control over the instrument lights. A good air intake shutter on the scuttle serves a heater of apparently rather limited power which also was unwilling to pass very much completely unheated air into the test car. The two-speed windscreen wiper motor operated through a flexible drive which was far from silent when the two blades were dragging on a drying screen. Combination of flashing direction indicators with the rear stop lights is an economy which we dislike on any car. All these are points which could be improved at moderate cost.

Fundamentally, this is a car which has few superiors in respect of providing smooth, quiet and comfortable travel for five people, yet which has speed and acceleration of the most remarkable order. The Jaguar used virtually no engine oil, and only a mild degree of restraint in using the available performance is necessary to record a fuel consumption better than 20 m.p.g., the engine being happy on every British or Belgian brand of Premium-quality petrol which we tried. No sports car designed as an amusing but impractical plaything for the wealthy, it is a first-class express carriage which will be invaluable to the many men for whom, literally time is money.

Specification

Engine
Cylinders	6
Bore	83 mm.
Stroke	106 mm.
Cubic capacity	3,442 c.c.
Piston area	50.4 sq. in.
Valves	2 overhead camshafts
Compression ratio	8/1 (optional 7/1)
Carburetters	2 S.U. horizontal, with automatic choke
Fuel pump	S.U. electrical
Ignition timing control	Centrifugal and vacuum
Oil filter	Tecalemit full-flow
Max. power	210 b.h.p. gross (190 net)
at	5,500 r.p.m.
Piston speed at max. b.h.p.	3,830 ft./min.

Transmission (Borg-Warner automatic)
Clutch: Hydraulic torque converter operating in 1st and 2nd gears, automatic plate clutch for top gear.
Top gear	3.54
2nd gear	5.3
	(with torque conversion, 10.95)
1st gear	8.16
	(with torque conversion, 17.6)
Reverse	7.08
	(with torque conversion, 15.2)
Propeller shaft	Hardy Spicer divided
Final drive	Hypoid bevel
Top gear m.p.h. at 1,000 r.p.m.	21.9
Top gear m.p.h. at 1,000 ft./min. piston speed	31.5

Chassis
Brakes: Lockheed Brakemaster hydraulic self-adjusting, with vacuum servo assistance.
Brake drum internal diameter... 11¼ in.
Friction lining area ... 159 sq. in.
Suspension:
 Front ... Independent by wishbones and torsion bars
 Rear ... Rigid axle with cantilever leaf springs and radius arms
Shock absorbers ... Telescopic
Steering gear ... Burman, re-circulating ball type
Tyres ... Dunlop Road Speed (with tubes), 6.40—15

Coachwork and Equipment

Starting handle	None
Battery mounting	Behind engine
Jack	Pillar type
Jacking points	Two each side

Standard tool kit: Jack, wheelbrace, grease gun, adjustable spanner, pliers, tyre pressure gauge, screwdriver, box spanner, 4 open-ended spanners, valve timing gauge, sparking plug spanner, brake bleeder tube and key, tyre valve extractor.
Exterior lights: 2 headlamps, 2 sidelamps/flashers, 2 stop/tail/flasher lamps, number plate lamp, reversing lamp.
Number of electrical fuses ... 2
Direction indicators Flashers, self-cancelling
Windscreen wipers ... 2-blade, 2-speed electrical, self-parking
Windscreen washers ... Trico vacuum type
Sun visors ... 2
Instruments: Speedometer with decimal trip distance recorder, tachometer, oil pressure gauge, coolant thermometer, ammeter, fuel contents gauge, clock.
Warning lights: Dynamo charge, low fuel level, direction indicators, headlamp main beam.

Locks:
 With ignition key Either front door, ignition
 With other key ... Luggage locker and glove box
Glove lockers ... 2 on facia panel (one open, one with lockable lid)
Map pockets ... 4 in doors
Parcel shelves ... None
Ashtrays ... 1 on facia, 2 in rear armrests
Cigar lighters ... 1 on facia
Interior lights: 2 in rear compartment, with courtesy switches (also luggage locker light)
Interior heater: Fresh air type with screen de-misters, taking air from intake on scuttle
Car radio ... Radiomobile as optional extra
Extras available ... None
Upholstery material ... Leather
Floor covering ... Pile carpets with felt underlay
Exterior colours standardized: 11 exterior colours (6 leather colours).
Alternative body styles ... None

Maintenance

Sump ... 11 pints plus 2 pints in filter, S.A.E. 30 summer, S.A.E. 20 winter
Automatic gearbox... 15 pints, automatic transmission fluid type A
Rear axle ... 3½ pints, S.A.E. 90 hypoid gear oil
Steering gear lubricant S.A.E. 140 gear oil
Cooling system capacity 22 pints (2 drain taps)
Chassis lubrication: By grease gun every 2,500 miles to 11 points, and by oil gun to 2 points.
Ignition timing ... 2° before t.d.c. static
Contact-breaker gap ... 0.014-0.016 in.
Sparking plug type ... Champion N8B
Sparking plug gap ... 0.025 in.
Valve timing: Inlet opens 15° b.t.d.c., and closes 57° a.b.d.c.; exhaust opens 57° b.b.d.c., and closes 15° a.t.d.c.

Tappet clearances (cold):
Inlet	0.004 in.
Exhaust	0.006 in.
Front wheel toe-in	0–⅟₁₆ in.
Camber angle	½° to 1° positive
Castor angle	½° to 1° negative
Steering swivel pin inclination	7°

Tyre pressures:
Front	25 lb.
Rear	22 lb.

(For fast cruising: Front 31 lb., rear 28 lb.;
Brake fluid ... Lockheed (S.A.E. 70 R2)
Battery type and capacity 12-volt, 51 amp.hr.

JAGUAR 2.4

by Walt Woron

photos by Bob D'Olivo

drivescription '56

THE STATEMENT, "With all the inherent quality of its marque" on the 2.4 Jaguar brochure pretty well summates our overall impression of this new car—a scaled-down version of the Mark VII. It's a quality product, assembled with the usual British and Jaguar care, and one that you might buy in preference to others largely because of this factor. The instrument panel is tastefully done in walnut, the instruments and controls are finely finished, the upholstery shouts that it's leather, the engine is a polished work of art, the trunk compartment is finished in the same high quality, tho of different and less expensive materials.

Other reasons you might want a 2.4 would be its sensible size (107⅜-inch wheelbase, 15 feet overall length), that allows extreme maneuverability in town, and its XK-like road-holding ability that stamps it as a good touring car.

The new model has combined the looks of its **bigger** sisters, using the grille of the XK-140 sports models with the general outline of the Mark VII sedan

Doors open wide on a conventional interior with step-down frame, extra-fine leather, scant rear knee room

Adjustable steering wheel and sporty driving position give unique feel. Gear lever throw is too long

Polished overhead-cam engine uses same-length block as larger Jaguars, but stroke is just under 3 inches

Trunk is mammoth by British standards, average by ours. Seldom-used spare is handier out of the way

The "Special Equipment" model (deluxe version selling for $3795) we borrowed from West Coast Jaguar distributor Chuck Hornburg had just been pulled out of the press showing and therefore had too little mileage to allow us to get performance figures. A quick check of its available power, its weight, and its gear ratio would lead us to the assumption that it will go from a standstill to 60 mph in 15-16 seconds. If you're not satisfied with this performance, you can always fit the D Jag head to the 2.4 block, replace the twin Solex carbs with S.U.s, and with a few other modifications, probably up the horsepower to somewhere around 150 from its present 112. This would give you a weight/power ratio of 18½ to 1 instead of the current 25 to 1.

Fuel economy should be a couple of miles per gallon better than the 21.4 tank mileage we got with this tight engine.

It would also be appreciably better when the car is equipped with the Laycock-de Normanville overdrive, optional at $160.

The 2.4 is simple enough to drive, requiring only that you not be averse to shifting thru 4 forward gears. Space around the pendulum pedals is adequate except that the clutch is too close to the large dimmer switch button. The steering wheel is mounted on a telescopic column and can be adjusted to your own desire. The key and choke are a fairly long reach since they are still located for right-hand drive, as are all the instruments; it would seem a fairly simple matter to make a more legible setup by transposing the instruments and controls, i.e., tach for speedometer. Openings are the same size.

This new Jag, suspended independently in front with coils and solidly in the rear with semi-elliptics, holds the road well with a minimum of body lean. It requires corrective action down a straight road only when there are side wind gusts of high intensity. The steering wheel is generally rock steady, except on the roughest of roads. Maintaining a cruising speed of 60-70 mph on the open road is pleasurable and safe; the servo brakes stop you smoothly and quickly.

Sitting on the wide-backed, semi-bucket, leather-covered seats, you ride firmly, with a lack of pitching about, have an insensibility to road noises, and experience no wallowing after hitting bad dips or bumps. There's lot of legroom for front seat passengers, not so much for those in the rear seat. Armrests are positioned so you can use them without contortions.

The 2.4 Jag is a solid approach to the philosophy of furnishing a quality product with no particular gimmicks to excite the flames of desire in the heart of the aficionado. Price-wise, it competes with cars like the Olds 98 and Buick Roadmaster, which may or may not be in its favor, depending on how much metal you want for your dollar. The 2.4 gives a great deal of craftsmanship per pound per dollar.

Lines of new car were immediate success with all who saw it because of cleanness, 57½-inch height. Designers refused to follow trend to small tires

31

THE FIRST 2,000

1,000 on the clock . . . the Jaguar poses for its picture on Rannoch Moor.

By Pamela Smith

Lunch from a well-packed boot. The "larder" door opens downwards so that food can be packed upright on the shelves.

"WHAT, *that?*" I used to wail in despair, peering after a small, neat silhouette disappearing rapidly round a corner. "But there won't be *room* . . ."

"Bigger than it looks," John would say, dreamy-eyed, and I let the subject drop. Motorists' wives are a resigned tribe, and at least I'm not expected to go to race meetings. Yet. "As long as it doesn't reek of petrol," I said, "That's all I ask."

I still wondered how my husband—ever the bright optimist when someone else does the donkey work—proposed to pack our usual loads of tents, camping gear, food and passengers, plus varying consignments of lighting equipment, suitcases, large colour cameras, etc., into that sleek, smart, unquestionably debonair little thing which was pointed out to me as a 2.4. We'd gone through a nasty phase when he'd hankered after something called a Gran Turismo—so having escaped with a boot and four proper seats I was inclined to count my blessings. There had also been much crony discussion about chassis v. integral construction, and a shocking litter of leaflets, books and technical specifications all over the house. Eventually, he decided to move with the times, and announced that integral construction and winking indicators were apparently here to stay . . . nevertheless, it was surprising how few cars measured up to our requirements—high performance and reliability, neat and light but with four seats and room for that mountain of gear which dogs me all over Europe.

Other people have a leisurely time running in a new car. You can see them any Sunday afternoon, toddling along at a steady 25, off to visit Auntie Mabel and Uncle George in Croydon, sporting a "running in, please pass" notice on the back to show that they really know what they are at. Our Jag arrived on Monday. "Put your hat on," said John, that evening, "We've got 500 miles to do before dawn on Friday." I know the angles on this running in business by now, so I said quickly that there wouldn't be time to cook dinner, and we'd better eat out . . .

There's always *some* reason why the first 500 miles have to be done in a hurry—this time it was a trip to Scotland which had long been

Sunset over the mountains of Sutherland.

THE FIRST 2,000

booked for the following week. On each of those intervening nights—and they were all wet—I would find myself round midnight at Aylesbury or Dunstable or some such place, nodding sleepily on the homeward run while the road glistened with rain and the windscreen streamed . . . only half awake, it occurred to me that the Jaguar was aptly named—it rode through the blackness like a great purring cat, and I dozed right off and dreamed that it was snarling because it had run out of petrol. I woke up in the rain outside the garage with another hundred on the clock and my husband enthusing about the road holding on wet roads in spite of the lack of weight on the back wheels . . .

First Impressions

We were both impressed by the interior finish of the car, and the boot at first inspection looked promising, but I was biding my time over that. John crowed happily over the tool kit, laid out like the Crown Jewels in a flat cushioned case kept under the spare wheel cover, although I could visualize some bitter feelings if it had to be extricated from a laden boot in the rain.

It says much for the attractions of Aylesbury, Tring, etc., that we had 498 on the clock by Friday morning, and the car went in for its first service, together with a note of faults, consisting of difficulty in engaging top gear, sticky throttle linkage, and the awkward positioning of the wiper switch, a small round switch on the far side of the dashboard and difficult to turn. John had this changed to a large lighting switch, much easier to locate and turn. The throttle linkage was oiled and adjusted, and the garage forecast that the top gear engagement would loosen up in the next thousand miles.

On Tuesday night we loaded the car for its real try-out, the run to Scotland. We packed in the dark, and were grateful for the built-in light in the boot. I was ready for trouble. As soon as anyone begins packing a car, it realizes what is going on and does an Alice in Wonderland trick. It shrinks. Mood on these occasions habitually starts with calm, builds up to optimism, deteriorates rapidly and ends with despair while we both cram things into odd corners, furtively encroaching on each other's territory in the hope of not being found out until it's too late. We had cut down as much as possible to accommodate a passenger, but nevertheless I was agreeably surprised to see the ease with which this compact car swallowed a load comprising two tents, three double sleeping bags, air beds, cooking stove, caravan water carrier, rucksacks, bags, three stores boxes, and a loose assortment of rugs, cushions, books, boots, cameras, coats and so on. And there was still an adequate space to pop our passenger in next morning. . . . Main grumble, in fact, was the glove pocket, as twee a cranny as ever I saw, and the dim interior lighting which needs stepping up with a front light for map reading.

Our companions on these trips are selected from a very short list—one requirement is that he or she shall be an incurable optimist. Ferdi, gazing serenely at the rain as we headed off at 8.15 next morning, remarked that she was sure we would have fine weather, and m'mmm . . . wasn't the car smart? We beamed upon her fondly.

John was anxious to avoid the first bit of A1, where traffic is often heavy by day and one runs the risk of languishing for hours behind the Biggest Boiler in the World, so we wiggled up through Aylesbury and Leighton Buzzard on to A50. This road, however, seemed nearly as lorry laden as A1, and the stench of Diesel fumes was appalling. We rejoined A1 at Bawtry from the new road 6097, a three-lane road where we had noted with approval the surfacing of the centre lane in red, a good psychological trick often used on the Continent.

Doncaster being jammed with race crowds, we abandoned our plan to lunch there and ate bread and cheese from the stores instead. The afternoon passed uneventfully up to Scotch Corner, across the Pennines and over the Border via Carlisle. Knowing all too well the dangers of leaving dinner too late in Great Britain, I got busy with the map at six o'clock to see where we could feed. John's favourite reading when he is depressed is the French "Guide Michelin." It makes him feel better just to know that all that wonderful food really does exist for the motorist, even if it's on the other side of the Channel. However, after totting up the miles I told him that if he kept it steady we could make Dinwoodie Lodge for seven. After that, we zoomed onward with a purposeful air. Dinwoodie Lodge, just past Lockerbie on A74, seems to maintain a very good standard, and we enjoyed an excellent meal. Sentiments towards the car at the end of the day were very favourable. Seating, both for driver and passengers, was extremely comfortable. No broken spines, and *no* petrol fumes. We drove on for a few hours after dinner through a clear twilight with a half moon rising, then through a starry night past Stirling and Callander. At 11, we stopped by Loch Lubnaig, put up the small tent by torchlight, fell into it and snored for the rest of the night.

By a happy coincidence, the Jag's 1000th mile came up on the clock as we took the sweeping bend on the climb up Black Mount, and soon we were on the superb straight stretch of road over Rannoch Moor. On this stretch, it is easy to get up to 100 m.p.h., but with a regretful sigh John kept to a careful 60-70, still keeping 500 revs in hand ... More cunning timing had us at Fort William for breakfast at the Hotel Alexandra, and then away through heavy rain into the single track country past Invergarry, where motoring really starts ... Arrogantly used to having this part of the world to ourselves, we were surprised to find lots of other people about, and we sorted ourselves out in the passing bays all the way up through Glengarry and Kintail. Not being burdened by hotel bookings, we took ourselves off up a side turning and camped for a night on the coast, in a deserted little bay with only a few cows and fishing boats for company. We used the little tent which is quickly erected and ideal for one-night camps. Next day we made for Strome Ferry and the North-West.

Ferry Wait

At Ballachulish Ferry, we know that it is quicker to go round the loch if there are a dozen cars in the queue, but it's a long way round from Strome via Inverness ... we sat for an hour watching vehicles chugging over two by two and wished we hadn't wasted time on a leisurely breakfast. Once over the ferry, the traffic thinned perceptibly and the sun came out. Here in the north west, the land is a curious mixture of lush vegetation and barren, rolling hills. We stopped for a picnic lunch, and when we were packing up again the boot catch jammed. Interesting sight, husband unpacking the boot to get that natty tool kit out, with a big rain cloud whipping smartly towards us. . . .

"That genuine charcoal flavour..." You can burn toast twice as quickly over a wood fire.

Calm morning at the camp site overlooking Loch Osgaig.

The wild, rocky coastline of Enard Bay.

THE FIRST 2,000

John had in mind to find a spot which would enable us to explore the coast and also climb the odd mountain or two if he could talk the women into it, and we turned off A835 and drove along the shore of Loch Lurgainn by the side of Stac Polly. We began to notice the prevalence of women drivers in this part of the world—they certainly outnumbered the men and ranged from a pretty girl who expertly backed a Rover some fifty yards up a hill to a passing point for us, to spunky elderly ladies touring in parties. These not unnaturally preferred to sit tight and let us do the scraping past with half an inch to spare. "I bet this poor car didn't know what it was in for," I murmured, wondering where my stomach had gone on a bend. John said something about bringing it up the right way, and this being better for it than a lot of cold starts in town. I yelled out news of a good camping site, and an hour later we were comfortably installed in the big tent.

Heavy weather came up, and it rained and blew all that night. We were glad of the big, comfy tent with the flysheet which will withstand continual rain for days. By morning, the weather had blown itself out and the loch was calm and still. It stayed that way for four days. We climbed Stac Polly in clear sunshine and enjoyed the superb panorama of hills and islands stretching out to the coast. We walked on the shore among the gaunt cliffs and unexpected sandy bays. We found a deserted village of ruined "black houses," uncannily still and silent apart from the waves and the sea birds. . . . We drove round some of the coast roads, and John found the Jaguar's lightness of steering good for the narrow twists and bends. From a passenger's angle, the ride was very smooth over the bumps, too—John murmured something about cantilever half elliptics, but I just said it was comfortable. He had been using 100 octane petrol as far as Fort William but there was not much available farther north, so he stuck to premium. This produced a little pinking, but was not too bad apart from the last load which started us on the road home. We have had many debates about that petrol. John affirms that it must have been either neat paraffin or hilly-billy hooch. At any rate, the Jaguar spat disgustedly all the way to Perth.

We took the quicker route back down A9 across the Cairngorms, rejoined our previous road and camped the final night near Lockerbie. The last day of a journey back from the wide open spaces is never very pleasant . . . the passengers dozed and the driver adjusted himself to a return to city life by considering his list of points needing attention on the car . . . brakes to be checked to cut out an occasional squeal at low speeds, an occasional irritating noise from the rear suspension, and things definitely still not quite right in the gear box. Apart from that, it was coming along very well indeed. With something like 1,700 on the clock, the engine was beginning to loosen up noticeably. It was most impressive to watch the way the rev counter flicked round with the surge of acceleration.

As we headed back through Yorkshire on this last day towards London, one ever-recurring problem presented itself—can a grimy party in camping clothes have the nerve to go to a good spot for lunch, even if it's really hungry? I sprayed John with lavender water and sent him in first. "I'm a bona fide traveller," he told the head waiter at Punch's Hotel, Doncaster, "Can we have some lunch?" They were very nice—they didn't even raise their eyebrows. One day we will go back looking very respectable, with me wearing my best dress and my pearls. . . . On this occasion, only one of us could afford to look complacent. After ten days of roughing it with all that gear, the car was by far the most presentable member of the party.

Jaguar 3.4

TESTING A SPORTS SEDAN WITH A HAPPY COMPROMISE

AN MT RESEARCH REPORT by Otto Zipper

THE 3.4 JAGUAR four-door sedan is a welcome import. It blends the luxuries and spaciousness of a full sedan with a real sports car feeling. This happy combination is not exactly new, but price-wise and performance-wise it's about the nearest thing extant to what a true sports sedan should be.

The 3.4 (an appropriate name since it's powered by the 3.4-liter XK engine) is externally identical to the 2.4 Jaguar introduced to the U.S. last year. But the one-liter (61 cubic inches) increase in the engine makes a vast difference in its appeal to the U.S. buyer. Not that it seriously challenges, in actual size or horsepower, the newer domestic products, but its performance is definitely pleasing, and our advice to "hot" car owners who have previously dusted off the 2.4 is — look before you tromp. The 3.4 can give a number of domestic '57 products a very bad time. Actual times show 0 to 60 mph in 10.7 seconds and the standing ¼-mile in 17.7 seconds at 78 mph. That's quite commendable for a relatively small engine of 210 cubic inches pulling a four-door sedan weighing 3280 pounds. Top speed is well over 110 mph.

The double overhead cam, six-cylinder engine, proven over many years, has been so successfully refined that it is remarkably

PHOTOS BY BOB D'OLIVO

THE ADDITION of 61 cubic inches is evident here, the 3.4 engine completely filling underhood space. Double overhead cam 6 gives over 110 mph speed.

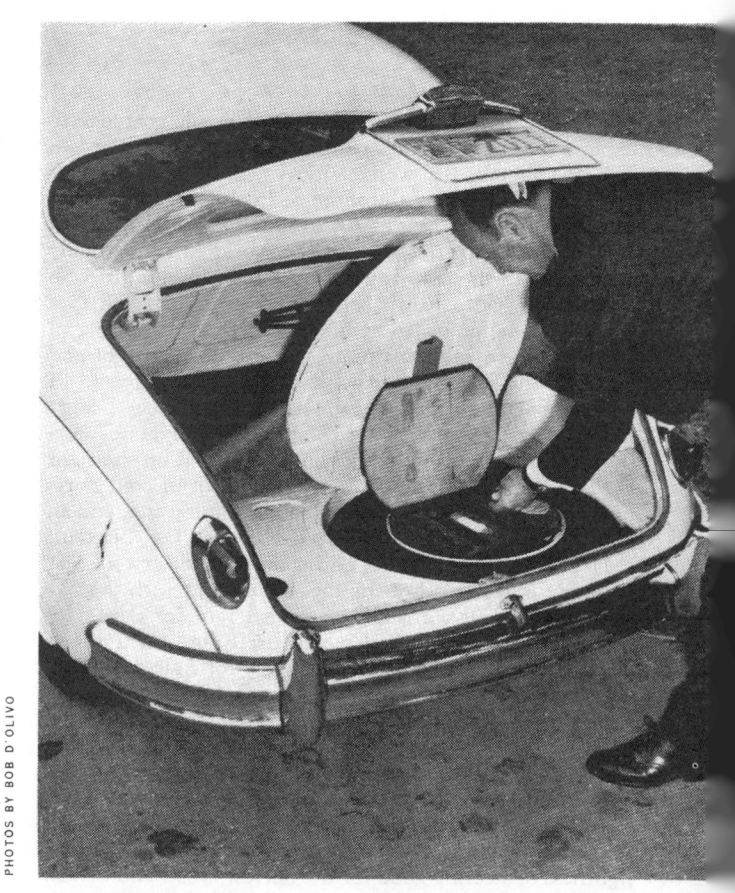

A DELUXE TOUCH is added with this easily accessible tool kit, built into center of spare tire. This well-designed feature allows full use of trunk compartment for luggage.

smooth and quiet. It's the same engine as used in the Mark VIII, but has an 8 to 1 compression ratio, which ups the power to 210 hp at 5500 rpm. Minor cam and port modifications have raised the torque in the 3.4 engine to 216 pounds-feet from the Mark VIII's 203. This increase is felt mostly in the low speed range, resulting in very pliable high gear driving. Even with the overdrive engaged, flexibility is good from 40 mph up. The optional Laycock-de Normanville overdrive operates only in fourth gear, with engagement being made simply by flipping a toggle switch conveniently located on the dash a finger's length away from your left hand on the steering wheel.

Normal shifting of the floor-mounted gearshift is pleasant, except for one point that could stand improvement — the lever needs a more positive stop for the entrance into the reverse gate. Shifting up is no problem, but a down-shift from third to second might find you in reverse gate instead.

An automatic Borg-Warner torque converter transmission is available on the 3.4 for an additional $250 over the f.o.b. Los Angeles price of $4445. A bench type front seat is standard installation with the automatic shift. Actually, about everything

COCKPIT VIEW shows convenient, uncluttered instrument panel set in center of dashboard, tachometer on left, speedometer on right.

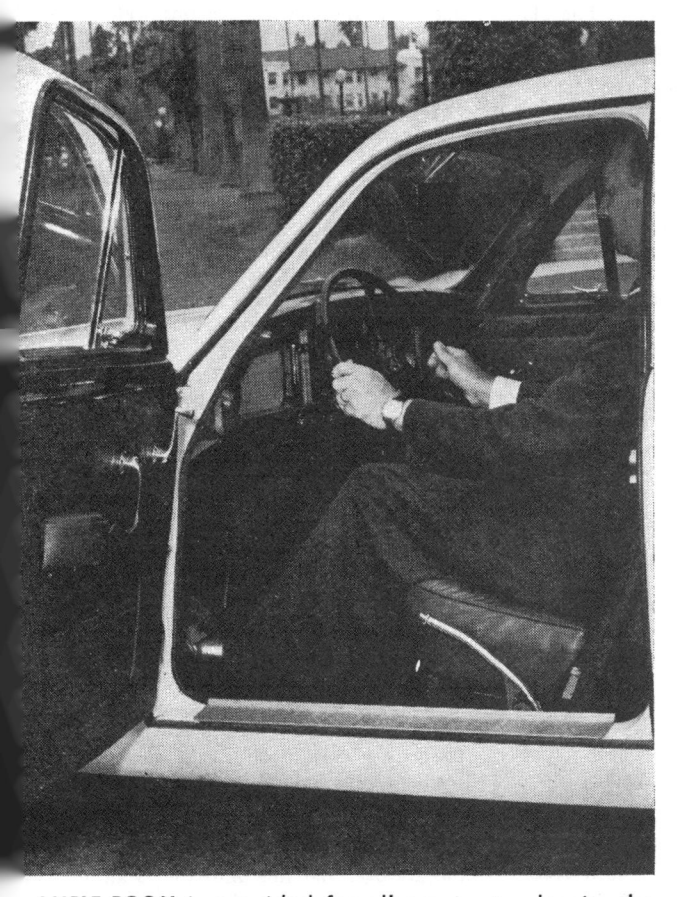

AMPLE ROOM is provided for all occupants, despite the Jaguar's small dimensions. The driver does not have to be a contortionist to get in or out of it gracefully.

LONG-LEGGED rear passengers may wish for a little more room, but the leather covered foam rubber seats are most comfortable. Arm rests are on all four doors.

Jaguar 3.4

found on the car is standard. At the base price the following "accessories" are included: overdrive, heater, defrosters, dual exhausts, directional signals, and power brakes.

The car has, as a matter of fact, a large number of desirable features. Take the dimensions, for instance. It's an inch shorter than a Thunderbird, has an inch less wheelbase than a Rambler, is only four inches wider than a Metropolitan, stands only as high as the new Chrysler line, yet amply accommodates five persons and has 13.5 cubic feet of luggage space. Vision is splendid in all directions. You can safely pass on narrow streets or park at an angle to the curb without having your derriere jut too far into the street. You can even parallel park in one space! And a U-turn is easy on the narrowest of streets.

Workmanship, as could be expected, is superb. The polished walnut of the instrument panel and window framings is beautiful. One should take care, however, to keep the wood conditioned to prevent cracking. The glove leather upholstery and heavy floor carpeting are in conservatively excellent taste, and the detailing shows the imprint of fine craftsmen. Map pockets and arm rests are in all four doors and there are two glove compartments, the driver's being doorless. Windows and doors are so well fitted there are no rattles or wind noise.

The instrument panel is located dead center, rev counter being left, speedometer, right. Jaguar Cars Ltd. shows proper concern for the limitations of their engine, but the driver will have to do some neck stretching to be sure he's conforming to the speed limits.

The seating position is very comfortable, contained and an asset to driving efficiency. It hardly seems to matter what size or shape you may be; the leather covered, foam rubber seats almost mold themselves to you.

As might be deduced from our opening paragraph, handling comes close to that of a sports car. While it is true that lean is apparent in sharp turns at speed, accompanied by tire squeal, the 3.4 still very definitely shows its sports heritage. It must be remembered that a comfortably riding four-door sedan cannot handle as well as an all-out sports car. The steering is precise, fairly quick, and needs no correcting action at speed. If the wheel is whipped from side to side, recovery is instantaneous. The ride itself is soft but firm and extremely comfortable.

There are many would-be sports car drivers who, for personal or business reasons, cannot own an out-and-out sports car. The Jaguar 3.4 should become a happy solution. Wives or customers who might balk at a two-seater, close-fitting roadster will in all likelihood comment favorably on the Jag. It has the satisfying performance and handling characteristics of the real article, yet it also has four doors, ample head, leg and luggage space. It's perfectly easy to get in and out of and should be impressive enough for anyone.

Specifications

ENGINE: Double overhead cam 6. Bore 3.27 in. Stroke 4.17 in. Stroke/bore ratio 1.28:1. Compression ratio 8:1. Displacement 210 cu. in. Advertised bhp 210 @ 5500 rpm. Bhp per cu. in. 1.0. Piston speed @ max. bhp 3830 ft. per min.

TRANSMISSION: 4 forward speeds, all synchronized. Overall ratios: 11.2, 6.56, 4.55, 3.77. Rear axle ratio 3.77:1. 10-inch dry plate clutch Borg-Warner automatic transmission. Rear axle ratio 3.54:1.

CHASSIS: Integral body chassis construction. Independent front suspension with semi-trailing wishbones, coil springs, tube shocks. Trailing link rear suspension by semi-elliptic rear springs, tube shocks. 6.40 x 15 tires. Servo-assist hydraulic brakes. Ball-type steering gear, with 33.5-ft. turning circle, 3.5 turns lock-to-lock.

DIMENSIONS: Wheelbase 107 in., overall length 180 in., overall height 57.5 in., overall width 66.7 in., ground clearance 7 in., front tread 54.6 in., rear tread 50 in., weight 2700 lbs.

PERFORMANCE: Max. speed 110+ mph. Acceleration: from standing start to 45 mph 7.2 secs., to 60 10.7 secs., 1/4-mile 17.7 secs. and 78 mph, 30-50 mph 4.1 secs., 45-60 3.6 secs., 50-80 10.1 secs.

PRICES (F.O.B. port of entry): $4530 for automatic transmission model, $4445 for overdrive model.

2.4 Jag

(Continued from page 17)

rear window looks narrow at a glance, but its generous width tells the whole story in the mirror, and the only snag in visibility is the old-fashioned width of all the body pillars.

Interior ventilation is easily regulated and draft-free, American-fashion, and the window winders have a ratio of one turn, top to bottom. Other fresh air can be scooped in at the cowl and either let in directly or sent through the heater, which has two outlets under the dash. The manual recommends the use of the heater blower under all circumstances, but forward ram action seems to shove air in pretty well when under way.

Interior room for the family is matched by the capacity of the trunk, which is cleanly laid out with the major jacking tools clipped well forward. The spare drops into a covered depression in the floor, next to the gas tank, and in its hub is stored a nicely fitted tool kit. That jack, by the way, is easy but infinitely boring to operate, since it's designed around an extremely shallow screw thread. Two Dzus fasteners under each rear door liberate the fender skirts.

An imaginative venture into integral construction yielded dividends to Jaguar in the form of a stiff, solid structure with minimum weight, and the few chances taken here were counterbalanced by the use of a much-reworked version of the well-tried series of six-cylinder engine. The old XK100 four was considered in the planning stages, but they wisely decided that the six would come closer to Jaguar standards of silence and smoothness. That the choice was good is reflected in SCI's upcoming Tech Report on this ruggedly precise powerplant, which reveals many of its potentialities even in humble 2.4 tune. Proper use of the Start control brings it to life instantly from cold, and the warm-up idle is smoothly inaudible. Accelerator pumps in the Solexes make up for the plumber's nightmare of manifolding to provide the instant throttle response that's expected in a Jag, and this rev-ready eagerness stays in there all the way to the top making you wish there weren't a peg on the tach. Within the visible range of figures there's no shadow of protest or vibration, and the red sector at 5500 is more for decoration than anything else. Peak power, after all, comes in at 5720!

We expected a hood full of revs, but we weren't prepared for unusually good low-speed torque for a 150-cubic-incher. A glance at the torque curve and its 2000 rpm peak told the story, though the 2.4's road behavior was more than convincing enough. It does like to wind, though, and it's a shame that the gearbox occasionally dampens its ardor. The cover of the basically standard Jaguar box has been redesigned to give a mechanically more direct lever control, which somehow manages to have a very vague and rubbery feel. It improves on acquaintance, but the pattern remains widespread and the knob still evades the wildly groping hand at crucial moments.

Jaguar's machinelike whine remains in the lower ratios, and is accompanied by slight dog clutch protest if the synchromesh is rushed at all. The indirect gearing is closer to that of the first XK's than it is to the close ratios used in the present 3.5 liter machines, and as a result you're all done a little sooner in the gears than you would like to be to use that engine to the full.

When you get on good terms with the gear train, the 2.4 responds with strong and steady acceleration that will carry it quickly to an easy cruising speed of 80 or so, which is still within the common piston speed limit of 2500 feet per minute. In its present trim, the 2.4 Jag is a nice balance between the paired factors of power and roadholding. The word now seems to be that the big 3.5 engine will find its way under this hood for the American market, which will dump 51 more pounds where they aren't needed and apply more power when it can't be fully used. We'd much rather see this fascinating short-stroke six developed further, with perhaps an optional C-Type head and close-ratio gearbox if more suds are demanded. Those plus stiffer shocks all around would push the 2.4 over the line into the Gran Turismo class and enable it to surprise many a sport car. Even now it's one of the most satisfying small sedans around. — K.E.L.

ROAD IMPRESSIONS

MIKE HAWTHORN'S JAGUAR 3.4

AS World Champion driver, with a prosperous motor business behind him, Mike Hawthorn has the pick of the world's cars for his personal transport. He is British concessionaire for Ferrari—and might well be expected to use a 250 GT in view of his connections with the Italian firm—and the fact that he chooses to run a Jaguar 3.4 is a tribute to British motor engineering, and in particular to the firm with which Hawthorn has enjoyed some of his most spectacular successes in sports car racing.

In addition to using his 3.4 for long distance journeys, both in the British Isles and on the Continent, Mike has raced it at Silverstone—at the Daily Express International Trophy meetings—in 1957 and 1958. Now the 3.4 is a very quick car in standard form, but the performance of VDU 881 on these occasions suggested that some ameliorative treatment might have been carried out at the Tourist Trophy garage. During conversation with Mike Hawthorn, round about Motor Show time, it was arranged to borrow the 3.4 for a few days to form some impressions of its performance and handling. I had previously gathered, from talking to "Lofty" England that the car was "virtually standard," and indeed closer inspection revealed that there were only two SU carburetters under the bonnet. The fact that these are 2 inch instruments prompted a non-motorist friend to enquire into the purpose of "those oil drum things alongside the engine."

To take advantage of the general availability of 100 octane petrol, Mike's car also has high compression pistons, giving a ratio of 9 to 1, while the exhaust system, with its twin tail pipes, was specially made at the Tourist Trophy garage.

To allow utilisation of the full power output of the modified engine a competition

Engine compartment of the Hawthorn Jaguar.

VDU 881, driven by Hawthorn, leads Tommy Sopwith's Jaguar at Silverstone. **3.4**-owners who would like to have their cars modified to the same standard should get in touch with the Tourist Trophy Garage, Farnham.

clutch is used, and this accommodates full throttle upward gear changes without a trace of slip; it is also remarkably free of vice in heavy London traffic, thanks to the wonderful low speed flexibility of the Jaguar engine, and copes admirably with a series of traffic-light starts in second gear. To the interior of the 3.4, which is obviously designed—except for the location of the tachometer at the far end of the dashboard—for the keen driver, Mike has made no change other than fitting an accurate 160 mph speedometer; he grew tired of going "off the clock" with the standard one.

A most important modification, one which many people would not even notice, is the fitting of special rear wheels, which increase the rear track by two inches and have a very considerable effect on the car's roadholding. In conjunction with this the suspension has been stiffened up—by the use of harder front springs, an extra leaf in the rear springs and competition shock absorbers all round—and although some body roll is still experienced during vigorous cornering it is quite predictable and does not have any undue influence on the car's handling characteristics.

To accommodate the increased track the rear wheel spats have been restyled—a personal touch whereby Mike can be sure of always finding his own car, even after a Jaguar Drivers' Club party.

The only other major departure from standard on the Hawthorn Jaguar is the fitting of a crown wheel and pinion which give a final drive ratio of 4.05 to 1 (standard is 3.77 to 1). Mike's reason for this is simple; he doesn't like being out-accelerated by "Kraut cars", referring to a well-known species with an unusual method of door opening.

I collected the Hawthorn Jaguar on a December Saturday afternoon. Unfortunately, to reach my destination I had to travel almost the whole length of the tortuous, ill-defined country lane known as A25, and in the prevalent traffic conditions there was little point in having a high performance motor car. Subsequently an almost complete *impasse* at Strood necessitated a 25-mile diversion before anything like the full performance of the 3.4 could be used in safety—truly, the frustrations of the fast car owner are almost greater than those of the impecunious enthusiast who hasn't a car at all.

When it is given a chance, however, the Hawthorn Jaguar really goes! The low axle ratio gives it tremendous acceleration, especially in second gear, and instead of the forward surge abating as frontal area asserts itself the car hurtles onwards in third and top gears until, at around 5000 rpm in top, a flick of the overdrive control "switches off" a thousand rpm and settles the car into its best, long-striding cruising gait. At this pace the Jaguar is acceptably quiet to its occupants; wind noise is kept at a low level and the engine gives no indication of stress.

Christmas traffic, patches of mist, and an unpleasantly wet road made the compilation of performance figures too hazardous to be worthwhile. Maximum speed is something in excess of 120 mph (Mike has seen over 130 mph), but at this velocity the 3.4. became a little unstable, due, perhaps, to the undulating road surface.

For the rest, Hawthorn's Jaguar is more or less like any other 3.4. The driving position is good, with the pedals designed to allow simultaneous actuation of brake and accelerator for downward gearchanges. The gearchange itself is very good by saloon car standards and the minor controls and instruments give the interior a prosperous, well-bred atmosphere. The lights-switch deserves a special mention for the pleasant and positive way in which it works.

In respect of roadholding, the 3.4 can be driven round corners extremely fast (vide the Silverstone photograph above) but it really needs a Hawthorn to get the best out of it. For ordinary mortals the combination of a wet road, spinning rear wheels and a sliding tail can become somewhat overawing, particularly as the relatively low-geared steering calls for rather a lot of wheel twirling when correction is required. Nevertheless, the tail of the Jaguar can be made to "go" a long way before one gets the feeling that all is lost, and for normal fast road use the car is obviously extremely safe.

The brakes—servo-assisted Dunlop discs all round with special Mintex pads—are absolutely superb, and slow the car in a straight line time after time; only light pressure is required and the pedal has a firm and responsive feel which greatly promotes confidence, particularly at night when, despite the excellent Marchal headlamps (another Hawthorn extra) and Raydyot spot and fog lamps, there is a definite limit to visibility, and corners and other hazards loom up alarmingly quickly.

Spacious and comfortable, with ample leg- and head-room in the rear seats, and a very reasonable luggage compartment at the rear, Mike Hawthorn's personal Jaguar can be summed up as a sports/racing limousine. Somehow I don't expect Mike to have much trouble with that little bubble-car with the irritating horn—even if the driver does manage to get it out of second gear! D. P.

1958 CARS

IN view of the introduction of three new Jaguar types since the beginning of this year, it will come as no surprise that, apart from one small detail point, all Jaguar models now current will go forward into the 1958 season. That, however, does not imply even a temporary lull in progress because the manufacturers have taken the notable step of offering disc brakes as optional equipment on both the 2.4 and 3.4-litre models.

In addition, Borg-Warner automatic transmission has been made available on the 2.4-litre car so that, apart from the two sports-racing models (the export-only XKSS and the D-type) all Jaguar models can now be obtained with the choice of three transmission variants —synchromesh gearbox, synchromesh gearbox with Laycock de Normanville overdrive, or Borg-Warner fully automatic transmission.

The small detail change mentioned above concerns the front grille on the 2.4-litre model which is now of the type introduced for the 3.4-litre car. It is 2½ in. wider, with 16 narrow vertical bars in place of the 8 wider bars used previously. From the front, these two models are now indistinguishable. Indeed, a keen eye is needed to tell them apart from any angle as the only difference from a side view lies in the wheel spats, which are of the cut-away type on the 3.4 model (and on the 2.4 as well when disc brakes and wire wheels are fitted); from the rear, the only clue is the dual exhaust system on the larger-engined model.

The Dunlop disc brakes now offered on these models are identical with those already fitted as standard to the XK150 introduced last May and furnish yet another example of how racing can, and does, improve the ordinary car. The Jaguar company was the first to adopt this type of brake for sports-car racing and, having immediately proved their worth at Le Mans and elsewhere, subsequently adopted them on the D-type, which was the first car in series production to be so equipped.

On the 2.4 and 3.4-litre models they are supplied only in conjunction with wire wheels with knock-on hubs, and cannot be fitted with the standard pressed-steel wheels. The extra cost, including purchase tax in each case, is £36 15s. for the brakes and £52 10s. for the wheels, making a total of £99 5s.

As on the XK150, the discs are of 12 in. diameter front and rear and in each case the callipers carry a single pair of opposed pads, each 2¼ in. in diameter, giving a total lining area of 31.8 sq. in. (In passing, and for the benefit of readers not *au fait* with this type of brake, it should be stressed that this figure is not in any way comparable with the lining area desirable on drum brakes, which is very much higher.) Appropriate distribution of braking effect as between front and rear is obtained by suitable disposition of the pads in relation to the hub centre, the wiped circle of the disc being of greater diameter at the front.

In both cases, the pads are disposed behind the hubs whilst, at the rear, an additional pair of manually-operated pads is provided for the hand brake, which is of the standard pull-up type mounted at floor level to the right of the driver's seat. The main pads are, of course, hydraulically-operated and are self-adjusting, whilst

NEXT YEAR'S

The 2.4-litre saloon is now to be fitted with this wider radiator grille, as used on the 3.4, and the two cars are thus virtually indistinguishable externally. Borg-Warner automatic transmission is now optionally available on the 2.4, controlled by the facia panel quadrant seen below.

operation is assisted by a Lockheed servo of the suspended-vacuum type.

The other new option on the 2.4-litre model, the Borg-Warner automatic transmission system, is identical with that already available on the 3.4-litre car, including the Jaguar-designed selector lever on the base of the instrument panel and the special switch which enables the transmission to be maintained in the "intermediate" range when desired. A lower rear axle ratio of 4.27:1 is used to suit the output characteristics of the smaller

When disc brakes are fitted, wire wheels are obligatory, as seen in this picture of a 3.4 model. With wire wheels both 2.4 and 3.4 have the cutaway rear wheel spats normally fitted only to the 3.4.

JAGUARS

Both 2.4 and 3.4 saloons can now be supplied with disc brakes fitted to all wheels. This drawing shows the Dunlop 12-in. disc with callipers carrying a single pair of pads; inset is the rear layout, with separate handbrake linkage.

Disc Brakes and Wire Wheels Optional on 2.4 and 3.4-litre Models. Automatic Transmission Now Available on All Except Sports-Racing Types

use on a high-performance British car of unitary construction of body and chassis. Particularly notable, too, are its cantilever rear springs which relieve the rear of the body structure of high loadings. The engine is a 2,483 c.c. short-stroke edition of the 3½-litre six-cylinder, twin o.h. camshaft Jaguar engine used in the larger models. It develops 112 b.h.p. at 5,750 r.p.m. and, with a car weight (dry) of only 26 cwt. offers exceptional performance for a car in the 2½-litre saloon class.

The 3.4-litre model is a direct development, being, in fact, a larger engined edition. Introduced in March this year, it has virtually the same 3,442 c.c. o.h. camshaft engine as the well-known Mark VIII model, developing 210 b.h.p. gross and giving a maximum mean speed, when tested by *The Motor* in April, of 119.8 m.p.h. and a rest to 100 m.p.h. acceleration figure of 30.3 sec.—and all with an old-gentleman standard of quietness and comfort.

For those who want even more performance, there are the XK150 coupé models (drop-head and hard-top) which come in the sports rather than touring category and have a maximum speed in the neighbourhood of 130 m.p.h. Descended from the original XK120 models, these cars also have the well-known twin o.h.c. Jaguar engine.

Where room and luxury are the prime considerations, there is the Mark VIII saloon which appeared at the Motor Show last year and, with improved equipment and appearance, carries on the tradition set by the highly-successful Mark VII.

Finally, there are the two sports-racing models—the D-type with its glorious Le Mans history and the XKSS which appeared as an export-only model last January and is, in effect, a D-type "civilized" for normal road motoring by the provision of a full-width curved windscreen, folding hood, luggage grid and bumpers.

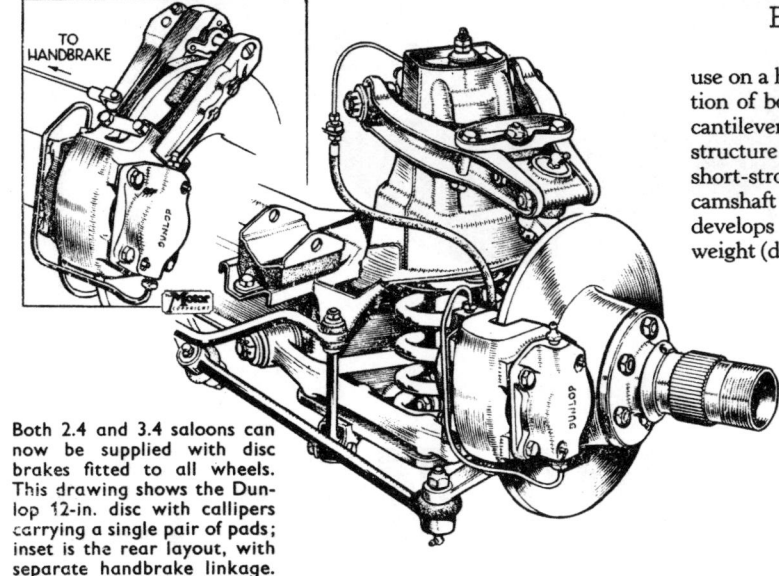

engine, giving overall ratios (including torque converter effects) of: Top, 4.27:1; intermediate range, 6.14:1 to 13.2:1; and low, 9.86:1 to 21.2:1. The extra price, including purchase tax, is £180.

So much for the features which are new for 1958. To put them in perspective, here is a brief review of the complete Jaguar range, which is unique in the fact that every model is capable of exceeding 100 m.p.h. in spite of the comfort, docility and general refinement in both performance and finish of the saloon models.

Smallest in the range is the 2.4-litre saloon which made its first appearance at the London Motor Show two years ago, when it created enormous interest by reason of a number of unconventional features including the first

Largest car in the Jaguar range is the Mk. VIII, introduced last year as a development of the very popular Mk. VII.

After 7,000 Miles the

Much has been written about the smoothness, tractability and dynamic performance of the graceful 2.4 litre Jaguar. Ever wondered what one'd be like after living through a year's hard driving?

IN Adelaide recently, "Wheels" was offered the opportunity of testing briefly a 2.4 litre Jaguar owned by Murray Jonas, and currently (at time of writing) in the hands of Grand Garage Ltd., for sale.

Lest it be thought that Murray was getting rid of his Jag because he didn't particularly like it, let us hasten to point out that he already has another, new one — as well as an Austin-Healey, in which he does most of his starker style motoring!

The green Jaguar in question had done an extremely hard 7,000 odd miles, much of it in competition work and we were interested to discover where, if at all, the car was beginning to show its age. Are these cars as durable as they appear . . . or will a few thousand miles' punishment quickly skim the cream off them? We thought we'd like to test one and find out.

Rodger Yates, from the garage, came along with us for the ride; mainly to show us the way around a few lesser-known by-roads where the Jaguar could be unleashed and made to run.

Immediately it was apparent that that smooth, smooth o.h.c. engine had lost none of its silkiness and verve! It was, in fact, difficult to detect the engine's presence when idling, and that "shove in the back" came with undiminished fury following any serious depression of the accelerator — particularly in the intermediate gears.

But, most of all, this small version of the Jaguar is a refined car. All controls feel taut and solidly mounted. The gears still change with that firm, crisp snick; the steering is notably lacking in either slop or indecision. One sits deeply in high, tailored seats, feeling immediately at home and an integral part of the car.

Easy access . . .

Entry is surprisingly easy for such a small, low car (scarcely longer overall than a Hillman Minx): one simply drops straight down into the seat and steps straight up out of it once again, due, we feel, to Jaguar's extremely cunning method of cutting the doors well up into the roofline.

Vision is excellent in all directions, and so silky is the 2.4 that parking it or manoeuvring it in tight spaces is simplicity. The windscreen pillars are thick and substantial, but there is so little interference with vision that one one wonders whether too much has in fact been made of their "elim-

One sits deeply in high, tailored seats; feels at home and part of the car.

2.4 Still Rates Tops!

ination" in other makes.

On the road, the car gave us no disappointment. That perfectly balanced feel was still there; the trace of understeer being just sufficient to permit of a genuine fast drift being induced, should one choose to engage in that sort of thing. Not a rattle marred the car's ghostly silence, which is of such an order that one invariably finds oneself cruising at 70 to 75 m.p.h. without realising that one has even exceeded fifty. Normal fifty mile an hour bends become seventy m.p.h. bends in this car, and there is always power in abundance to get you out of any tight spot you may happen to get into. The power brakes were sure, smooth, and positive; and semi-emergency applications of the pedal from speeds in the vicinity of ninety m.p.h. evoked no mis-

◀

One finds oneself cruising at 70 to 80 without realising that the speed has gone above 50. At all speeds the taut little car behaved delightfully.

behaviour from the Jag.

Off the clock . . .

Revs are on tap at all speeds, and we did on one occasion (this was a non-overdrive car) manage to wind the tachometer needle clean off the dial on one downhill stretch in high gear. The speedo., at this time was showing about 105/107 m.p.h.; and although at least two seconds elapsed after we had taken our foot off before the tach. needle came back on to the dial once again, that beautiful, tractable motor seemed to be not in the least concerned.

And certainly the motor *is* tractable. The car could be — and was — moved off from a standing start in high gear several times with only momentary slipping of the clutch, and with the engine barely ticking over. Top gear acceleration is strong and constant right throughout the range, 30-50 m.p.h. taking 7.5 seconds (5.2 sec. in 3rd); and 40-60 m.p.h., 8.5 seconds (6 sec. in 3rd).

The clutch, however, was showing some signs of its previous hard usage,

and open-throttle changes through the two lower ratios produced some slip — therefore our 0-50 m.p.h. times, averaging 8.0 sec., could probably have been improved on when the car was new. Conversely, the same run, using second gear only, took a bare 9.9 sec., which is not mucking about for a car of any size.

We wished keenly that we did *not* have a 'plane to catch, so that our acquaintance with this truly delightful motor car could have been substantially prolonged — but it was not to be. Final chapter, a 75 m.p.h. rush along straight roads to the airport with Rodger Yates at the wheel, left an impression seldom superseded in any car of our acquaintance of sheer smooth effortless power.

There was only — maybe — one stronger memory.

The pride felt in stepping casually out of "one's" Jag, bidding the "chauffeur" farewell, and walking nonchalantly away towards the aircraft without even a backward glance. ●

43

modern MOTOR ROAD TEST

JAGUAR 3.4

I HAVE just come across an entirely new concept in motoring and I'm still in the clouds; I have road-tested and fallen in love with Coventry's incredible new product—the Jaguar 3.4.

This surely must be Britain's finest car in any class, particularly if measured by that old-fashioned yardstick, "value for money."

The quality the Jaguar people offer the motorist at such relatively low cost has always amazed me. Who else but Jaguar can supply a full five-seater saloon with a body and interior finish that rivals the best of the English coachbuilders, a most capacious boot, a moderate thirst and a performance that will take you from an ambling 30 m.p.h. to a cruising speed of 100 m.p.h. in roughly 25 seconds? I say "cruising" because the engine is turning only at 4500 r.p.m.; you still have another 1000 safe revs to lift the speed to a maximum of more than 120 m.p.h.

All this for £2873 (and until recently the price was £2740, a rise in duty taking care of the additional 130-odd pounds)! Any comparable Continental car—and I can't think of any off-hand—would cost more than twice as much. The Americans don't make this sort of machinery.

My version of a motorist's dream was supplied by Bryson Industries, the N.S.W. distributors. It was ducoed light grey, with dark blue leather seating and trim. The uninitiated might mistake the 3.4 for a 2.4—stationary, of course—for apart from the 3.4's twin pipes, emerging slickly behind the off-rear wheel, and the cutaway in the rear wheel-spats, the appearance of the two cars is identical. The lines of the body are rounded, compact, and somehow convey that here is no ordinary automobile.

Luxury Appointments

The overall height is only 4ft. 9½in. and length barely 15ft.—6in. lower and 18in. shorter than the Mark VIII. Both the 3.4 and the Mark VIII use the famous "Blue Top" engine, of which more later.

The 3.4 is 6½cwt. lighter, and remember the big one is no slouch. Apart from an array of lighting, every item of which I used and found perfect, and the bare-fanged crouching Jaguar mascot, no ornamentation mars the 3.4's contours. No fins, flaps or rudders despite the high speeds at which the car can travel—even when we are led to believe that these deformities really play an important stabilising role!

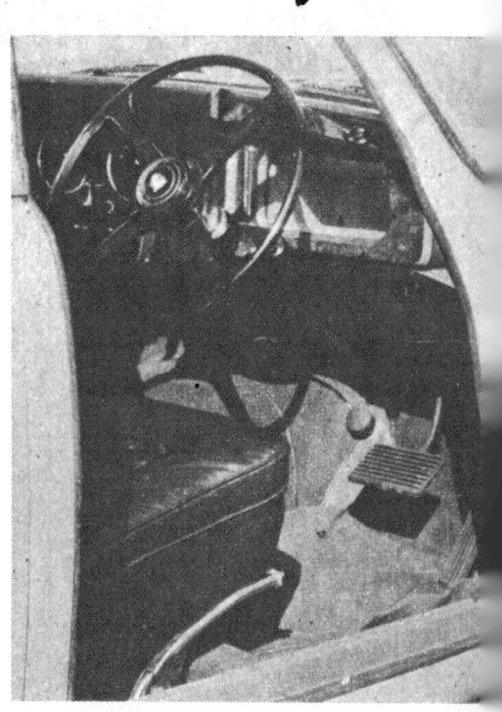

HANDSOME "office"—but it's time they put instruments over the wheel.

An air of luxury pervades the interior. The deep-pile carpets, superb hide upholstery over foam-rubber seating, polished walnut instrument panelling and interior garnishings are extremely tasteful and comfortable. It jarred to find the full instrumentation still centrally placed. Criticising this motoring masterpiece for such an oversight seems like telling your beloved that her new hat is really most unbecoming.

Vision is mostly improved on previous Jags, and so is the driving position. This is the first Jaguar I've honestly felt at home in—the others have given me a faintly submerged and claustrophobic sensation. The plain black four-spoke telescopic wheel is pleasant to handle and not too large, though I thought it tended towards the horizontal a trifle. The "pull-up" handbrake comes readily to the driver's right hand and is most effective.

Enthusiasts' Extras

The test car was fitted with automatic transmission, and this naturally was a disappointment to me. Mr.

MAIN SPECIFICATIONS

ENGINE: 6-cylinder, twin o.h.c. (high-lift cams); bore 83mm., stroke 106mm., capacity 3442 c.c.; compression ratio 8:1; maximum b.h.p. 210 at 5500 r.p.m.; max. torque 215 ft./lb. at 3000; twin SU type HD6 carburettors, twin exhaust system; twin SU electric fuel pumps, 12v. ignition.

TRANSMISSION: Automatic gearbox; rear-axle ratio 3.54:1 or 22 m.p.h. per 1000 r.p.m. (Manual box as above; overdrive model rear-axle ratio 3.77:1 with 2.93 overdrive or 26 m.p.h. per 1000 r.p.m.)

SUSPENSION: Independent in front, by semi-trailing wishbones and coil springs; trailing-link type with cantilever semi-elliptic leaf springs and radius arms at rear; telescopic shock-absorbers all round.

BRAKES: Lockheed servo-assisted self-adjusting hydraulics; lining area 159 sq. in.

WHEELS: Bolt-on discs with Dunlop 6.40 x 15in. Road Speed tyres.

STEERING: Recirculating-ball type with 17in. adjustable wheels; 4¼ turns lock-to-lock; turning circle 36ft.

WEIGHT: As tested, 29cwt.; distribution 58/42 p.c.

FUEL TANK: 12 gallons.

DIMENSIONS: Length 15ft. 1in.; width 5ft. 7in.; height 4ft. 9in.; wheelbase 8ft. 11 3-8in.; track, front 4ft. 6 5-8in., rear 4ft. 2 1-8in.; ground clearance 7in.

PERFORMANCE ON TEST

CONDITIONS: Fine, warm, windy; two occupants; premium fuel.
MAXIMUM SPEED: 120 m.p.h.
STANDING QUARTER-MILE: 18.2s.
MAXIMUM IN GEARS: Low 43 m.p.h.; intermediate 81 m.p.h.
ACCELERATION in "drive" range: 0-30, 4.5s.; 0-40, 6.5s.; 0-50, 9s.; 0-60, 11.5s.; 0-70, 15s.; 0-80, 18.5s.
SPEEDO: Accurate at 60 m.p.h.
CONSUMPTION: 19.5 m.p.g.
AUTOMATIC BOX: Changes up — low to inter., light throttle 12 m.p.h., full throttle 38; inter. to direct, light throttle 24 m.p.h., full throttle 60, kick-down 78.

Changes down — direct to inter., closed throttle 16 m.p.h.; inter. to low, closed throttle 3 m.p.h.; direct to inter., kick-down, below 70 m.p.h.

PRICE: £2873 including tax

HAS THE LOT!

Mark VIII engine and 2.4-type body combine to make this car the pick of the present Jaguar stable, enthuses David McKay

POWER to spare: Mk. VIII engine churns out the full 210 b.h.p., though 3.4 weighs 6½cwt. less than its big brother.

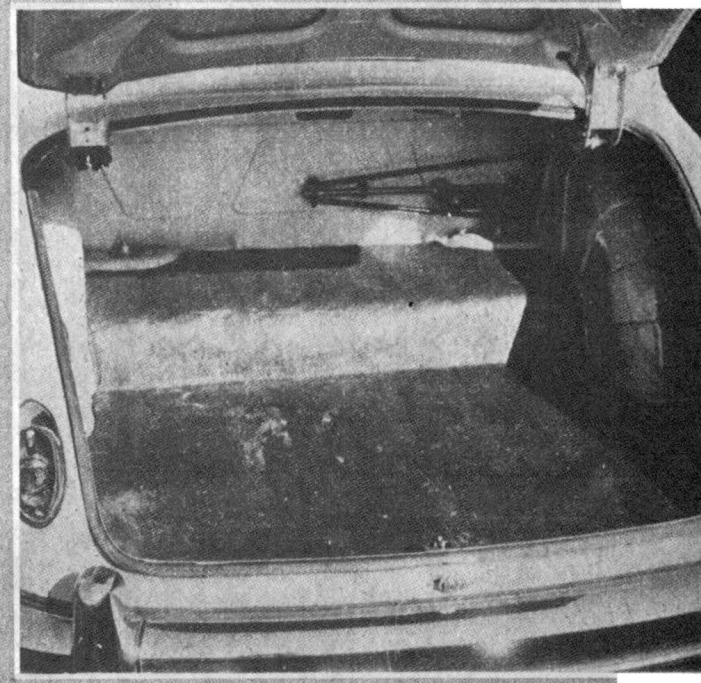

BOOT is roomy, uncluttered, with spare tucked so well out of the way that it's hard to see in photo.

JAGUAR 3.4 HAS THE LOT!

Cook, of Bryson's, tells me orders are about 50/50 for standard and automatic 3.4s, so it's good to know that at least half the owners apparently appreciate the car Jaguar have put into their lucky hands.

Incidentally, a couple of 3.4s have been delivered with disc brakes and wire wheels. Should these ever find their way to a local circuit, the days of Holden and Customline domination will be numbered.

The "mods" available to the enthusiast come under the heading of "performance equipment" and have been developed from the racing Blue Top engine. These include special pistons to raise compression from the normal 8:1 to 9:1, lead-bronze bearings (not advised for general touring, as the greater clearances make the engine run less quietly), a limited-slip differential with a choice of several axle ratios, a competition clutch, higher-geared steering, disc brakes and wire wheels.

The old C-type head is no longer in production, being superseded by the improved B-type unit, which gives the same b.h.p. but better torque in the lower rev ranges. The B-type head has larger valves for better breathing at high revs but retains the narrower Mark VII porting, which gives good gas velocity in the middle and lower revs, and therefore good torque.

The six-cylinder twin-overhead-camshaft 3.4-litre engine must be the best fitted to any production passenger car today. The direct result of many years' racing, the Blue Top version is the logical step up from the old 190 b.h.p. power plant. Its 210 b.h.p. is delivered at 5500 revs and a massive 215 ft./lb. of torque comes in at 3000, or just under 70 m.p.h., giving that terrific urge that usually is associated only with thirsty five-litre V8 engines.

The Jaguar engine is renowned for its long life—Le Mans results over the past few years testify to that—and its ability to run thousands of miles without attention is well known to Australian countrymen. Bill Pitt, driving a Mark VIII in last year's Round Australia Trial, forcefully proved this when he finished hard on the heels of the all-conquering VWs in a contest quite foreign to the marque.

Town and Country

After taking over the car we pottered around to test its docility. Here the automatic transmission was faultless and the small turning circle a wonderful help in making rapid U-turns when navigating some of our new one-way streets the wrong way.

The 4½ turns lock-to-lock was a surprise but proved no hindrance later, though I was busy round Druitt, which I lapped in 2min. 2sec. Also, I did think the throttle pedal needed rather high pressures. Despite the heat of the day, the engine ran cool in traffic and its only sign of impatience was to run-on once or twice.

Away from town we were treated to motoring par excellence — smoothly and in silence we cruised around 80, with an occasional dab of throttle bringing up the ton in a most matter-of-fact manner. Our time from the bottom to the crest of the test hill was 2½min. dead.

The ride was very comfortable, best described as a cross between the better Continental and American styles. The 3.4 has a definite feel of solidity. Its 29cwt., fully fuelled, is apparent—yet it never feels big or cumbersome.

The Lockheed servo-assisted self-adjusting brakes are smooth and powerful, needing only light pressures. I was rather disappointed to find the same lining area of 159 sq. in. on both 2.4 and 3.4 models, despite the latter's increase in weight and the additional 100 b.h.p. (not five, 10 or even 50 additional horses, but 100!).

The manual thoughtfully gives tyre pressures for normal, touring and full-chat speeds, these being respectively 25/22, 31/28, and 34/31 lb., the front pressures being higher to compensate for the 16cwt. carried forward against the 13cwt. aft. The new Dunlop Road Speed tyres are 6.40 x 15in., and while noisy on sharp bends at slow speeds, become more silent as they do their important work in the higher registers.

Having the poor grazier in mind, we subjected the 3.4 to the dust bowl and horror section. Some road shock was transmitted through the steering wheel, but this was remarkably little considering our speed over the holes. The ride was 100 percent and the all-steel body was rattle-free and dustproof—even the boot was dustless, and, believe me, we really put down a dustscreen behind us. So full marks to Jaguar for thinking of Australian conditions.

What, No Low?

Before we got to Kurrajong, the "hold in the intermediate range" switch, which is a feature of this automatic transmission, became inoperative and we were stuck in that range. Without "low," which would have been a help through the hairpins, the 3.4 swept up the hillclimb in 1min. 41sec., which is a good sports-car time.

Descending Kurrajong, three percent fade was recorded on the brake meter. The run home could have been unpleasant in almost any other car stuck in one gear, but intermediate range gives more than 80 m.p.h., so we were hardly inconvenienced. After a burst in maximum revs the box changed up, and on a traffic-free road we experienced the full 210 b.h.p. giving their best in direct drive.

Once among the lesser machinery, we slipped back into intermediate and used that during the night run over the mountain circuit. Handicapped without low, we still averaged an easy 52 m.p.h., and I estimate we could have made fastest time to date with the full range of ratios.

At the end of the test the m.p.g. was checked. The result: 19.5. Normal m.p.g. could be assessed around 23, which, with the smallish tank (12 gallons), doesn't give a long range.

That evening we did another 50 miles over the usual test mileage, though the hour was late—it was pure joy and I hated parting with the car. One compensation, however, is that this has been one of the easiest road tests to write about, and it is pleasant to be able to praise so wholeheartedly an English car. ● ● ●

CAR STORY

The first in a series of case histories of cars bought and owned by the publishers for at least a year or 20,000 miles

The 3·4 Jaguar

Painted in this journal's blue and white colours the 3·4 was a visitor to many European circuits last year. Here it is on the way to Monza for the Italian Grand Prix

THE 3·4 Jaguar is a fine car in many ways, as will be seen from the following history, but it is unique in Britain, and perhaps in the world, in what it offers in terms of performance, accommodation and quality of finish for its initial cost. Acceleration and maximum speed is superior to all but the more exotic sports cars, it accommodates four adults and their luggage for a fortnight in comfort, fittings and finish are in the de-luxe class and yet the British price of the basic car was only £1,672 including purchase tax in 1958. It is now almost £100 less than this.

132 EBH was bought in May 1958 in anticipation of a hard season's motoring on the continent covering motoring events and making other business calls, with steady daily duty in and around London between times. A conventional gearbox was preferred to the automatic version (bucket seats therefore being fitted) and the optional extras of disc brakes, overdrive and Lucas Le Mans headlamps were specified. After some consideration it was decided against wire wheels and against any modifications to the engine such as lead bronze bearings.

The standard medium blue, with navy blue carpets and leather upholstery made up the initial colour scheme, but arrangements were made to have the top and wheels painted white as soon as the car was delivered. This was with the dual purpose of keeping the interior cool and lightening the rear part of the car which was felt to look rather heavy above the rear wheels. Moreover, the colour scheme now matched the colours of Autocourse and Sporting Motorist!

Apart from the items mentioned above, and the later fitting of a Cicca wind horn for continental use, the car was perfectly standard, as it would be bought from any authorised agent, and it has remained thus for the first year of its life.

The running-in of the car was very carefully undertaken, in the belief that this period is highly significant on the later performance of the car. Revolutions were kept within the limits of 2,000 for the first 1,000 miles and 2,500 for the subsequent 1,500 miles, and the engine was as lightly loaded as possible at all times, which is probably of more importance. The running-in period was in no way irksome, since the overdrive allowed cruising speeds up to 70 m.p.h. within the rev. limits.

The 3·4 takes quite a lot of getting used to. Most people, for instance, find the positioning of the pedals rather unusual. Of the suspended variety, they are cocked at too high an angle for comfortable manipulation. Moreover it is impossible for any normally shaped person to lift the right foot directly off the accelerator and on to the brake in an up-and-down movement with the steering wheel at an arms length position because there is insufficient clearance for the legs. It is necessary to practise a sort of "knees splayed" position which causes the brake and clutch to be depressed at their edges. Positioning of a steering wheel on its telescopic control is, of course, largely a matter of personal preference, but it must be pointed out that if the wheel is brought as far from the dashboard as possible, allowing free manipulation of the legs, it is not possible to reach either the facia-mounted overdrive switch or the steering column-mounted indicators without taking the hand from the steering wheel. A great improvement would be the mounting of the latter *above* the telescopic adjustment so that it stays with the wheel.

Another disconcerting item was the long stretch to the gear lever which is of a pleasantly short remote control variety. It is not possible for a 5 ft. 10 in. driver sitting relaxed at the wheel to reach the gear lever in bottom gear without bending slightly forward.

Offsetting these criticisms were innumerable highly pleasing features. The quality of finish was impeccable. The paintwork was good, the doors shut with a solid 'clonk', the front seats were firm yet comfortable (and were later found to be much more than merely this), the engine had immense pull and an almost turbine smoothness even thus early, and, perhaps most remarkable of all, the brakes were a recurring revelation.

Running-in proceeded smoothly and 500 miles were soon on the clock. A shower or two of rain raised a few misgivings about wet weather handling, but the enforced sparing use of throttle avoided any excitement at this stage.

At the first servicing only three faults had evidenced themselves: the nearside front door was not shutting completely and required re-setting of the catch; the windscreen wipers had moments of refusing to work; and the passenger's front seat had developed a rattle in its adjusting mechanism. Seat and door were quickly fixed but a mystified mechanic could see nothing wrong with the wipers (they worked alright for him, *of course*). Willy nilly, they were left untouched.

By mid-June 132 EBH was fully run-in and the first opportunity to try the car before embarking on a continental trip to take in the European Grand Prix at Spa and the *Vingt-Quatre Heures du Mans* came when a flying visit had to be paid to Blackpool for a conference which began at 9.30 a.m. on a Monday. In order not to break up the week-end and also to have (hopefully) a

A weakness in the chromium plating (noted on other cars of the marque) lies behind the over-riders. Or is it just lazy cleaning?

traffic-free run, a start was made at approximately 5.30 a.m. Traffic proved considerably denser than expected, particularly coming the opposite way in the form of lorries, which made overtaking a difficult matter. Even so, 198 miles to the mouth of the Mersey Tunnel were covered in 4 hours 5 minutes, whereafter 9 o'clock Liverpool rush-hour traffic caused arrival at Blackpool to be five minutes late. This journey made a great impression, not because it was done at any sensational speed but because the truly remarkable character of the 3·4 was made plain. The speed with which the car could reach any desired cruising speed without noise or fuss—without even a gear-change if desired—made nonsense of minor checks and hazards. As surely as the silent surge of power took you rushing up to the eighties and beyond, the disc brakes would bring you down as if a giant hand were winding you in on a line, without the slightest pull or fade and with the lightest pedal pressure.

The seats, too, demonstrated their virtues. Although they look ordinary enough, their support for long-distance driving is quite remarkable and it is unheard-of to feel any back-ache after driving 200 miles at a stretch. Yet many much more luxurious seats will give back-ache after a quick spin round the block!

The handling of the 3·4 had taken a good deal of getting used to. There is a basic incorrigible (so far as we can envisage) weakness in the design of the car which is by no means confined to this marque. This is the distribution of too much weight on the front wheels, caused by the forward-positioning of the engine. In the case of the 3·4 the problem is probably aggravated by the fact that the heavier engine goes where the lighter 2·4 unit does in the smaller-engined car. The forward emphasis of the weight allied to the tremendous power—about 160 b.h.p. per ton in road trim—make for a "tail-happy" motorcar in the wet. It has been suggested that the cantilever-type rear springing plays a part, but we have found no reason to believe this. Indeed, for the purposes that this car has been used this type of springing has proved itself generally superior to half-elliptics.

A contributory factor may lie in the narrow rear track (it is three inches less than the front). The late Mike Hawthorn had, by fitting special wheels, effectively increased the rear track of his car. When we drove his car, in the wet, shortly before his tragic accident, we were most impressed by the improvement in rear-end 'stick' that had resulted. It is opportune here, therefore, to pass a few comments on Hawthorn's car, because there have been a lot of half-baked opinions expressed in the press and elsewhere about its contribution to the accident, with no sound basis of knowledge.

Hawthorn's 3·4 had received a great deal of attention and was undoubtedly faster than the standard vehicle. Attention had not been confined to the engine, however, and the ingenious fitting of non-standard rear wheels of a slightly dished pattern had resulted in an effective increase of the rear track by some two inches. When we tried Mike Hawthorn's 3·4 we had visited him in our own 132 EBH, and reaction was therefore particularly sensitive. There is no doubt that the modifications had notably improved the road-holding and in particular the rear-wheel traction under power. A point mentioned by Mike was noted, and subsequent events give a possible significance to this. The throttle pedal was decidedly sticky, and Mike indicated that a bit of care was required in its use. We found it embarrassingly 'on or off' in relation to the power of the car, and inevitably the thought occurs that perhaps just at the wrong moment on that tragic day on the Guildford by-pass the throttle was just a little more sticky than hitherto.

What must be made clear at this stage, in view of the bar-room chat, is that the 3·4 Jaguar—even in standard form—is light at the rear but not *viciously so*. By this we mean that any experienced driver—and we presume that no-one would argue Mike Hawthorn's status—will know precisely when the tail will break away. Indeed, on a recent trip to Goodwood with a colleague he remarked that correction for a slight tail-wag at the exit to a sharp corner had apparently been made *before* the slide occurred. In effect this was more or less the case. The *extent* of the correction was of course not decided but it had been anticipated that some such action would be required under the prevailing circumstances. We have always found that the 3·4 was thus predictably easy to control, with the characteristics only altering appreciably under full load conditions. And this brings us back to the second phase in the first year's life of 132 EBH.

After the Liverpool trip there was an almost immediate embarkation for the Continent via the excellent Lydd-Le Touquet air service of Silver Cities. As always, the car was aboard the aircraft almost before the handbrake had been applied in the airport car park; shortly afterwards the editor and photographer were safely aboard and some 30 minutes later all three of us were happily established on French soil, awaiting the arrival of Edward Eves who had somehow managed to miss the 'plane at Lydd after driving down from the Midlands.

After a belated lunch in the excellent but expensive airport restaurant, Ted Eves duly arrived and a late afternoon start was made for Belgium and Spa. The new motor-road to Liege gave the first taste of what the 3·4 could do under sustained high speed conditions. At any cruising speed attainable on this stretch of road which was between 110 and 120 m.p.h. (indicated) there seemed to be no more effort required than for 40 m.p.h. The engine was not obtrusive and the wind noise was comparatively slight. Exhaust, too, was completely subdued. Spa was reached in time for a late evening meal and a brief reconnoitre of the town. A tour of the Spa Francorchamps circuit made the following day, before practising for the European Grand Prix began, was illuminating. On the back part of the course speeds between 90 and 110 m.p.h. can be kept up for considerable periods quite

Heavy heels made short work of the matting beneath the pedals, and a suggested improvement is stronger material for this strategic position

As beautifully finished now as the day it left the showroom, the Jaguar engine seems second to none for quality externally

comfortably in a touring car. Yet you certainly seem to be getting along just as fast as you would wish. When it is considered that the record lap for this circuit is around 130 m.p.h. it will be seen that the straight stretches must be negotiated by a grand prix car at something between 130 and 160 m.p.h. in order to make up for the slower parts of the circuit. It is not our purpose to discuss the 1958 Belgium Grand Prix, but it is a sad thought that the three musketeers of the then-Ferrari team are now departed.

From the European Grand Prix at Spa the itinerary was to Brussels for a brief visit to the World Fair and thence to Le Mans in order to attend practising for the 24-hours race. A lack of accommodation in Brussels necessitated a journey to Ostend in order to find beds for the night. By this time the occupants of the car had increased to four, the newcomers being Peter Easton and his newly-wedded wife, complete with (modest) honeymoon baggage. Thus laden, the 3·4 put up one of the most impressive performances of its career and covered the length of the Jabbeke road in traffic conditions that were no more than reasonable at an average speed in excess of 80 m.p.h. with the speedometer registering "headlamps". (i.e. an indicated 125-130 m.p.h.) for minutes at a time.

Another very pleasant run from the Belgian coast to Le Mans demonstrated that the rear suspension of the 3·4 was equal to the very poor surface encountered on some of the secondary French main roads, even up to 100 m.p.h. cruising speed. The famous 24-hour race passed without any incident relative to the car, except that on a journey from Le Mans to La Chatre late at night, and in pouring rain, the windscreen wipers chose to give up the ghost. It was raining so hard that there was no question of getting out to have a look at the motor, so an obliging passing motorist was used as pathfinder and the party regained home safely. The next morning, strange to relate, the windscreen wipers worked perfectly and have done so ever since!

After Le Mans the 3·4 came home and continued with routine London work until mid-July, when it embarked on a further intensive three-weeks motoring on the Continent, during which time a further 3,500 miles were covered. During this period it gave no trouble whatsoever, but the indications were that the shock absorbers were growing tired of a rather arduous life. At 9,000 miles, therefore, the routine servicing was undertaken and new shock absorbers were fitted all round. In view of the fact that a further Continental trip was imminent the Dunlop RS4 Road Speed tyres which had been fitted were removed, although not completely worn out, and 15 × 6·40 Michelin X-tyres were fitted, with the manufacturers interested to know how things fared with them.

It is opportune here to mention that the RS4 tyres were highly satisfactory, particularly at very high speeds. They constantly gave the impression that the car was becoming more stable and more in contact with the road as the speed mounted. Against them it could be said that they tended to squeal unduly on sharp corners, and rear-wheel traction, particularly in the wet, left something to be desired. Conversely, the 'X'-tyres, while improving rear-wheel traction and completely eliminating the tyre-squeal on corners, caused the car to wander slightly at high speeds from 90 m.p.h. upwards. This can be corrected, Michelin inform us, by an alteration to the castor angle of the front wheels. The 'X'-tyres, as is well known, also cause the steering to become rather heavy at low speeds and they also have to be inflated to nearly 40 lbs. in this size for very high speed driving, in order to prevent overheating. This of course results in a hard ride for passengers at low speed.

Another trip of nearly 4,000 miles on the Continent followed, taking in visits to various German motorcar factories on the way south, then to the Italian Grand Prix at Monza, followed by a holiday trip to Rome and the south, with some more business calls for good measure.

The 3·4's Continental season had thus been run and there were something like 14,000 miles on the clock in the course of four months. Since then, the closed season has asked no more than a mere 200 or 300 miles' trip to the north of England. To date the car has covered 24,000 miles and is eleven months old.

During this period nothing was done to the car barring routine maintenance and the items mentioned. Prior to acceleration tests checking performance an engine check-up was made in which carburettor settings, ignition timing etc. were attended to and the first replacement set of sparking plugs fitted. At about the same time the brake pedal quite suddenly became flabby and braking deteriorated to a considerable extent. A leaking servo pump

49

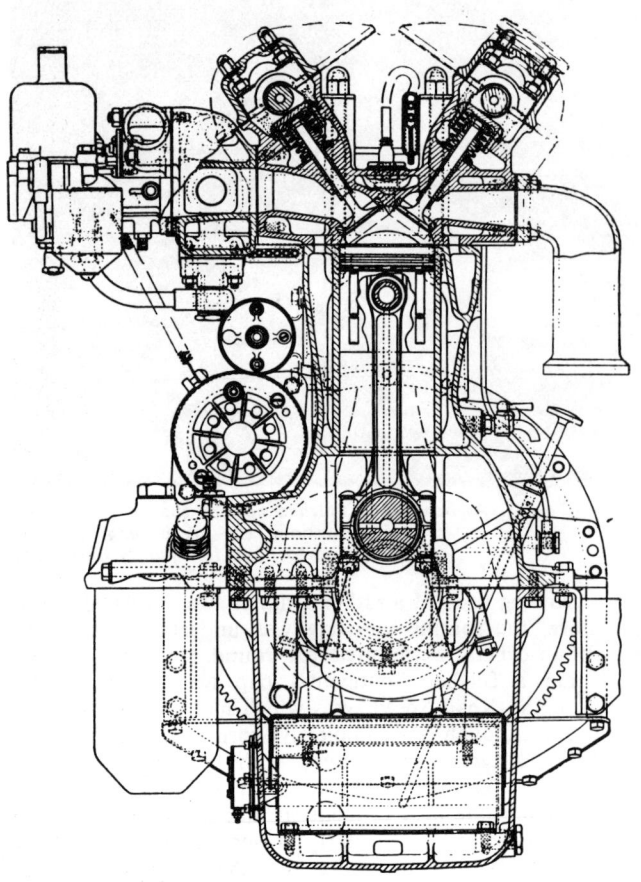

Internal arrangement of the 3·4 litre engine which has proved unfailingly reliable, unburstable, economical and very good at holding tune

diaphragm was diagnosed and rectification made.

It is not only fair but necessary to say that the brakes have functioned faultlessly apart from this. They are immensely powerful, have never been made to fade and do not suffer from the squeak when cold of some of the earlier models. The handbrake requires more than average attention to keep in adjustment.

After the first settling-in period the car has been a constant joy to drive. Starting has always been instantaneous from the coldest to the hottest day and—again unlike some earlier models—the automatic choke now has a gremlin who knows when to switch off. The handling of the car is not unimpeachable. It rolls too much on corners and, even with the 'X'-tyres, is skittish in the wet. Against this, the ride is very comfortable for both front and back seat passengers and the driver automatically adjusts himself to going easily round corners and using the formidable acceleration to regain cruising speed.

The interior appointments are comfortable without being lavish, and they have remained in very good order without special attention. The only criticism here lies in the mat provided beneath the pedals. Made of some plastic material, it is not nearly hard-wearing enough. A small hole appeared where the heel of the right foot rests after about 12,000 miles and it is now gaping, with a corresponding hole where the heel of the left foot goes.

Externally, the paintwork has remained very good indeed, but the chromium has been patchy. For some curious reason the backs of the over-riders on the bumpers have rusted badly; although the bumpers, tail pipes and window surrounds remain good. Also slightly rusted are the wheel dishes and the chromium strip on the side of the bodywork. Except on Continental trips the car has been kept in a garage every night but has been in the open for the best part of nearly every day.

In all, the 3·4 is a wonderful car for those who want tremendous performance with saloon car comfort and accommodation. It is not the most viceless car in the world to drive and it should be treated with respect by the experienced. Those who have not served a good apprenticeship on fast motor cars would be well advised to buy a 2·4 first!

Whatever its shortcomings, however, it is the most reliable high-performance car yet owned by the publishers, and the memory of its effortless and silent cruising at over 100 m.p.h. stays in the mind and is a constant yardstick for perfection when essaying all but the élite of the world's sports cars.

132 EBH JAGUAR 3·4. Condensed details

Average fuel consumption over 25,000 miles: 22·4 m.p.g.

Mechanical attention beyond normal servicing:

at 500 miles: nearside front door adjusted
passenger front seat tightened
traffic indicator adjusted

at 9,500 miles: new shock absorbers all round

at 2,350 miles: new diaphragm for servo pump for brakes

at 2,400 miles: new sparking plugs

Wear and tear: No bodywork damage; rusting on certain chrome parts; wearing out of drivers foot mat.

Looking sleeker than its predecessor, the Mark II Jaguar benefits greatly from the larger windscreen and from the wider rear window

Jaguar Widens its Range

NEW Mk. II SALOONS WITH 2.4, 3.4 or 3.8 LITRE ENGINES

ALREADY 1959 is being talked of as a vintage year for new cars, and undoubtedly well up the quality list will come the new Mark II Jaguars. These are included in a price range now containing ten basic models. On them variations afforded by choice of transmissions, disc or drum brakes for the standard 2.4 and 3.4 saloons, XK150 series body styles and a wide range of exterior and interior colour schemes, provide the potential customer with great variety from which to select.

The large and luxurious 3.8-litre Mk. IX saloon, the 2.4 and 3.4 standard saloons and the 3.4 XK150 and 150S continue as before, and the Mk. II 2.4, 3.4 and 3.8 saloons, with major body and mechanical changes, are additions; also the XK150 range is extended by using two versions of the 3.8-litre engine.

Basically the specification of the Mk. II Jaguars resembles that of the standard models. The well-known twin-overhead camshaft engine, coupled to a four-speed synchromesh or Borg-Warner automatic gear box, is installed in a four-door, all-steel, integral body and chassis structure. There is coil spring independent front suspension, and long cantilever half-elliptic leaf springs at the rear. Alterations to front suspension and widening of the rear track are intended to give the new cars superior suspension qualities and freedom from roll on corners.

Roll centre of the front suspension has been raised to 3.25in above ground level; on the standard cars it is 0.75in below ground level. The bottom section of the vertical link carrying each stub axle has been sloped downwards, lowering the centre for the bottom suspension ball joints by 1.68in, and by repositioning the top suspension joint by 0.5in, the king-pin axis angle is reduced from 7.5deg to 4.25deg. Ground clearance remains at 7.0in.

Rear track, previously 4ft 2.1in, is now 4ft 5.4in. No alterations are made to the rear spring centres, the extra width being provided by using a new and stronger axle casing, of similar design to the previous unit. Hypoid bevel gearing is used, with a choice of ratios to suit the size of engine and type of transmission specified. Road noise is practically eliminated by extensive use of rubber bushing between the springs and body structure.

Extra power is given to the 2.4-litre engine used in the Mk. II cars by fitting the B-type cylinder head. This has larger valves, inlet of 1.5in dia and exhaust of 1.375in dia, both using a 45deg angle seat; standard valve sizes are 1.375in and 1.25in respectively. This head, developed from the famous D-type engine, allows improved breathing and produces increased maximum b.h.p. and torque, from 112 to 120 b.h.p., and 140 to 144 lb ft. The respective engine speeds of 5,750 and 2,000 r.p.m. are unchanged. Compression ratio can be 8.0 to 1 or 7.0 to 1.

The 2.4-litre engine is the only Jaguar power unit using Solex carburettors—the twin B32 PB1 downdraught type; they breathe through a cast alloy manifold connected to a circular oil bath air cleaner which has a single, forward-facing inlet pipe. A dual exhaust system employs two silencers, with tail pipes terminating on the left below the rear bumper.

An owner who requires more performance can have a Mk. II with either the 210 b.h.p 3.4-litre engine or the 220 b.h.p. 3.8-litre unit, which in this car should give

Below: An ashtray with a hinged lid is recessed into the armrest of each rear door, which is also fitted with a map pocket. Right: A great deal of thought has gone into the layout of the Mark II instrument panel. All dials are clearly visible to the driver and the switches are laid out in a practical manner

Above: Apart from major body changes, Mark II cars may be distinguished from the rear by the wider track and new position of the exhaust tail pipes. Right: Twin Solex carburettors and a round oil bath air cleaner with a single air intake tube distinguish the 2.4-litre engines

The 3.8-litre S type engine has three S.U. HD6 carburettors. The overdrive is an extra

Jaguar Range . . .

These gross power curves for the current Jaguar engines make interesting comparisons, particularly in respect of the breathing capacities as gauged from the b.m.e.p. curves

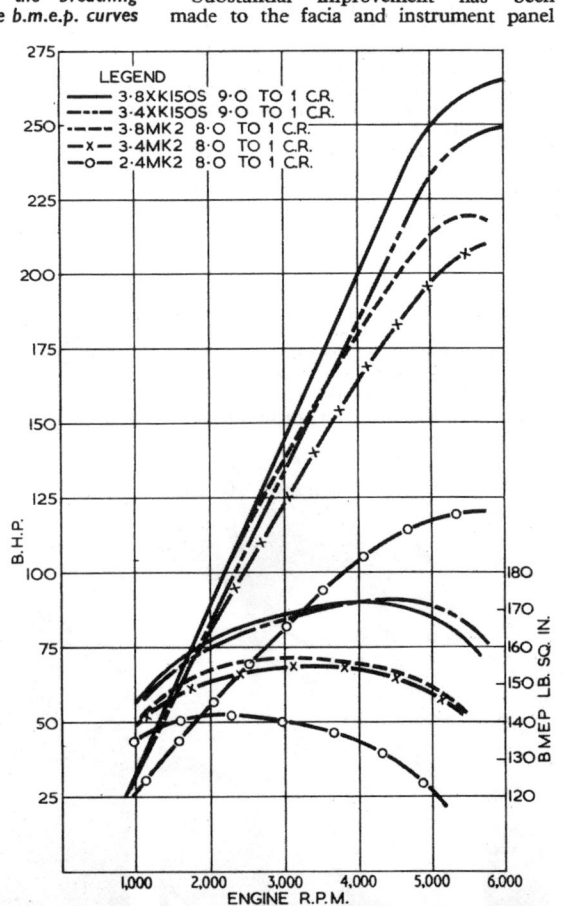

a truly *gran turismo* performance. Both have twin S.U. carburettors with automatic choke control, and 10in dia clutches.

Dunlop 12in dia disc brakes—standard equipment—are fitted at front and rear, and servo assisted; they have a swept area of no less than 540 sq in. Mounted on the right side of the facia panel is a warning light which indicates when the hand brake is applied, and when the level of the brake hydraulic fluid requires topping up.

It is a rare production car in which the driver can find all the instruments and equipment he desires. Many items are often classed as extras, and others the owner has to supply himself. Jaguar design engineers have sifted customers' demands and added the more practical ideas to their own, and the result is a body so fitted that it is difficult to fault.

Exhaustive laboratory tests at the Motor Industry Research Association establishment have assisted in the design of the deep, wide windscreen and wrap-round rear window, for improved driving visibility. The large screen has pillars of dimensions which offer considerably less obstruction than before. Plated frames are used for all the windows and the roof guttering.

Rear side window styling resembles that used on the Mk. IX saloon, as the rear opening ventilators merge into the roof corner panels. Not only does this provide the rear seat occupants with a better view but also it allows easier entry to the compartment. A new-pattern door lock has a smooth exterior handle, and a push-button action designed to avoid freezing-up in extreme temperatures.

To accommodate the extra width of the rear track, the panels behind the rear door opening follow the lines of the body to the wheel arches, where they then curve round to merge into the rear wing line. The overall width of the car, 5ft 6.75in, is the same as before, as are other overall dimensions of the car which affect garaging.

Although the Mk. II models bear a strong resemblance to their fellows in their exterior lines, many changes have taken place in the interior. A new, more comfortable and practical shape has been given to the separate front seats and to that in the rear compartment. The front seat back rests are designed to hold the occupants with firm and comfortable support; all seats have Dunlopillo foam rubber over spring cases, and are covered with leather.

A curved padding over each rear wheel arch, an extremely deep cushion, well-shaped back rest and folding central arm rest give maximum relaxation for occupants of the rear seat, who have foot room under the front seats. Folding tables are recessed in the back of the latter.

Considerable thought has been given to provision of stowage for maps, cameras and personal items. The facia locker is illuminated when the lid is opened; there is a pocket in each locker and a shelf behind the rear seat. A very practical size of ash tray is provided for the front compartment, and one is recessed into each door arm rest. All have hinged covers to prevent ash flying about the car. Other facilities for all occupants include four interior lamps, one in each rear quarter panel and on each centre pillar. All light up when any door is opened.

Substantial improvement has been made to the facia and instrument panel

LEGEND
— 3·8XK150S 9·0 TO 1 C.R.
--- 3·4XK150S 9·0 TO 1 C.R.
-·- 3·8MK2 8·0 TO 1 C.R.
-x- 3·4MK2 8·0 TO 1 C.R.
-o- 2·4MK2 8·0 TO 1 C.R.

52

Mark II points. Left: A shallow table is fitted into the back of each front seat, between which is the outlet for the rear compartment ventilation system. Right: All the windows have plated surrounds and a moulding extends the length of the car at waist rail level. The new pattern door locks have very smooth handles with a push-button action designed to avoid the risk of freezing-up in cold weather

layout, and the use of lacquered wood is continued for panel and window fillets. Immediately behind the new-pattern steering wheel, and in full view of the driver, is an electrically recording engine tachometer and a matching speedometer. In the centre panel are dials for oil pressure, water temperature, fuel gauge and ammeter.

These instruments, and the comprehensive row of toggle switches beneath them, are well illuminated after dark. In addition to the passenger compartment illumination, a separate map light is fitted beneath the top rail of the facia. Two-speed screen wipers, screen washer, electric clock and cigar lighter with an illuminated holder are standard equipment. Two recessed sun vizors are provided for the front compartment; that for the passenger is fitted with a mirror.

Drivers of fast cars have become increasingly aware that flashing head lamps give more warning than the average horn. This facility is provided on the Mk. II Jaguars by an extra position on the steering-column-mounted switch of the turn indicators. A two-spoke steering wheel carries a half-ring for the twin horns.

The standard four-speed transmission may be supplemented by a Laycock-de Normanville overdrive or, alternatively, a fully automatic Borg-Warner transmission may be specified. In each case the overdrive or automatic control lever is placed on the left of the steering column. A small segmental dial on the upper part of the column indicates overdrive engaged or automatic gear positions selected, and this is illuminated when the side lamps are switched on.

Earlier model Jaguars have been criticized for the inadequacy of the heating and de-misting equipment. The Mk. II cars have a 3.9kw unit which draws fresh air through a manually controlled vent forward of the windscreen. Air from here is ducted to the floor area of the front compartment, and to the windscreen. A two-speed fan boosts air flow.

From the front compartment, a duct above the propeller shaft tunnel carries more air to a grilled vent which has an outlet at floor level between the two front seats, controlled by means of a flap valve. The console beneath the facia, which carries the vertically operating main controls for the heating system, also has provision for a radio and loudspeaker.

In equipment and bodywork the Mk. II Jaguars are identical; to allow the extra power of the 3.8-litre engine to be used fully, the cars fitted with this unit also have a Powr-Lok limited slip differential which enables rapid acceleration to be achieved without wheelspin. This type of differential is also standard equipment on the 3.8-litre XK150 and XK150S two-seaters. As fitted to the XK150, the 3.8-litre unit is the same as that used for some

PRICES								
	Basic £		Purchase Tax £ s d			U.K. Total £ s d		
Mk. IX saloon ...	1,329		554	17	6	1,883	17	6
2.4 standard	1,019		425	14	2	1,444	14	2
3.4 standard	1,114		465	5	10	1,579	5	10
2.4 Mk. II	1,082		451	19	2	1,533	19	2
3.4 Mk. II	1,177		491	10	10	1,668	10	10
3.8 Mk. II	1,255		524	0	10	1,779	0	10
3.4 XK150 coupé ...	1,175		490	14	2	1,665	14	2
3.4 XK150S coupé ...	1,457		608	4	2	2,065	4	2
3.8 XK150 coupé ...	1,370		571	19	2	1,941	19	2
3.8 XK150S coupé...	1,535		640	14	2	2,175	14	2

time in the Mk. IX saloon and now in the Mk. II cars. It has a high rate of torque compared with the 3.4 engine, 240 lb ft against 215 lb ft at 3,000 r.p.m.; when fitted with the straight port Gold Top head and three S.U. carburettors, it produces 260 lb ft at 4,000 r.p.m. and 265 b.h.p. gross at 5,500 r.p.m.

The XK models are available with a choice of closed, open or convertible bodywork, ranging upwards in price from the standard 3.4-litre coupé. The same varied specification of alternative transmission systems applies to these models as to the saloons, with left-hand drive available for all types. Prices given are for the basic models only, and are unchanged.

The current standard 2.4 and 3.4 models correspond with the former special equipment models, disc brakes are extra.

SPECIFICATION

ENGINE

	2.4	3.4	3.8
No. of cylinders	6 in line	6 in line	6 in line
Bore and stroke	83 x 76.5mm (3.268 x 3.012in)	83 x 106mm (3.268 x 4.173in)	87 x 106mm (3.425 x 4.173in)
Displacement	2,483 c.c. (151.50 cu in)	3,442 c.c. (209.96 cu in)	3,781 c.c. (230.65 cu in)
Valve position		Overhead, twin camshafts	
Compression ratio	8 to 1 (alternative 7 to 1)	8 to 1 (optional 7 or 9 to 1)	
Max. b.h.p. (gross)	120 at 5,750 r.p.m.	210 at 5,500 r.p.m.	220 at 5,500 r.p.m.
Max. b.m.e.p.	142lb sq in at 2,000 r.p.m.	155lb sq in at 3,000 r.p.m.	155lb sq in at 3,000 r.p.m.
Max. torque	144lb ft at 2,000 r.p.m.	215lb ft at 3,000 r.p.m.	240lb ft at 3,000 r.p.m.
Carburettor	Twin Solex B.32 PBI downdraught	Twin S.U. HD6	Twin S.U. HD6
Fuel pump		S.U. Electric	
Tank capacity		12 Imp. gallons (54.5 litres)	
Oil Sump capacity		13 pints (7.4 litres)	
Oil filter		Tecalemit full-flow	
Cooling system		Pump, fan and thermostat, pressurized	
Battery		12 volt, 51 ampère hour	

TRANSMISSION

	2.4	3.4 and 3.8
Clutch	s.d.p., 9in dia	s.d.p., 10in dia
Gear box	Four speeds, synchromesh on 2nd, 3rd and top; floor gear lever	
Overall ratios	top 4.27	top 3.54
	3rd 5.48	3rd 4.54
	2nd 7.94	2nd 6.58
	1st 14.42	1st 11.95
	reverse 14.42	reverse 11.95
Final drive	Hypoid; Ratio 4.27 to 1 (manual and automatic transmissions); 4.55 to 1 with overdrive	Hypoid; Ratio 3.54 to 1 (manual and automatic transmissions); 3.77 to 1 with overdrive

CHASSIS

Brakes	Dunlop disc, bridge type, front and rear; vacuum servo assisted
Disc dia	12in
Suspension: Front	Independent, coil springs and wishbones; anti-roll bar
Rear	Live axle, cantilever leaf springs
Dampers	Telescopic hydraulic
Wheels	Bolt-on disc. Wire centre lock optional
Tyre size	6.40-15in
Steering	Recirculating ball
Steering wheel	Two-spoke, 17in dia Telescopically adjustable column
Turns, lock to lock	4.3

DIMENSIONS

Wheelbase	8ft 11.38in (272.7cm)
Track (with disc wheels): Front	4ft 7in (139.7cm)
Rear	4ft 5.38in (135.6cm)
(with wire wheels): Front	4ft 7.5in (141cm)
Rear	4ft 6.13in (137.5cm)
Overall length	15ft 0.75in (459.1cm)
Overall width	5ft 6.75in (169.5cm)
Overall height	4ft 9.5in (146.1cm)
Ground clearance	7in (17.8cm)
Turning circle	33ft 6in (10.21m)
Kerb weight	2.4—3,204lb (28.5 cwt) 3.4 and 3.8—3,288lb (29.25 cwt)

PERFORMANCE DATA

2.4		3.4 and 3.8	
Top gear m.p.h. Manual and at 1,000 r.p.m. automatic	18.1	Top gear m.p.h. Manual and at 1,000 r.p.m. automatic	21.9
Overdrive	21.8	Overdrive	26.45
Top	17.0	Top	20.6
Torque lb ft per cu in engine capacity	0.95	Torque lb ft per cu in engine capacity 3.4	1.02
		3.8	1.04
Brake surface area swept by linings	540 sq in	Brake surface area swept by linings	540 sq in
Weight distribution (dry)	57% front, 43% rear	Weight distribution (dry)	57% front, 43% rear

1960 CARS

IMPROVED 2·4 and 3·4

Mark 2 Version with Wider Rear Track, Modified Front Suspension and Replanned Interior. Alternative 3.8-litre Engine Available in 3.4-litre and XK150 Models. Mark IX Unchanged

JAGUAR innovations for the coming season centre mainly round the 2.4- and 3.4-litre saloons, which now appear additionally in Mark 2 form with considerable improvements in appearance, user amenities and general handling. In addition, the 3.8-litre engine is now available as an alternative to the 3.4-litre unit in both this Mark 2 model and in the XK150 range. Otherwise the XK150 models are unchanged, and so is the big Mark IX saloon.

The mechanical alterations in the Mark 2 models concern the front suspension, which has been modified to give a higher roll centre, and the rear axle which has been increased 3¼ in. in track; in addition, the 2.4-litre engine now has the B-type head. In external appearance, the cars have been improved by a much greater window area, whilst minor modifications have been made to the front and to such items as door handles, lamps, beadings and so on. Inside, the interior trim has been completely revised and the facia board redesigned, whilst an added refinement lies in the provision for rear-compartment heating.

* * *

To deal first with power unit developments, the cylinder head fitted to the 2.4-litre Mark 2 6-cylinder engine embodies the same principles which were introduced originally for the Mark VIII model, that is to say the valves have been increased in diameter and now have 45-degree seats (instead of 30-degree) and convex faces, but are used in conjunction with the same diameter ports as hitherto. The resulting freer gas flow gives an increase in maximum output from 112 b.h.p. to 120 b.h.p. at an unchanged peak speed (5,750 r.p.m.) but the fact that the port diameters have not been increased maintains the gas speed at low and medium r.p.m. so that low-speed pulling powers are not affected. A single exhaust system is retained.

The 3.4-litre engine remains unchanged, but the exhaust system has been shortened in the Mark 2 edition and the twin, paired pipes are now directed straight to the rear on the left side and not angled towards the kerb.

As an alternative to the 3.4-litre engine in the new Mark 2 saloon, buyers are given the option of a 3.8-litre unit at an extra charge. This engine is identical to that used in the Mark IX except for minor installation modifications. It develops 220 b.h.p. at 5,500 r.p.m. as opposed to the 210 b.h.p. (at the same speed) of the 3.4-litre unit. The rest of the specification, including gear ratios, remains unchanged.

The final power unit innovation for 1960 consists in offering the 3.8-litre engine in the XK150 range. Here the larger engine is available in

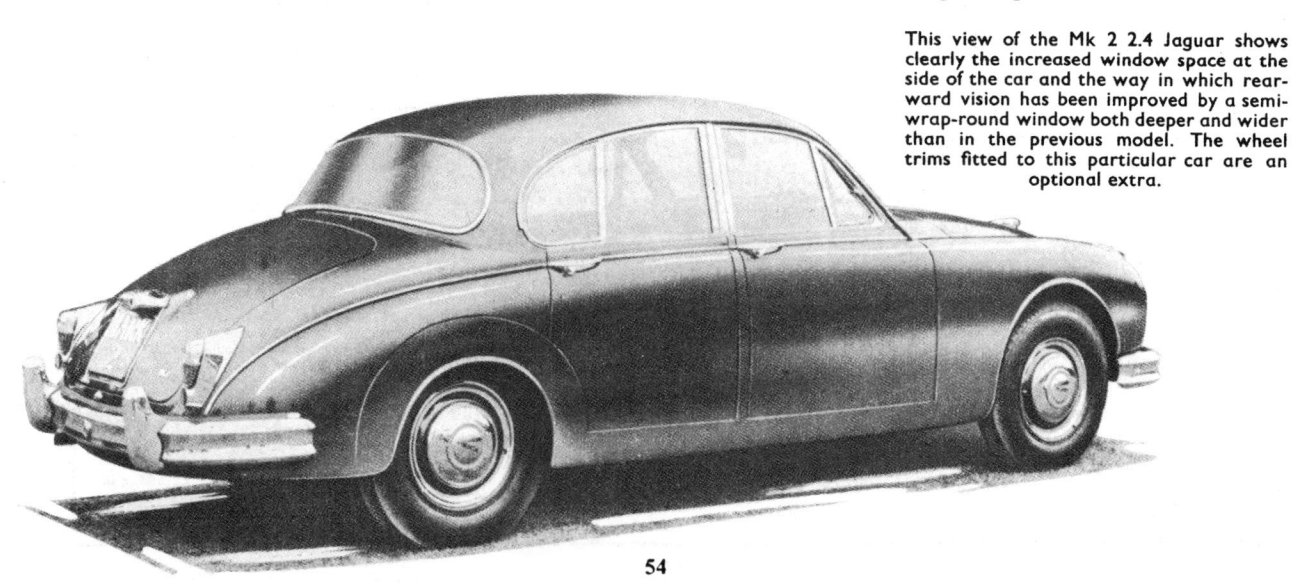

Externally there is little to distinguish the 3.8-litre engine from the 3.4-litre except that slightly larger carburetters are fitted.

This view of the Mk 2 2.4 Jaguar shows clearly the increased window space at the side of the car and the way in which rearward vision has been improved by a semi-wrap-round window both deeper and wider than in the previous model. The wheel trims fitted to this particular car are an optional extra.

54

JAGUARS

Forward visibility has been achieved by making the windscreen not only deeper but more curved to produce a semi-wrap-round effect. A further distinguishing feature of the Mark 2 models, of which the 3.4 is here illustrated, is a more pronounced centre rib to the radiator grille, semi-recessed fog lamps and small faired side lamps at the top of the wings.

A new design of front seats with deeper squabs and improved door arm rests are among the modifications which have been made to the latest Mk 2 Jaguar saloons.

standard (220 b.h.p.) form or to "S" type specification with straight-port head and three carburetters. In the latter form, the output is 265 b.h.p. at 5,500 r.p.m.—the highest output ever offered by Jaguar on a production car.

In the interests of better handling, minor but important alterations have been made at both ends of the Mark 2 chassis. At the rear, the modification consists of extending the rear axle casing outwards by $1\frac{5}{8}$ in. on each side to give a $3\frac{1}{4}$-in. increase in rear track, thus considerably reducing the crab-track effect on standard 2.4-litre and 3.4-litre models.

At the front, coil and wishbone independent front suspension is used as before, but the roll tendency has been reduced by subtle changes in the geometry of the wishbones which have had the effect of raising the roll centre from $\frac{3}{4}$ in. below ground level to $3\frac{1}{4}$ in. above.

In external appearance, the new models follow the generally familiar lines, but the numerous changes have both æsthetic and practical merit. Most notable amongst them is the change in window area. At the front, the screen has been increased in depth by 1 in. to $16\frac{1}{2}$ in., whilst the glass has simultaneously been given a greater curvature to produce a semi-wrap-round effect.

An even greater improvement in vision is to be found at the sides, where the doors have been modified so that the main pressings end at waist level, only narrow chromium frames being extended above. The effect has been to reduce the blind spots caused both by the screen pillars and the posts between the doors. In addition, the rear door windows have been extended backwards beyond the normal door opening to provide very large ventilating panels for extractor purposes. Some idea of how effective these changes are can be gathered from the fact that the front door windows are now $1\frac{1}{4}$ in. deeper and 3 in. wider than before, whilst the rear side windows have been increased in maximum depth by $1\frac{1}{2}$ in. and in effective width by no less than $6\frac{1}{2}$ in. Completing this campaign for better vision is a semi wrap-round rear window which is 3 in. deeper and 7 in. wider.

At the front, the Mark 2 models can be distinguished by a modified grille with a more pronounced centre rib, by the partial recessing of the formerly separate fog lamps into spaces previously occupied by horn grilles and by the fitting of small faired side-lamps on the tops of the wings, with separate flasher units in the positions formerly occupied by the side lamps.

At the sides, the chromium-framed windows are offset by chrome finishers on the rain guttering and top of the door waist rails, whilst revised push-button door handles of a neater and more waterproof type have been fitted. The rear wing shape has been slightly modified to accommodate the wider track, and cutaway spats are used. Changes visible from the rear include the larger rear window, new tail lamp clusters and a deeper apron below the rear bumper to conceal the petrol tank.

As already indicated, the interior has virtually been completely re-furnished. Most noticeable is an entirely different design of facia board in which the two main dials for the speedometer and rev. counter are directly in front of the driver, whose view of them is quite unobstructed

55

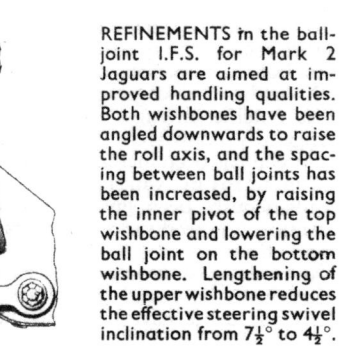

IMPROVED 2·4

DRIVER CONVENIENCE has been very carefully studied in the new Mark 2 Jaguar models. The example shown is the 2.4-litre which differs from the 3.4 edition only in having a manual choke and warning light. Note the two-spoke wheel giving a clear view of the speedometer and revolution counter and the separate dials for the water thermometer, oil gauge, fuel gauge and ammeter, with smaller switches of the toggle type below. The latter are identified on a panel illuminated at night. An automatic transmission model is shown, but on overdrive examples the overdrive switch takes the place of the automatic transmission selector lever. Note also the brake-fluid/handbrake warning light and the new "tower" for the heater and, when fitted, radio.

through a new two-spoke steering wheel with a half horn ring.

In the centre of the board is a panel carrying separate small dials for the ammeter, fuel gauge and water thermometer, with the main lighting switch located in the centre of the row. Horizontally disposed below them are the minor switches, which are all of the easily operated toggle type and clearly labelled on a small illuminated panel beneath; also included in this row are the ignition switch and starter button, with an illuminated cigar lighter between. All the dials have clear white markings on black faces and, in the case of the main speedometer and r.p.m. dials, the figures are translucent and illuminated (dim or bright) from behind. Except the speedometer, all the instruments are electrically operated.

On the driver's side of the two main dials are the choke control (2.4-litre only, as the 3.4-litre engine has an automatic choke), and a notable innovation in the form of a warning light to indicate if the fluid level in the hydraulic brake reservoir is getting low. Because a warning light of this type would normally come into use only at rare intervals, the manufacturers have taken the sensible step of linking it also to the handbrake so that it lights up whenever the latter is applied with the ignition switched on. Thus the bulb is frequently tested and the arrangement serves additionally as a memory jog if a driver forgetfully starts off with the hand brake applied.

The adjustable steering column is surrounded by a binnacle with provision for an indicator segment if automatic transmission is fitted, and this is flanked by warning arrows for the direction indicators.

Particular praise must be accorded to Jaguar for introducing a fingertip headlamp flasher. Actually, the steering-column direction indicator switch is used, pressure of the fingers towards the steering wheel rim serving to flash the headlights in the Continental manner. A similar lever on the left of the column operates the overdrive, when fitted.

Below the new facia is a central "tower" which incorporates a radio, when fitted, together with controls for the heater. The latter is of higher output (3.9 kilowatts rating) and a welcome refinement is the provision of ducting to the rear compartment; the outlet is on the propeller shaft tunnel between the front seats, with directional flutes to deflect the air towards the floor on each side of the car. Also on the propeller shaft tunnel, below the "tower," is an outsize in ashtrays.

The front seats are new, with deeper squabs and neat flush-fitting folding tables in their backs. The door trim

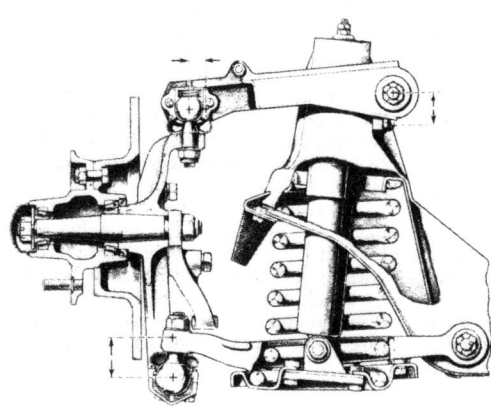

Small folding picnic tables for the use of rear compartment passengers are recessed into the backs of the re-designed front seats; also clearly shown here are the improved arm rests and capacious door lockers.

REFINEMENTS in the ball-joint I.F.S. for Mark 2 Jaguars are aimed at improved handling qualities. Both wishbones have been angled downwards to raise the roll axis, and the spacing between ball joints has been increased, by raising the inner pivot of the top wishbone and lowering the ball joint on the bottom wishbone. Lengthening of the upper wishbone reduces the effective steering swivel inclination from $7\frac{1}{2}°$ to $4\frac{1}{2}°$.

and 3·4 JAGUARS

has also been revised and, at both front and rear, armrests of new shape are so arranged that they give the necessary elbow support without in any way restricting width across the seat cushions. Other details of interest include flush-fitting sun visors and repositioned catches for the ventilating panels on the leading edges of the front doors; these catches are now situated on the vertical portion of the frame where they are not only easier to use, but are virtually thief-proof.

Finally, particular attention has been paid to interior illumination and lights are now fitted both above the centre door pillars and in the rear quarters with, of course, courtesy switches on all doors. At the front there is also a map reading light, whilst the interior of the glove box is illuminated.

As will have been gathered, these new 2.4-, 3.4- and 3.8-litre Mark 2 models are in addition to the existing standard versions of the 2.4- and 3.4-litre saloons. On the Mark 2 cars, disc brakes (an extra on the existing standard range) are normal equipment; and, all Jaguars excepting the "S" series XK150 models, can be had with normal synchromesh gearbox, with synchromesh gearbox and overdrive, or with Borg-Warner fully automatic transmission. A Powr-Lok limited-slip differential is available as an extra on the 3.4-litre XK150 "S", but is standardized on both the normal and "S" versions of the 3.8-litre XK150 models, and also on the 3.8-litre Mark 2 models.

Including transmission options, but excluding the effect of disc brakes, wire wheels and Powr-Lok differential where these are offered as extras, the total Jaguar range now totals no fewer than 45 separate and distinct models.

Continuing unchanged both in the standard and in "S" type is the 3.4-litre XK150 seen here in open form. The model is also available in 3.8 litre versions.

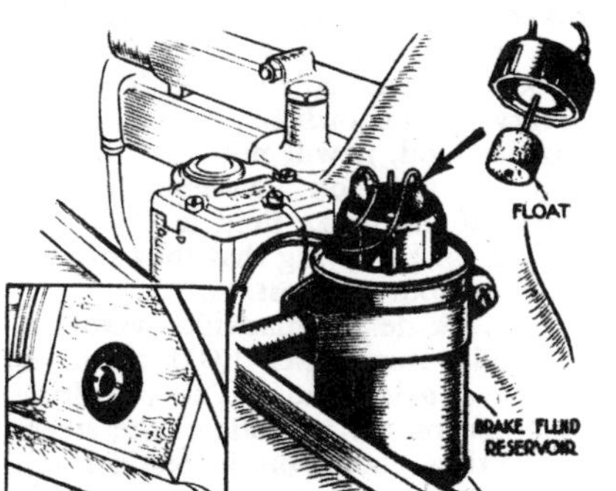

SAFETY PRECAUTION. A brake fluid level indicator is a new feature of the Mark 2 models. A float is used in the reservoir and an indicator light is placed on the driver's side of the facia board. The warning light also serves to indicate when the handbrake is applied (see text).

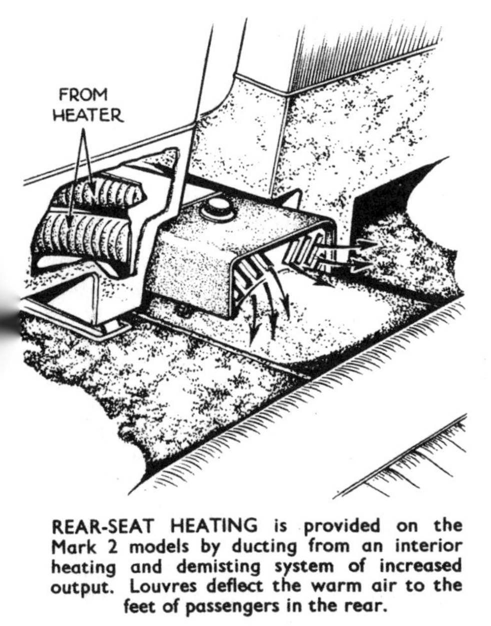

REAR-SEAT HEATING is provided on the Mark 2 models by ducting from an interior heating and demisting system of increased output. Louvres deflect the warm air to the feet of passengers in the rear.

1960 JAGUAR PROGRAMME SUMMARIZED

2.4-LITRE STANDARD

Continued unchanged.

2.4-LITRE MARK 2

ENGINE—Dimensions: Cylinders, 6; bore, 83 mm.; stroke, 76.5 mm.; cubic capacity, 2,483 c.c.; piston area, 50.4 sq. in.; valves, overhead (twin o.h. camshafts); compression ratio, 8 : 1 (7 : 1 optional). **Performance:** Max. b.h.p.; 120 gross at 5,750 r.p.m.; b.h.p. per sq. in. piston area, 2.48. **Details:** Carburetters, two Solex downdraught; ignition timing control, centrifugal and vacuum; plugs, Champion N5; fuel pump, S.U. electric; fuel tank capacity, 12 gallons; cooling system capacity, 20 pints; engine oil capacity, 11 pints; oil filter, Tecalemit full-flow.

TRANSMISSION—Clutch, Borg and Beck 9 in. s.d.p.; overall gear ratios: top, 4.27 (s/m); 3rd, 5.48 (s/m); 2nd, 7.94 (s/m); 1st, 14.42; rev., 14.42; also available with Borg-Warner automatic transmission, or Laycock-de Normanville overdrive; prop. shaft, Hardy Spicer, open; final drive, hypoid bevel.

CHASSIS DETAILS—Brakes, Dunlop disc (servo assisted); disc diameter, 12 in.; suspension: front, independent (coil); rear, cantilever; shock absorbers, Girling telescopic hydraulic; steering gear, Burman re-circulating ball; wheels, bolt-on disc (wire centre lock optional extra); tyre size, 6.40—15 tubeless (tubed with wire wheels).

DIMENSIONS—Wheelbase, 8 ft. 11⅜ in.; track: front, 4 ft. 7 in. (4 ft. 7½ in. with wire wheels); rear, 4 ft. 5⅜ in. (4 ft. 6⅛ in. with wire wheels); overall length, 15 ft. 0¾ in.; overall width, 5 ft. 6¾ in.; overall height, 4 ft. 9½ in.; ground clearance, 7 in.; turning circle, 33½ ft. dry weight, 27¾ cwt. (with o/d, 28 cwt.; with auto-trans, 28½ cwt.).

PERFORMANCE DATA—Top gear m.p.h. per 1,000 r.p.m. 18.1; top gear m.p.h. per 1,000 ft./min. piston speed, 36.0.

3.4-LITRE STANDARD

Continued unchanged.

3.4/3.8-LITRE MARK 2

ENGINE—3.4-litre: Dimensions: Cylinders, 6; bore, 83 mm.; stroke, 106 mm.; cubic capacity, 3,442 c.c.; piston area, 50.4 sq. in.; valves, overhead (twin o.h. camshafts); compression ratio, 8 : 1 (7 : 1 or 9 : 1 optional). **Performance:** Max. b.h.p.; 210 gross at 5,500 r.p.m.; b.h.p. per sq. in. piston area, 4.17

3.8-litre: Dimensions: Cylinders, 6; bore, 87 mm.; stroke, 106 mm.; cubic capacity, 3,781 c.c.; piston area, 55.3 sq. in.; valves, overhead (twin o.h. camshafts); compression ratio, 8 : 1 (7 : 1 or 9 : 1 optional). **Performance:** Max. b.h.p., 220 gross at 5,500 r.p.m.; b.h.p. per sq. in. piston area, 3.98.

Details: Carburetters, two S.U. HD6; ignition timing control, centrifugal and vacuum; plugs, Champion N5; fuel pump, S.U. electric; fuel tank capacity, 12 gal.; cooling system capacity, 22 pints; engine oil capacity, 11 pints; oil filter, Tecalemit full-flow.

TRANSMISSION: Clutch, Borg and Beck 10 in. s.d.p.; overall gear ratios: top, 3.54 (s/m); 3rd, 4.54 (s/m); 2nd, 6.58 (s/m); 1st, 11.95; rev., 11.95; also available with Borg-Warner automatic transmission or Laycock-de Normanville overdrive; prop. shaft, Hardy Spicer, open; final drive, hypoid bevel.

CHASSIS DETAILS: Brakes, Dunlop, disc servo assisted; disc diameter, 12 in.; suspension, front, independent (coil); rear, semi-elliptic; shock absorbers, Girling hydraulic telescopic; steering gear, Burman re-circulating ball; wheels, bolt-on disc (wire centre-lock optional extra); tyre size, 6.40-15 (with tubes).

DIMENSIONS: Wheelbase, 8 ft. 11⅜ in., track: front, 4 ft. 7 in. (4 ft. 7½ in. with wire wheels); rear, 4 ft. 5⅜ in. (4 ft. 6⅛ in. with wire wheels); overall length, 15 ft. 0¾ in.; overall width, 5 ft. 6¾ in.; overall height, 4 ft. 9½ in.; ground clearance, 7 in.; turning circle, 33½ ft.; dry weight, 28¼ cwt. (with o/d, 28½ cwt.; with auto trans. 29 cwt.).

PERFORMANCE DATA: Top gear m.p.h. per 1,000 r.p.m. 21.6; top gear m.p.h. per 1,000 ft./min. piston speed, 31.4.

MARK IX

Continued unchanged.

XK150 (3.4-LITRE)

Continued unchanged (including "S" type).

XK150 (3.8-LITRE)

Specification as 3.4-litre XK150 except:—

STANDARD ENGINE: Dimensions: Cylinders 6; bore, 87 mm.; stroke, 106 mm.; cubic capacity, 3,781 c.c.; piston area, 55.3 sq. in.; valves, overhead (twin o.h. camshafts); compression ratio, 8 : 1 (7 : 1 or 9 : 1 optional). **Performance:** Max. b.h.p., 220 gross at 5,500 r.p.m.; b.h.p. per sq. in. piston area, 3.98.

"S" TYPE ENGINE: Dimensions as above. Compression ratio, 9 : 1; max. b.h.p. 265 gross at 5,500 r.p.m.; b.h.p. per sq. in. piston area, 4.97; carburetters, three S.U. HD8; fuel pumps, two S.U. electric; overdrive standard.

Limited-slip differential standard on these models.

The Autocar ROAD TESTS

1762

Jaguar 3.8 Mk 2 OVERDRIVE

Always a good looker, the Jaguar in Mk. 2 style with larger windows and slim pillars, is a much improved car

VERY few cars indeed set out to offer so much as the 3.8-litre Mk. 2 Jaguar, and none can match it in terms of value for money. In one compact car an owner has *Gran Turismo* performance, town carriage manners and luxurious family appointments.

The model in its latest Mk. 2 version as tested is the outcome of several years of development, since it was first introduced (with 2.4-litre engine) in September 1955. The changes made for 1960* without doubt represent together the greatest improvement so far achieved between a Jaguar model and its predecessor—short of a wholly new design.

Externally, the rather wider (by 3¼in) rear track is quickly noticed; it is not a very significant design change but it plays its part, with other minor improvements, in providing a more stable ride with increased resistance to roll when cornering. The 3.8-litre engine, which has an 87mm stroke compared with 83mm for the 3.4-litre unit, has lost none of its sweetness nor flexibility in producing an extra 10 b.h.p., and 240 lb ft torque instead of 215—at 3,000 r.p.m. What is more, this big twin-cam engine apparently has no objection to revving momentarily to its extreme limit of 6,000 r.p.m. This figure was touched or approached on several occasions during the test without the engine losing tune or showing any sign of distress. The normal limit—and there is little advantage in exceeding it—is 5,500 r.p.m.

When one examines the appearance and interior arrangements, marked improvements are apparent at once. The larger windows and screen, with slim chromium frames, brighten the car both inside and out. A new and much superior instrument layout has been introduced for the driver, and a central leather-trimmed console carrying the radio installation and heater control quadrants now merges with the transmission hump, and has a large covered ashtray buried in it. The steering wheel, too, is of pleasing new design, having two safety-flat spokes beneath a horn half-ring, and a rim that is comfortable to grasp. Matched tumbler switches are identified on a transparent strip beneath them, which has internal-glow illumination at night.

No one will have doubted that the performance of this 3.8 saloon would prove exceptional; a standing quarter-mile in under 16.5 seconds is matched by a 0-100 m.p.h. figure of 25.1 seconds and a maximum of 125 m.p.h. At the other end of the scale this four- occasional five-seater saloon will also glide silently in heavy traffic, snatch-free in top gear, down to 14 m.p.h. But the aspects of the performance which our drivers most appreciated were the smooth, silent cruising up to 100 m.p.h. in overdrive top on auto-routes at home and on the Continent, and the splendid acceleration for quick, safe overtaking, in direct top or third.

It is proper to turn attention directly from speed to brakes. The Dunlop 12in discs, as the recorded data indicate, are very powerful indeed. The pedal pressures, up to a maximum of 95 lb, are well matched against retardation which itself reaches almost the theoretical maximum. Repeated use of the brakes from high speeds caused no loss of stopping power, unevenness or increase in pedal pressures. The front brakes did feel "cobbley" when they became exceptionally hot, but recovered as they cooled.

The hand brake, a good big lever on the outside of the driver's seat, is satisfactory for all ordinary uses, though like most disc systems of its kind, it will not hold the car on a 1-in-4 test gradient. A red "on" warning light is provided which also signals if the brake fluid level is low. The brakes may be summed up by saying that they proved completely dependable throughout the test.

The car tested has a Laycock de Normanville overdrive, for top gear only, giving 26.4 m.p.h. per 1,000 r.p.m. in conjunction with a 3.77 to 1 axle ratio. The overdrive switch—a short lever mounted on the left side of the steering column, needs heavier spring loading on the test car; the brush of a sleeve will move it. Such is the power available that overdrive serves as a high top for nearly all out-of-town driving, and progress is the more silent and restful as a result.

Maximum speed is attained in overdrive, normal peak revs being reached in direct top at about 115 m.p.h. For smooth engagement of direct top from overdrive, plenty of throttle is desirable, and from top to overdrive, also some throttle.

No changes have been made in the familiar Jaguar gear box; the intermediate ratios are well chosen, and quiet in use. Leisurely changes can be made very sweetly, but the synchromesh is weak and the lever movements are rather long; when the box became hot the lever movement between bottom and second was stiff.

In the course of performance measurements a clutch naturally comes in for some heavy treatment; that on the 3.8 was very good indeed. Never was there any slip or roughness in engagement in spite of the very considerable torque transmitted. The pedal load is not unduly heavy, but a shorter travel would be appreciated.

Overall fuel consumption, including extensive performance testing at the Motor Industry Research Association track and on the Continent, was 15.7 m.p.g. for 1,690 miles. A brisk run of 221 miles in England returned exactly 17 m.p.g.; another of 349 miles, on both sides of the Channel, 18.9 m.p.g. A spell of gentle driving on British roads gave 21 m.p.g. The load was mainly two passengers and some luggage, but a full load was carried at times. Ordinary premium fuel was used without any signs of pinking. In France and Belgium the best available fuel suited the car well enough. Accommodation of the fuel tank has been something of a problem. Holding just over 12 gallons, it is smaller than that of the original 2.4 model. Total oil consumption during the test was 4.5 pints.

Suspensions seldom have to be designed for such a wide variety of speeds and conditions as that provided by the 3.8 Jaguar. Few owners will be other than well satisfied, for

---

* The Mk. 2 improvements were described in detail in *The Autocar* of 2 October, 1959,

there is sufficient firmness for fast driving, yet the ride is smooth and comfortable over all but the worst road surfaces. The pitching which a hump or ridge will induce is damped at once, and rolling movements are firmly restrained. From a following car, the suspension may be observed absorbing road irregularities with a whole variety of movements not transmitted to the passengers.

An important improvement in the Mk. 2 is the complete elimination of the rear axle hop or tramping which handicapped some earlier models when using full acceleration from a standstill. This may be largely the result of the fitting, as a standard feature, of a Powr-Lok limited-slip differential. For optimum get-away from a standstill, between 1,500 and 2,000 r.p.m. seemed best for clutch engagement.

Very little wheel noise or thump can be sensed in the car, and no reaction is felt at the steering wheel (the column of which now includes three universal joints). Over rough roads a slight body shake was noted at the front door pillars. With the windows closed, the interior silence is remarkable, particularly when overdrive is in use; an occasional moaning may be heard from the back axle.

As a result of careful design the quarter-lights may be opened even at high speed without causing more than very slight wind roar. They are spring-loaded on the test car to stay at about a 40 deg angle, which is the most silent and draught-free position, and there are also 90 deg and 130 deg positions. The exhaust note is very restrained, being scarcely audible inside the car.

There are three recommended pairs of tyre pressures, of which the normal is 25 lb sq in (1.76 kg sq cm) front and 22 lb sq in (1.55 kg sq cm) back. During the test the opinion was formed that the car handled better for general driving with an all-round increase of 3-4 lb sq in cold; this makes very little difference to the ride comfort. For fast cruising a 31/28 setting is recommended by the manufacturers, and maximum speed figures were obtained at 34/31—also according to the handbook. The handling of the car does not seem to be affected appreciably by the extent of the load carried, within normal limits.

Pronounced understeer is a characteristic of this Jaguar—the wide slip angle on the front tyres is clearly seen by an observer. The steering is better suited to the low and moderate speed sections of the car's wide range. Rather than comment upon what is a quite well-known design, it may be more helpful to discuss the feel and functioning. For manœuvring and parking the steering is rather heavy, and sufficiently low geared to require a fair amount of wheel-winding. Ordinary guidance in a traffic stream requires only light and gentle movements, while with more lock for sharp bends the heaviness increases. As speeds rise over about 50 m.p.h. there is increasing need to take bends early and to hold the car tight into them, otherwise it may swing wider than intended. There is quite powerful self-centring and this, no doubt, is in part responsible for the good line that the car holds on straight roads at all speeds. Up to about 100 m.p.h. side winds have little effect on the car; above this speed (and strong side winds were experienced at times during the testing) quick and delicate corrections were needed to hold course.

Tastefully trimmed and upholstered in hide and with polished veneers, the 3.8 interior has also a full complement of equipment and comforts. There are arm rests and pockets on all doors

Since there is sufficient power to spin the rear wheels quite readily in third gear on wet roads, and because the weight distribution is markedly in favour of the front, care has to be taken to use only light throttle when coming out of bends or away from corners; experienced drivers likely to be attracted by the 3.8 will adopt this technique instinctively. Incidentally, the stiff and notchy accelerator linkage of some earlier models has been redesigned; it is now smooth, light and progressive in its action.

Should the back end of the car slide, lifting the accelerator foot is usually enough to check it at once. Here, however, the low-geared steering—5 turns lock to lock with rather slow response around the mid-sector—is at a disadvantage, and it is difficult to apply quickly enough opposite helm to correct a skid at once. Worn tyres can have the effect of slowing the steering response. The turning circle between kerbs of just under 36ft is good in relation to the car's dimensions. On certain dry road surfaces there is some tyre squeal, but it is seldom at all obtrusive.

This Jaguar is one of the comparatively few cars in which any size of driver should be able to make himself comfortable. The seat has a long slide, the rails rising gently to their front ends, and the steering column has a convenient rake and a long adjustment. Obviously great care has been taken with the arrangement of hand controls and instruments. The pedals are not quite up to this high

Left: the neat, smart and convenient new panel arrangement and steering wheel. Right: a large and handsome engine in a small space. Most components likely to need routine attention have been raised to accessible positions

Jaguar 3.8 Mk. 2 . . .

Wide-track rear wheels and full-width window characterize the latest Jaguar model from this view, taken at Boom while the car was being tested in Belgium

standard, for there is a discrepancy in the levels of accelerator and brake pedals, and it is scarcely possible to heel/toe, though frequently desirable to do so. Certain drivers found the accelerator pedal (no longer of the organ type) and its angle of movement a little inconvenient.

Increased screen and window area have, of course, improved all-round vision. The scuttle no longer seems high, and the steering wheel rim is below the immediate horizon of bonnet top and front wings. Since the screen pillars and the sealing strips for the glass are thin, there is very little blind area. A wide new rear-view mirror, top mounted, takes advantage of the full-width rear window, and does not interrupt the forward view. Two sun vizors are neatly flush-fitted into the roof. They are effective, but when extended remain rather too close to the forehead.

In wet weather, vision is less good since the area swept by the driver's wiper blade is too small and stops well short of the bottom rail. The linkage produces a clicking noise. A toughened glass screen is normally fitted to cars delivered in the home market; owners may specify a laminated screen at an extra cost of £7 8s 6d inclusive of P.T.

At night, maximum speed must obviously be restricted, but the head lamps are powerful. As delivered, the setting was rather too low for maximum range with the result that cut-off, when dipped, was to the close, foreign standard. A head lamp semaphore switch is provided, operated by lifting the turn indicator lever on the steering column.

It is unusual these days to find side lamps which are fitted as separate units faired into the tops of the front wings, but there is much to be said for this arrangement. On the Mark 2 Jaguars the lamps are tiny and carry red "tell-tales" in their tops. A tall driver can just see the top of the near-side lamp from his seat.

Jaguars fit a pair of bright, semi-built-in auxiliary lamps. These are particularly good for winding secondary roads but seem rather too strong for straight main roads, since oncoming drivers regularly show disapproval. No fog was encountered to test them in those conditions. A bright reversing lamp is switched on with the engagement of reverse gear.

The excellent new instrument and control layout has already been mentioned. Among the supra-mundane

Auxiliary lamps are now flush-fitted in front of the horn grilles, so tidying up the frontal appearance. Diminutive side lamps are neatly faired into the wing brows; they have "tell-tales" on top

equipment are the cigarette lighter and handsome covered ashtray (two more neat, covered ashtrays are let into the rear arm rests); small polished veneer tables folding from the backs of the front seats; rear-seat heating duct; a map light and a separate blue light in the dash locker; a large tachometer paired with the speedometer; smaller matching dials for water temperature, oil pressure and ammeter in addition to the fuel gauge. Their illumination at night is clear and subdued, and no reflections were seen in the screen. There is no positive fuel reserve, but an amber lamp warns of the need to refill. The horns have a penetrating, high-frequency note.

Extensive use is made of hide leather and fine veneer in the interior trimming. The floor is carpeted, and reinforced under the driver's heels. The roof is cloth-trimmed, to meet customers' demand and Jaguar standards of silence, we learn. Large and comfortable seats are fitted; at the back there is a central folding arm rest. The front pair with advantage might have a little more curvature to give greater cornering location to the occupants.

If the front seats are set in their rearmost positions, knee-room for the back passengers is restricted. It is very easy to get in or out of all the doors, and sill height is such that, although the car is quite low-built, they do not catch a high kerb. The doors carry pockets and arm rests, and courtesy lamps are lit when a door is opened. The courtesy-cum-interior lamps are not, perhaps, in the best position— beside the rear passengers at the points where they might wish to rest their heads.

The test car is fitted with an H.M.V. radio, neatly built in, together with its speaker, below the instrument panel. The tone was good and there was no distortion at high volume. The aerial mounted in the offside front wing can be retracted or extended with the aid of a winder mounted horizontally under the instrument panel on the driver's side.

A comprehensive heating and ventilation system is neatly built in. A scuttle ventilator may be opened to admit fresh air, with the aid of a lever inside the map-slot beneath the switch line-up. Selector levers, flanking the radio, give varying degrees of heat and rate of delivery to screen or car interior, or both. A two-speed fan is provided which operates very quietly. The central air duct to the rear seats has already been mentioned. Since the engine is controlled to run at about 90 deg C, a greater heat supply might be expected; it was just sufficient on the test car for comfort when the outside temperature was at freezing point. The car interior is well sealed against draughts and fumes, and no water entered it in torrential rain.

Much luggage can be stowed in the boot. Should it be necessary to change a wheel, jack and handle are found in clips in the boot, and the spare is beneath the carpet, in a hatch in the floor. The screwjack is a sturdy, triangulated affair, with a lifting bar to slide into any one of four sockets. The quite elaborate tool kit is carried in rubber in a circular box which fits snugly into the well of the spare wheel. If the car jack is used to change a rear wheel, the semi-spats need not be removed, since the wheel drops on the springs; but a garage jack, giving a central lift, leaves insufficient clearance. In any case only two quick-release screw fasteners need be undone to remove a spat.

Like the lid of the luggage boot, the bonnet lid, hinged at the rear, is spring loaded; a control in the car releases the first catch. The engine compartment is a sight for enthusiastic eyes, being tightly filled by the big, bright,

Jaguar 3.8 Mk. 2 ...

twin-cam engine. Parts likely to need attention are fairly accessible and fillers are easily reached—oil, water, brake fluid, windscreen washer and enclosed battery (through the hinge gap of the open bonnet if one wishes). The oil dip stick is easily withdrawn, but difficult to replace; level readings should not be taken until the engine has been standing for some minutes, owing to the slow drain-down.

The Mark 2 3.8 Jaguar was introduced in October last year with a special eye on the American market, and with automatic transmission much in mind. Such a model is even better suited to American roads and journeys and offers even better value than its predecessors. The manual change model tested is a very distinguished car; it calls for an experienced driver to take full advantage of its great potential.

JAGUAR 3.8 MK.2 OVERDRIVE SALOON

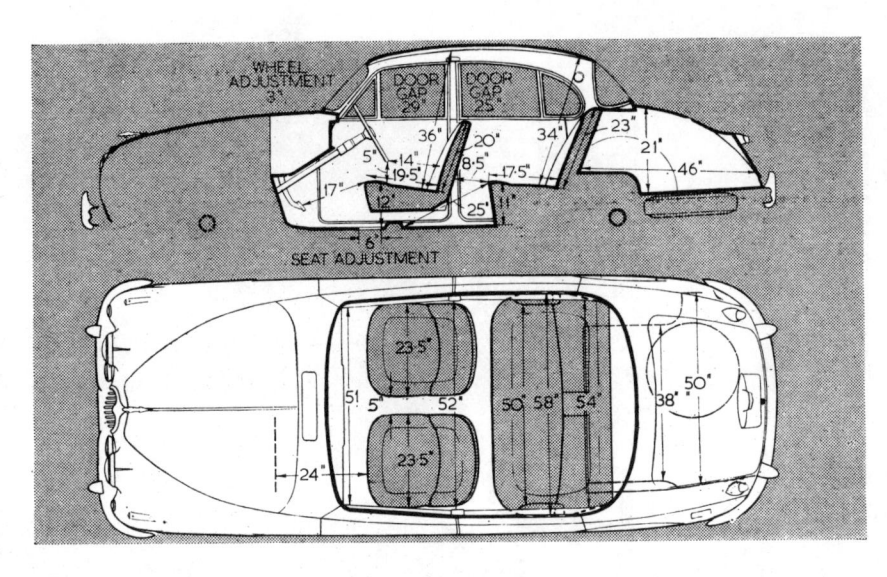

Scale ¼in. to 1ft. Driving seat in central position. Cushions uncompressed.

PERFORMANCE

ACCELERATION TIMES (mean):
Speed range, Gear Ratios and Time in Sec.:

M.P.H.	*2.933 to 1	3.77 to 1	4.836 to 1	7.012 to 1	12.731 to 1
10—30	—	6.7	5.1	3.1	2.5
20—40	—	6.0	4.7	3.4	—
30—50	8.5	6.1	4.9	3.6	—
40—60	7.8	5.7	4.2	—	—
50—70	8.6	5.9	4.9	—	—
60—80	9.3	6.3	5.3	—	—
70—90	9.5	7.7	6.5	—	—
80—100	12.7	9.7	—	—	—
90—110	19.2	15.5	—	—	—

*Overdrive

From rest through gears to:

30 m.p.h.	3.2 sec.
40 ,,	4.9 ,,
50 ,,	6.4 ,,
60 ,,	8.5 ,,
70 ,,	11.7 ,,
80 ,,	14.6 ,,
90 ,,	18.2 ,,
100 ,,	25.1 ,,
110 ,,	33.2 ,,

Standing quarter mile 16.3 sec.

MAXIMUM SPEEDS ON GEARS:

Gear		M.p.h.	K.p.h.
O.D.	(mean)	125.0	201.2
	(best)	126.0	202.8
Top		120.0	193.1
3rd		98.0	157.7
2nd		64.0	103.0
1st		35.0	56.3

SPEEDOMETER: 2 m.p.h. fast from 30—120 m.p.h.

TRACTIVE EFFORT (by Tapley meter):

	Pull (lb per ton)	Equivalent gradient
O.D.	270	1 in 8.2
Top	365	1 in 6.0
Third	455	1 in 4.9
Second	610	1 in 3.6

BRAKES (at 30 m.p.h. in neutral):

Pedal load in lb.	Retardation	Equiv. stopping distance in ft.
25	0.30g	100
50	0.49g	61
75	0.80g	38
100	0.96g	31.4

FUEL CONSUMPTION (at steady speeds):

	Direct Top	O.D. Top
30 m.p.h.	27.7	31.7
40 ,,	24.3	29.8
50 ,,	23.5	27.7
60 ,,	21.5	25.0
70 ,,	20.4	23.2
80 ,,	17.1	20.0
90 ,,	14.9	17.5
100 ,,	12.2	15.5

Overall fuel consumption for 1,690 miles, 15.7 m.p.g. (18.0 litres per 100 km.).
Approximate normal range 15—22 m.p.g. (18.8-12.8 litres per 100 km.).
Fuel: Premium grade.

TEST CONDITIONS: Weather: Dry, light wind—strong gusts.
Air temperature, 48 deg. F.

STEERING: Turning circle:
Between kerbs, L, 35ft 10in, R, 35ft 9in.
Between walls, L, 37ft 9in, R, 37ft 8in.
Turns of steering wheel from lock to lock, 5.

DATA

PRICE (basic), with saloon body, £1,300.
British purchase tax, £542 15s 10d.
Total (in Great Britain), £1,842 15s 10d.
Extras: H.M.V. radio £47 15s 2d.

ENGINE: Capacity, 3,781 c.c. (230.6 cu in)
Number of cylinders, 6.
Bore and stroke, 87 × 106mm (3.425 × 4.173in).
Valve gear, 2 overhead camshafts.
Compression ratio, 8 to 1.
B.h.p. 220 at 5,500 r.p.m. (B.h.p. per ton laden 133.7).
Torque, 240 lb ft at 3,000 r.p.m.
M.p.h. per 1,000 r.p.m. in top gear, 20.5; O.D. top 26.4.

WEIGHT: (With 5 gals fuel), 29.9 cwt (3,350 lb).
Weight distribution (per cent); F, 56.5; R, 43.5.
Laden as tested, 32.9 cwt (3,686 lb).
Lb per c.c. (laden), 0.97.

BRAKES: Type, Dunlop disc with quick-change pads.
Method of operation, hydraulic vacuum servo assisted.
Disc diameter: F, 11in; R, 11.375in.
Swept area: 495 sq in total (300 sq in per ton laden).

TYRES: 6.40—15in Dunlop Road Speed.
Pressures (lb sq in): F, 25; R, 22 (normal).
F, 31; R, 28 (fast driving).

TANK CAPACITY: 12 Imperial gallons.
Oil sump, 13 pints.
Cooling system, 20 pints.

DIMENSIONS: Wheelbase, 8ft 11.375in.
Track; F, 4ft 7in; R, 4ft 5.375in.
Length (overall), 15ft 0.75in.
Width, 5ft 6.75in.
Height, 4ft 9.5in.
Ground clearance, 7in.
Capacity of luggage space: 13 cu ft (approx).

ELECTRICAL SYSTEM: 12-volt; 67 ampere-hour battery.
Head lights, Double dip; 60-36 watt bulbs.

SUSPENSION: Front, Independent, wishbones, coil springs, telescopic dampers, anti-roll bar.
Rear, Cantilever, half elliptic springs, radius arms, Panhard rod, telescopic dampers.

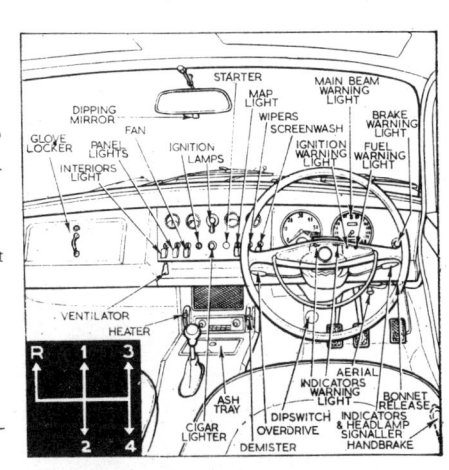

MARCH H

Below: At the top of the Jaun Pass in Switzerland, the snow lay firm and crisp and very uneven. It was only possible for the chainless Jaguar to achieve this route in winter by virtue of its Powr-Lok differential, a good deal of trials driving technique and some manual help from the passengers.

WHENEVER confronted by a long journey we try and make the motorcar fit the route. In fact it seldom works out and on occasions we have ended up by driving a 7-passenger limousine single-handed in the North of Scotland and sweltering in a 100-m.p.h. sports car in London traffic but, for the Geneva Show this year, we really planned the operation with care.

The idea was to drive a 2.4-litre Jaguar to Switzerland and back, incorporating this run as part of a road test, and we calculated that, as a mount, the Mark 2 version would give three of us an effortless and economic ride. Passages were booked, forms filled in, and then, at the very last moment, something went wrong in the Coventry area and we found ourselves heading for France in the latest 3.8 version of the Jaguar range with "normal" gearbox plus overdrive, and the certainty of a three-figure cruising speed and a maximum not far short of 130 m.p.h. To add to the general frenzy another plan had to be hatched in order to fit in Continental driving experience on a 6.2-litre V-8 360 b.h.p. Facel Vega, concerning which our regular readers will, last week, have seen a full road test report.

None of us really likes the Dover-Dunkirk night ferry, but there cannot be any more effective way of rising early than being thrown on to a semi-deserted French dockside at 4 o'clock in the morning, and this has a profound effect upon what is possible by nightfall. For those who this summer wish to follow in

by CHRISTOPHER JENNINGS, *Editor of The Motor*

Two Cars, Three Drivers and 580 b.h.p.

An impressive front to a very impressive car. The Facel Vega is the result of some fine blending of European and American ideas.

Quiet flows the Seine while a contented crew relax in the excellent Hostellerie de la Seine, Polisot, Aube.

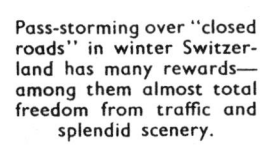

Pass-storming over "closed roads" in winter Switzerland has many rewards—among them almost total freedom from traffic and splendid scenery.

our footsteps, I have two words of advice. One concerns a wholly admirable little place called the White Horse Inn at Sandway, near Lenham, Maidstone, Kent, where they dined and wined us well at no great cost and where, when an urgent telephone call to London broke down, the landlord himself re-established communication with the exchange from a neighbour's house as part of the unsought obligation of a true host. The other piece of incidental intelligence concerns the cabins on the night ferry which, on the British boats, are adequate if they are on the outside of the ship and airless furnaces if they are not. The French boat, being much newer, does not have this trouble.

We had a little time in hand before our duties commenced at Geneva, so we set our sights for Lucerne and by lunchtime had covered 337 miles of the 450 or so involved. Here we made an error of judgment when we forgot to check the prices of meals at an attractive-looking hotel rightly endowed with Michelin's highest qualifications. The lunch was splendid but the bill sobered us for the next 48 hours! We agreed that at our time of life we should know better than to disregard the full implications of the gastronomic guide.

Our main reactions to the dash across Northern Europe were the great value of the green "come on" signs now exhibited by many large French long-distance lorries—the surprising fact that the Kleber Colombes tyre advertisement uses the English word "tubeless," and the admirable warning given at night by cyclists in Switzerland who, being licensed, have to carry a small number plate at the rear which is treated to reflect a strong white glow.

The Technical Editor and my wife were both brought up on a motoring diet of trials driving, and so it came about, inevitably, that if one route was considered easy and the other impossible, we always took the latter. On the Jaun

Only the fact that there was no other traffic in sight for miles justifies this remarkable picture taken through the windscreen by the driver of the Facel Vega as it pursued the 3.8 at resounding speed across France.

MARCH HARE

Pass in Switzerland, which was supposedly closed, the idea was to see if the limited-slip differential would prove a good substitute for chains on snow. Practising a few tricks which must have been acquired from a war-time tank driver, combined with skill and patience, Joseph Lowrey finally got us over the top, but not before I had come to discover what it is like to stand behind a bogged-down car when 200 h.p. takes possession of one back wheel. On this section our m.p.g. fell from a hitherto recorded figure of nearly 20 to just under 17.

Further desire to discover the worst at all costs took us up to Verbier in a heavy snowstorm. But this at least produced a magnificently simple lunch of the kind that never seems to happen outside good winter sports resorts. It also enabled me to discover a unique and highly attractive little locomotive from a bygone age.

Geneva was less interesting as a motor show than as a place to meet leaders of the automobile industry from all corners of the world. It was refreshing to find how many of our own top men had driven themselves from England, and some of them had had more than a fair share of ice and snow on the last leg over the Col de la Faucille. It was with regret that we could not find time to drive the 2100 Fiat married to the new Smiths automatic transmission, because our instincts told us that this might prove a very attractive combination. My wife and I did, however, motor far and fast in a Chrysler Valiant which coupled admirable steering and roadholding with seats so shiny that even the steering wheel proved insufficient as a hand grip. I also went for a rousing run with Lance Macklin in the little Facellia coupé, a vehicle which may one day challenge the Porsche at Le Mans. Perhaps the most interesting session of all arose from a discussion with an American who forecast that General Motors' next move would probably be to add two more cylinders to the Chevrolet air-cooled Corvair engine, thus making it a "flat eight," and then transplanting it so that the final result would be a front-wheel-drive Buick of about 3½ litres!

The run home was like something out of a chapter from the late Dornford Yates. With three of us to share two of the most desirable cars in the world and a glorious spring day, we were able to leave Geneva comfortably after breakfast, have a prolonged lunch at the delightful Hostellerie de la Seine at Polisot, Aube, and yet reach Le Touquet in time to unload our luggage at the only hotel which was open and then drive back towards our beloved Auberge de la Grenouillère near Montreuil for dinner. There we duly celebrated one of the great motoring days of our lives. The memory of the Facel Vega with its incomparable feeling of power hurtling down the tree-lined, deserted roads of France in pursuit of the Jaguar, with both speedometers far beyond the magic 100, will remain with me when much else is forgotten.

All over Switzerland the miniature tractor is coming into its own as a load-tower and machinery driver. But this was the first and only snow plough attachment which we encountered.

West of St. Maurice in the Rhone Valley, a filling station has preserved a delightful little locomotive named Herbert. Already the owner of one historic steam engine, the Editor was quite unable to resist a prolonged investigation. *Below:* Photographed by Ing. Dott. Carlo Felice Bianchi Anderloni of Carrozzeria Touring, this view of the sunburnt Technical Editor sitting in the Maserati at Geneva is somehow reminiscent of the Demon King!

GRACE . . . SPACE . . . PACE

With over 30 new luxury and safety features

THE NEW 2·4, 3·4 AND 3·8 LITRE

Mark 2 JAGUAR

MODELS

Already enjoying pride of place in the esteem of press and public alike for exceptional performance, Jaguar now introduce a new conception of safety and luxury in these lavishly equipped new models. 18% greater visibility is provided by new slim pillars, semi wrap around windscreen, greatly enlarged rear window and an overall increase in all window areas. The entirely new instrument panel layout has matched and grouped dials and switches as is customary in aircraft practice.

New comfort and safety features include interior heating carried to rear compartment, "brake-fluid-level" warning light, courtesy lights operated by all four doors, finger-tip controlled head lamp flasher independent of foot-operated dip switch and flush-fitting folded tables in rear of front seats. These and a score of other refinements recommend the Mark 2 models to all who demand the utmost in high performance motoring in a car of compact dimensions.

London Showrooms: 88 PICCADILLY W.1

The New Mark 2 Jaguars have Dunlop race-proved Disc Brakes **on all 4 wheels**

JAGUAR'S

Ten miles cruising at 100 m.p.h., newly furnished interior, terrific brakes and first class handling have sent **PETER COSTIGAN** *into raptures over the Mark 2 version of the 3.4-litre.*

IT'S hard to talk or write about Jaguar cars without a feeling of awe — or, at least, deep respect.

You can't help remembering that the Jaguar has been one of the greatest success stories in the history of the motor industry, has been one of its finest engineering achievements and has done more, perhaps, than any other make, to bring the thrills (and performance) of the race track to essential day-to-day transport.

Not every accelerator puncher has been able to afford a Jaguar, but its makers have brought the feeling of thoroughbred motoring within reach of a much larger section of the community than ever before.

Of course, heavy customs duty on all imported cars has not helped things much in this country, and the Jag is relatively much more expensive here than it is in Britain.

But still, anyone with a nose for motor cars follows Jaguar developments with great keeness and never more so than when this vigorous independent company released a modified series of their brilliantly successful 2.4 and 3.4 litre range at last year's London Motor Show.

The standard 2.4 and 3.4 litres remained (and will remain) in production but a new model, the 3.8 litre, has been added to the range. It has the same body as the other Mark 2 models, but is powered by the highly potent 3.8 litre engine and has a limited slip differential. But most of the 3.8 litres are going to the United States and we are not likely to see them for some time, and then only in small quantities.

The 2.4 and 3.4 litre Mark 2 models have started to arrive and are being snapped up eagerly by well-heeled Jaguar fans. It was rather a thrill then, to get the chance of road-testing the 3.4 litre Mark 2 as soon as Brylaw Motors in Melbourne had run one in. A surprise, too, because full-scale tests of Jaguars in this country have been hard to come by.

But before analysing the test, here are the major changes that have been made to the Mark 2 models:—

● BODYWORK. The windscreen is higher and wider, the pillars narrower. The door windows are also bigger and the rear quarter lights curve deeper into the metal corner. The rear window is much wider and higher and there is more chrome plating around all the glassware. The fog lights are set into the front bodywork, not sitting on the bumper bar as in the standard models, and the radiator grille ribs are heavier.

Inside, there are a number of changes, most important of which is the dashboard. The big speedo and rev. counter are now placed, side by side, right in front of the driver, not uselessly out of the way in the centre of the panel as in the standard model. The panel is occupied by the smaller gauges (ammeter, fuel, water temperature and oil pressure) and a row of magnificent black toggle switches that work with a healthy click. Below the centre panel is provision for a radio, the heater controls and, recessed into the transmission hump, the best car ashtray I've ever seen — big and just where the driver can use it without taking his eyes off the road.

The steering wheel is now a graceful two-spoker with a half-circle horn ring, instead of the former four-spoke, horn-button job. The heater now incorporates a separate outlet to the rear compartment. The bucket seats have been slightly redesigned for greater comfort and

The heart of the beast! Under that glossy exterior lurkes Jaguar's extremely willing 3.4-litre motor. It has twin overhead camshafts, two SU carburettors and the crankshaft runs in seven bearings.

GLOSSY NEW CLAWS

wheels FULL ROAD TEST

have flush-fitting tables on their backs.

● MECHANICAL AND CHASSIS: The most important mechanical feature of the Mark 2 models is the fitting of servo-assisted disc brakes on all wheels as standard equipment. (On the standard models they are an optional extra). Jaguar drivers who constantly got caught on tram lines will be pleased to know that the track has been widened on the Mark 2 range, by ⅜ in. at the front and by 3-4 in. at the back.

The engines are unchanged except for the addition of the 3.8 litre model to the range.

The car I tested was the top luxury model of the line — a 3.4 litre equipped with Borg Warner automatic transmission and disc brakes. A gearbox man to the end, I was a bit disappointed it wasn't the less expensive four-speed gearbox with electrically controlled Laycock overdrive. But most 3.4 litre Jags — both standard and Mark 2 — now being sold are automatics, and I was curious to see why the preference. Particularly as many of the automatic buyers, in Melbourne anyway, are among the most enthusiastic car drivers I know, including several racing drivers.

I learned why quickly. This particular automatic transmission, with a driver controlled intermediate gear hold is about the best automatic you can get anywhere. Power loss seems to have been minimised and the intermediate gear hold gives the driver almost as much control as the manual gearbox.

On the road, the Jag behaved with impeccable, thoroughbred manners. At one stage of the test, I found myself early in the morning on a near-Melbourne four-lane superhighway. I hadn't looked at the speedo for some time and had had only a couple of glances at the rev. counter, noting the engine was loping along at about the 4250 mark. I felt I had found its best fast cruising speed, and after 10 miles I glanced at the speedo to record for history what story it told — the needle was sitting as evenly on the 100 mark as the car was sitting on the road.

Smoothed out lines of the 3.4-litre give it new appeal. This is the Mark 2 model distinguished by its built-in fog lights, centre strip in the grille and more glassware.

JAGUAR'S NEW CLAWS

I have driven plenty of cars that had top speeds well over the ton, even two (the M.G. twin-cam and Repco's experimental car) that have a higher top speed than the Jag, although I think under the right circumstances, the Jag would outpace them both (see later) but I have never been in a road car that took 100 m.p.h. cruising so easily in its stride.

The car was absolutely stable, the steering behaved exactly as it did at half the speed, there was no lightening of the front end, no sensitivity to cross winds or irregularities in road surface. The engine note was no louder than a purr and wind noise was slight. There was so much power left that a jab on the accelerator would send it accelerating hard and without effort. Corners taken at this magnificent cruising speed disappeared as soon as they were seen. There was no drift, lurch or roll. The whole car gave an indelible impression of utter safety.

Now, a word about top speed. My best run over the flying quarter mile registered 7.7 seconds, giving the Jag a true top speed of 116.8 m.p.h. At that speed the rev. counter registered just over 5000 r.p.m. Recommended top engine speed is 5500 r.p.m. (which I reached several times in the lower ratios) and it will do 6000 revs easily. The car I tested had done 1800 miles and I felt had not yet fully settled in. Indeed, I was the first driver to take it over 70 m.p.h. Again, my test course had a comparatively short run up of about three miles. Given more miles, both on the speedo and run up, there is no question that 5500 r.p.m. would have been reached and the top speed would have been well in excess of 120 m.p.h.

WHEELS tester Peter Costigan was highly impressed by the wonderful interior layout, including sensible ash-tray. Note simple dials and superb toggle switches. This is the automatic model.

That would clinch the issue, but even my top figure makes the 3.4 Jaguar Mark 2 the fastest stock, standard saloon car available on the Australian market. It would also be among the fastest anywhere in the world.

Much bigger rear window, new tail lamps and a wider rear track make the Jag Mark 2 instantly recognisable from the back. There is also more chrome, more glass and better handling.

The steering qualities matched its brilliant performance. I threw it round all manner of corners, fast bitumen ones, slow gravel ones, pot-holed ones, even a devil's elbow or two and was unable to find a trace of either oversteer or understeer. You chose a line through a corner, pointed the Jag along the line and just forgot about the rest. The car always went exactly where it was told to go.

On rough country, the independent coil springs at the front and quarter elliptics at the rear gave the 3.4 litre a level comfortable ride, pleasantly on the firm side. The front end had been stiffened slightly on the Mark 2 and there is none of the standard model's tendency to bottom over deep holes and gutters.

Driving position will probably continue to be a matter for argument among enthusiasts. It is the same as the standard model 3.4 litre, but the seat is slightly higher, thanks to the heavier padding.

Many tall drivers complain their knees get a bit of a beating from the low dashboard and steering wheel. I'm 5 ft. 11 ins. tall and I found the driving spot quite good. It was a bit strange at first. The wheel was nearly a 45 degree incline, nothing like the vertical rake one comes to expect from sporting machinery and with a bit of imagination I could have been sitting at the wheel of a very glamorous bus.

But it didn't take long to get used to.

CONTINUED ON PAGE 77

Technical Details

SPECIFICATION

MAKE:
Jaguar 3.4 litre, Mark 2.

AVAILABILITY:
Small delay, depending on ship arrivals.

PRICE:
£2986.

ENGINE:
Cylinders, six; pattern, in line; valves, overhead with twin overhead camshafts; bore and stroke, 83 m.m. x 106 m.m.; compression ratio, 8 to 1; b.h.p., 210 at 5500 r.p.m. carburettors, twin SU H.D.6; fuel pump, S.U. electric; capacities, fuel tank 12 gals.

TRANSMISSION:
Type, Borg Warner Automatic with driver controlled intermediate gear hold. Gear ratios: low 17.6 — 8.16, intermediate 10.95 — 5.08, direct top 3.54. Hardy Spicer propellor shaft. Hypoid rear axle.

BODY AND CHASSIS:
All-steel integral body-chassis construction.

SUSPENSION:
Front, coil spring and semi-trailing wishbones; rear, cantilever quarter-elliptic springs; shock absorbers, telescopic.

BRAKES:
Type: Disc hydraulics with servo-assistance. Handbrake, mechanical on rear wheels.

STEERING:
Type: Burmann re-circulating ball.

ELECTRICAL EQUIPMENT:
Voltage, 12; 60 amp/hour battery. Standard features; the usual plus fog lamps, twin speed wipers, 4 courtesy lights, map light, glove box light.

WHEELS AND TYRES:
Type: Pressed steel disc. Tyre size, 6.40 x 15.

DIMENSIONS:
Wheelbase, 8' 11⅜"; Track, front 4 ft. 7 in., rear, 4 ft. 5⅝ in.; Overall length 15 ft. 0¾ in.; overall height, 4 ft. 9½ in.; overall width, 5 ft. 6¾ in.; ground clearance, 7 in.

PERFORMANCE

TOP SPEED:
116.8 m.p.h. (at 5000 r.p.m.)

MAXIMUM SPEED IN GEARS:
In low range, 51.4 m.p.h.
In intermediate range, 85.7 m.p.h.

SPEEDOMETER CALIBRATIONS:
Indicated 30 m.p.h., actual 30 m.p.h.; indic. 40, actual, 38; indic. 50, actual, 47.9; indic. 60, actual, 57.8; indic. 70, actual, 67.6; indic. 80, actual, 78.2; indic. 90, actual, 88.5; indic. 100, actual, 99; indic. 110, actual, 110.

ACCELERATION:
Standing quarter mile, average: 18.35 sec. best time, 18.1 sec.
Through gears: 0-30, 3.6 sec.; 0-40, 5.7 sec.; 0-50, 7.5 sec.; 0-60, 10.6 sec.; 0-70, 13.9 sec.; 0-80, 17.8 sec.; 0-90, 23.3 sec.; 0-100, 29.7 sec. Top gear: 20-40, 4.4 sec.; 40-60, 4.8 secs.

FUEL CONSUMPTION:
Over 162 hard driven test miles, 15.4 m.p.g.

WEIGHT:
30 cwt. with driver, test equipment.

TEST CONDITIONS:
Very hot, no wind. Surface: All performance figures recorded on level, bitumen-bonded gravel surface. Figures averaged from several runs in opposite directions.

ROAD TEST

JAGUAR 3.8 SEDAN

A family car that puts the sport in transportation

IN JUST 12 YEARS the British Jaguar has become virtually synonymous with the term "expensive foreign car" in the minds of the American public for, if you ask people you meet at random to name one or more imported cars, Jaguar always comes out on top of the list. All this has been done with a meager advertising budget and fewer than 5000 cars per year available. How was it done? The answer, obviously, lies in the product itself for, in the final analysis, a car of the Jaguar class is bound to be recognized as something out of the ordinary.

To obviate the "expensive" connotation, Jaguar introduced a lower priced compact, the 2.4 sedan of 1955/56. This car was not a success on the American market, so two years later the 3.4 was announced—a very high performance compact, with no less than 210 honest bhp. Despite a higher price tag, this car immediately caught on. Now, for 1960, we have a further refined and improved compact model from Jaguar, known as the 3.8.

Although the 3.8 is still more expensive, it offers so many advances over the previous 2.4 and 3.4 models that it is definitely worth every penny—provided only that one is looking for a car of this category: i.e., a high-quality, high-performance sedan of sensible size. As with the 3.4 sedan, the 3.8 is a very special type of automobile. It has even been described as a compact Cadillac —though the Jaguar people probably don't care much for this description. In fact, Jaguar calls this model a sports sedan and while we, as purists, object to this terminology, the type of performance illustrated by 0 to 100 honest mph in 25 sec certainly justifies the description "a sedan with sports car performance."

While there are innumerable chassis and body improvements, the most important change is undoubtedly the use of a larger engine; the actual piston displacement increase being from 3442 to 3781 cc, hence the designation 3.8 for the approximate size in liters.

The actual power output and concomitant top speed have not been increased appreciably. Instead, the designers have chosen to improve torque by 11% and the net result in terms of driver "feel" is very noticeable when driving or riding in the car. That the above impression is not imaginary is shown by comparing the Tapley pull readings in each gear, the net gain being almost exactly in line with the 11% increase in torque:

Gear	Ratio	Tapley Readings, lb/ton 3.4	3.8
od	2.94	210	230
4	3.78	280	300
3	4.84	360	400

With more torque, and pulling power, we find that acceleration times through the gears are also improved, as should be expected. In both our 1957 test and in this one, the driver used a degree of restraint and we feel that the figures we quote for either car can be duplicated by anyone, and, with more "vigorous" techniques, it should be possible to improve on our data slightly. Here is a quick comparison of acceleration times:

	3.4	3.8
0-30	3.4	3.0
0-60	10.5	9.2
0-100	27.5	25.1
SS 1/4	17.6	17.0

Top speed is pretty much academic. Our test car, which, incidentally, belongs to contributor Bill Corey and had 6000 miles on the odometer, has indicated 130 mph on an instrument which shows very little error in the upper speed ranges (due to tire expansion at 100 mph and up). In this case the car was equipped with the standard 3.78 axle, and with overdrive the effective ratio becomes 2.94. Theoretically, a speed of 146 mph would correspond to 5500 rpm (in overdrive), but the combination of power available and wind resistance is such that this figure is not attainable. In short, the speed of this car is far above any possible use, but the ability to cruise safely and comfortably at a continuous 100 mph is certainly there if road conditions permit. Corey says that he drove as above for hour after hour

A comfortable compartment for the head of a sporting family.

across Arizona and Nevada and got just a fraction under 20 mpg in the process.

Cruising speed brings up the question of comfort, and everyone who rides in the Jaguar expresses amazement over its comfortable ride. The car is particularly surprising to those people who own and drive nothing but the largest cars. The rear seat is especially luxurious, thanks to ample leg room, high seat backs and a useful center arm rest which folds out of the way when not required. A particular feature of the riding quality is that it seems so well controlled over rough and wallowing roads. In our opinion, few, if any, cars of this category can keep pace with the Jaguar when the going gets rough.

The new power steering is one of the few we have tried that we liked. While the average man driving the previous model felt no need for power steering, many women did object to a slight heaviness, which increased to a considerable force during parking. The new system eliminates this objection, and its particular virtue is that it is absolutely impossible to detect when the power comes on. Thus, the oft-encountered lumpiness at either side of straight ahead is not present and, as a matter of interest, several people drove the car without suspecting that it had power steering.

The "old" cars handled very well for sedans, this feature being particularly appreciated by a driver used to a standard American car. On the other hand, a driver used to any of the more popular sports cars would, generally, dislike the feel of the 3.4 models, particularly the very pronounced understeer when cornering hard and fast. Michelin X tires would completely transform the general feel of the 3.4 but, unfortunately, parking effort went even higher and the car's normally excellent high speed stability decreased. The 3.8 has new suspension geometry with roll center raised from ground level to a point 3.25 in. above. Power steering makes it difficult to evaluate this change because rim pull in a high speed bend is now very light. Nevertheless, the car definitely handles better than before and body roll appears to be less, as it should on theoretical grounds. The new model also turns in an appreciably smaller circle than before. Our only criticism of the steering would be that, with power, it could have been made closer to 4 turns lock to lock, rather than 5.

Our 1957 test car had the Jaguar C-type close ratio gears which, in our opinion, were just about perfect. However, for reasons known only to Jaguar, they have again "juggled" the ratios; 3rd and 2nd are close to the C-type, but first is the old creeper gear of earlier days. As before, first gear is not synchronized and the gears are rather noisy. The only advantage of the wide-ratio box appears to be that the synchronized 2nd gear is quiet and quite sufficient for normal starts. Thus, the provision of 5 speeds forward (via overdrive) seems superfluous and overall ratios of the order of 3.0, 4.0, 5.5 and 9.0 would do the job just as well, if not better. However, the 4-speed with overdrive model is primarily for the enthusiast and most of the cars coming over are equipped with the 3-speed-plus-converter type of automatic transmission. While this item takes all the fun out of driving a car such as this, we must admit that the performance is still fantastic and a driver used to, say, a four-passenger Thunderbird, will find the Jaguar a tremendous advance in terms of roominess, accuracy of steering and ability to cruise safely and comfortably at very high speed.

One might think that fuel economy in a $5000 car would be unimportant, but many owners of even more expensive cars are quite fussy about this item of the budget. Here the Jaguar scores heavily with never less than 17 mpg under the worst possible conditions. In our own experience, involving two 3.4's (a 1957 and a 1959) and a total of 34,000 miles, the day-to-day figure never went below 19.5 mpg. This is only 1.0 mpg worse than we are currently getting from a popular U.S. compact on our staff, with 130 *less* bhp! A consumption of 23 mpg is easily possible, and quite normal, using overdrive and cruising at around 70 mph. The automatic version will get 21 to 22 mpg under the same conditions, which is identical to the *best* figure we get from the previously mentioned U.S. compact, with stick shift.

Although the 3.8 is a refinement of the 2.4 and the 3.4, external and internal changes are quite extensive. Most important is the new "cab" with vastly improved visibility in all directions. The front corner posts are now very narrow and the rear window much larger. The rear tread has been widened from 50.1 to 53.4 in. and the optional wire wheels add another 0.7 in. to this. Thus, the former crab-track effect is greatly reduced and this is a change we approve. Formerly the car had a most annoying tendency to wobble about on car tracks or on certain types of lateral ridges in the road. Also, in our opinion, the general appearance from the rear is much improved by this change.

The interiors are absolutely sumptuous, even by previously high Jaguar standards. While last year's 3.4 model was very nicely finished (and is continued as the Mark I version, by the way), the interior of the new 3.8 Mark II model is completely re-designed throughout. The speedometer and new electric tachometer are now directly in front of the driver. The rest of the instruments, which, incidentally, are all-electric and very complete, are located in a central panel which is hinged and can be dropped for access in one minute, and without tools. At the bottom of this central panel is an imposing array of

The 3.8-liter dohc six fills the engine compartment.

Larger rear window aids driver vision, and safety.

switches, each labeled in translucent letters which glow dimly at night so that they are plainly visible at all times.

As before, the parking brake is located alongside and at the left of the driver, a position that has much to recommend it on cars with individual front seats. A light on the dash glows if the brake is not released and this same warning light is wired to show when brake fluid is needed. Linkage in the parking brake system has received a much needed modification—it will now hold the car on a moderate grade where, before, it was almost useless. Genuine leather and varnished wood trim are continued, of course, in Jaguar tradition.

Having now dwelt at some length on the favorable aspects of the Jaguar, we must also point out some of the bad. We receive quite a few complaints from Jaguar owners. These can be tabulated by frequency about as follows:
1. Speedometer and tachometer failure
2. Trouble with the brake booster
3. Minor electrical problems
4. Automatic choke doesn't work
5. Engine leaks oil
6. Timing chain rattles
7. Floor mats wear out quickly
8. Paint fades and blisters

Having had no small experience with these cars (the 2.4 and 3.4), we must agree that the above list is fairly typical of our own cars. It is interesting to note how these troubles have been corrected. The speedometer trouble was caused by oil coming up the cable. This was cured by a re-designed seal, which attaches to the gearbox drive outlet. The electric tachometer solves that problem, and also eliminates item 5 where oil leaked profusely out of the tach drive fitting. The brake booster trouble was caused by gasoline running down to it via the rubber vacuum line. This line now takes off from the top of the manifold, not the bottom (where liquid fuel accumulates during cold starts). Electrical problems are somewhat difficult to avoid on any modern car, owing to the complexity of the system. The hinged center panel mentioned above at least makes the most inaccessible and most complicated portion of the system easy to work on. Jaguar's automatic choke seems particularly allergic to dirt and sand. Since it is electrically controlled, the standard "fix" is to put a driver-controlled switch on the dash. This could be wired to the brake warning light to help the driver remember to turn it off. The Jaguar's timing chain layout provides a hydraulic back-lash device, plus an external adjustment for wear. This needs attention every 10,000 miles or so, and the owner's manual tells how to do it.

The external finish problem is not unknown on domestic cars, though we admit that Jaguar's white, in particular, shows surprising differences in color, from panel to panel, even when the cars are new. This is a quality control problem for the manufacturer, as are the front floor carpets. From the foregoing it is obvious that the company is making every effort to improve the cars.

Incidentally, the price of $4795 quoted in our data panel includes heater and overdrive, this car not being available in the U.S. without them. (Despite changes, the heater still is not satisfactory for zero weather operation.) Automatic transmission adds $100 to the price, power steering is $130 extra and wire wheels cost $117, or $278 if chrome plated. Thus, without tax and license the full list price, as tested, comes to $5243, with whitewall tires. Dealers can install a well designed proprietary air-conditioning unit for about $700.

Summed up, the Jaguar 3.8 Mark II sedan may not be everyone's cup of tea, but it is exactly the kind of car we have been advocating for many years: a compact, high-performance, high-quality family sedan for $5000.

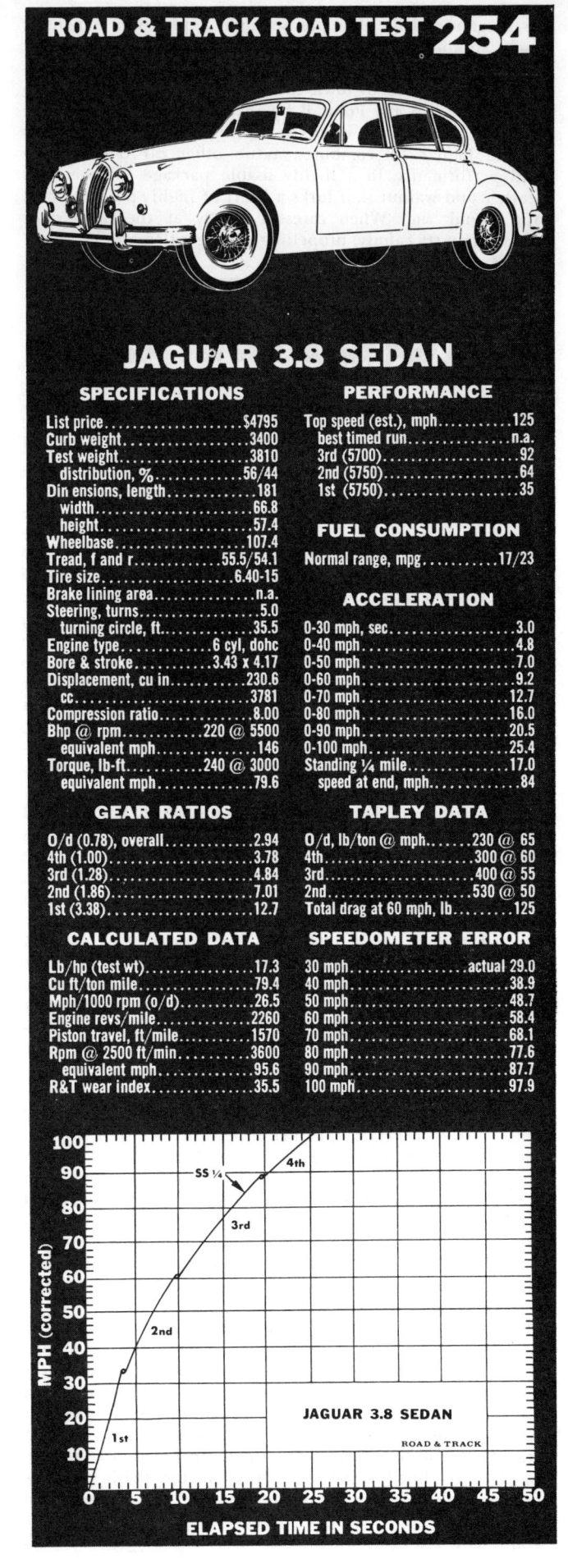

ROAD & TRACK ROAD TEST 254

JAGUAR 3.8 SEDAN

SPECIFICATIONS
List price	$4795
Curb weight	3400
Test weight	3810
distribution, %	56/44
Dimensions, length	181
width	66.8
height	57.4
Wheelbase	107.4
Tread, f and r	55.5/54.1
Tire size	6.40-15
Brake lining area	n.a.
Steering, turns	5.0
turning circle, ft	35.5
Engine type	6 cyl, dohc
Bore & stroke	3.43 x 4.17
Displacement, cu in	230.6
cc	3781
Compression ratio	8.00
Bhp @ rpm	220 @ 5500
equivalent mph	146
Torque, lb-ft	240 @ 3000
equivalent mph	79.6

GEAR RATIOS
O/d (0.78), overall	2.94
4th (1.00)	3.78
3rd (1.28)	4.84
2nd (1.86)	7.01
1st (3.38)	12.7

CALCULATED DATA
Lb/hp (test wt)	17.3
Cu ft/ton mile	79.4
Mph/1000 rpm (o/d)	26.5
Engine revs/mile	2260
Piston travel, ft/mile	1570
Rpm @ 2500 ft/min	3600
equivalent mph	95.6
R&T wear index	35.5

PERFORMANCE
Top speed (est.), mph	125
best timed run	n.a.
3rd (5700)	92
2nd (5750)	64
1st (5750)	35

FUEL CONSUMPTION
Normal range, mpg	17/23

ACCELERATION
0-30 mph, sec	3.0
0-40 mph	4.8
0-50 mph	7.0
0-60 mph	9.2
0-70 mph	12.7
0-80 mph	16.0
0-90 mph	20.5
0-100 mph	25.4
Standing ¼ mile	17.0
speed at end, mph	84

TAPLEY DATA
O/d, lb/ton @ mph	230 @ 65
4th	300 @ 60
3rd	400 @ 55
2nd	530 @ 50
Total drag at 60 mph, lb	125

SPEEDOMETER ERROR
30 mph	actual 29.0
40 mph	38.9
50 mph	48.7
60 mph	58.4
70 mph	67.9
80 mph	77.6
90 mph	87.7
100 mph	97.9

▶ Elegance adrift. The phrase might apply to an expensive yacht sulking quietly on calm seas, but it's also a succinct way of describing a Jaguar 3.8 being hurried home from the office, or back to town for the theatre or, in fact, from any Point A to any Point B by anyone who cares about the path in between and enjoys traversing it.

The Jaguar 3.8 combines, as no other car does, luxury and performance in a highly usable package. Beneath its leather and walnut skin lurks a heart of highly-polished steel and aluminum. When caressed gently at the accelerator pedal, it purrs calmly, propelling this "gentleman's carriage"

JAGUAR 3.8

in a smooth, dignified manner. But like a Douglas Fairbanks hero, when hard-pressed it reacts in a violent, soul-stirring fashion. Wheels spin, tires scream with rage and two tons of ironmongery and precision woodwork hurl themselves down the road.

The "Three-Point-Eight" is the improved successor to the popular, distinctive 3.4 sedan. The essential changes are the bored-out engine which gives a change in name, an increase in rear tread of 3.3 inches which adds measurably to ess-bend stability, and a thinning down of all window posts to improve visibility significantly. The latter two, with subtle variations in rear body contours, give rise to the added designation, "Mark 2". In England, it is possible to have your choice of either Mark 1 or 2 according to your preference of privacy or vision and either 3.4 or 3.8 liters of engine displacement. In the American market, however, we are limited to such choices as wire wheels or discs, Borg-Warner automatic or Moss-built four-speed manual transmission, and if the latter, with or without overdrive.

Everybody gets the six-cylinder engine which has powered so many Le Mans winners and also the Dunlop disc brakes which have stopped them, lap after lap, though both are somewhat detuned to make them more suitable for road use. In its brief history, Jaguar has scored two mighty firsts in the automotive sales scene. With the XK 120 it was the first to offer lots of power (160 bhp) for a price that many, if not exactly the masses, could afford. With the 3.4 it was years ahead of Detroit in offering a compact version of its full-sized sedan. (The 3.8's wheelbase is 12½ inches shorter than the Mark IX's.) They are still ahead in providing full-sized performance (though in races the big-inch, stark Lark V8's give them a bit of a run) and quite alone in this class in providing elegance inside as well as out.

Geometrically the 3.8 is a "compact car" in the wheelbase sense (107.4 inches) but in price, performance and furnishings it has little in common with the herd of new models available. It costs just under five thousand, which puts it on a par with the cheapest Cadillac. Its performance well and truly deserves the term "sporting". Top speed is some 125 mph and acceleration beats the Ace-Bristol or Austin-Healey.

As on the 3.4 predecessor, the dashboard and window sills are made of walnut, polished beautifully to a sparkling luster. An improvement is that the instruments are no longer clustered at the center which had suited the production-line problems of building both left- and right-hand drive cars. They are now spread out directly in front of the driver where they belong. Each instrument is circular with plain white figures on a flat background, a "functional" concept which is in effective and dramatic contrast to the extravagance of the walnut panelling. Included are a tachometer (red-lined at 6000), a speedometer (with ordinary and trip odometer), an ammeter, and fuel, oil pressure, and water temperature gauges. The zero on the fuel gauge contains a red light which warns that the 14½-gallon tank is down to one or two. A cigar lighter and a truly man-sized ash tray care for the smoker. The English push-button radio fitted was not in as good voice as the engine, a matter of tuning, we trust.

Each front seat has a full 6 inches adjustment fore and aft. Since the rails are inclined, the seats move up as they slide forward, on the reasonable presumption that short people have short legs and tall guys have long ones. The former will especially appreciate this feature as the rapid falling-away of the fender-line otherwise would tend to make accurate placement in tight traffic a touch ticklish, though the high-mounted parking lights do help to define the car's corners.

The front seats are separate but to term them bucket seats would be like calling the Queen's throne a chair. The deep foam rubber is covered with leather hide; our test car's were suede green to complement the richly dark-green exterior. Other paints available include three shades each of gray and red, two of blue, plus cream, black and, of course, British Racing Green.

Entry is easy and comfortable. Though you sit slightly askew with your legs a bit to the left and the 17-inch diameter steering wheel somewhat to your right, the position is very comfortable and its asymmetry is soon unnoticed. The wheel has two tapering spokes and a semi-circular horn ring. It is much nearer to horizontal than we are used to seeing in sedans, reminding us of the comfortable installations in Indy cars (of all things!).

To those well-acquainted with the 3.4, the improvement in visibility is striking. It was achieved by drastically slimming the windshield posts and the between-the-doors posts, in both cases by a full inch. The value of this sort of increase in visibility cannot be too highly stressed.

"Elegance adrift" might apply here. Jaguar's small sedan exhibited quite a bit of understeer both here and on SCI's 400-foot handling test circle.

There's logic too in the hoodless headlights and the finless fenders; their good aerodynamic form helped us get 19 mpg on a traffic-free, gentle run from New York City to Bridgehampton and return. If we'd used the full 225 horsepower frequently we would have paid for it, but it's pleasant to realize that unlike the gas-guzzling "power-packs" the 3.8 can be thrifty when you wish.

The car we tested was loaned to us by Mr. C. Gordon Bennet of Jaguar Cars Inc. Sporting three impressive badges (British Racing Driver's Club flanked by Road Racing Driver's Club and SCCA), his is one of the earliest 3.8's. Current production models intended for the United States feature a nine to one compression ratio, one ratio up on his car's. Though this raises both torque and horse-

CONTINUED ON PAGE 135

ROAD TEST

JAGUAR 3.8

Price as tested: $4890

Importer: Jaguar Cars Inc.
32 East 57th St.
New York 22, N. Y.

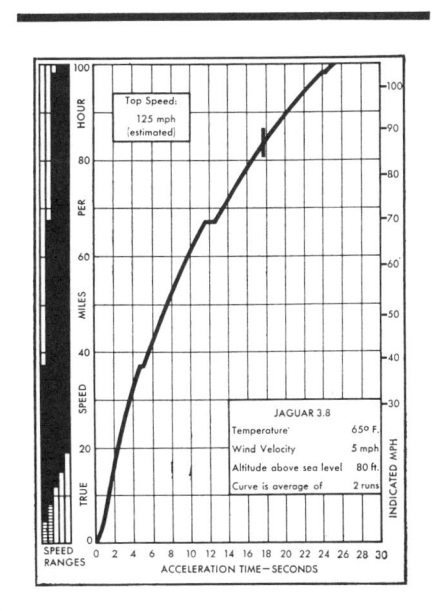

ENGINE:

```
Displacement ..................230.6 cu in, 3781 cc
Dimensions ....................Six cyl, 3.42 x 4.17 in
Compression Ratio ...................8.0 to one
Power (SAE) ..................225 bhp @ 5500 rpm
Torque ......................240 lb-ft @ 3000 rpm
Usable rpm Range ....................700-6000 rpm
Piston Speed ÷ √s/b
  @ rated power ......................3465 ft/min
Fuel Recommended....................Premium
Mileage ................................13-19 mpg
Range ................................190-275 miles
```

CHASSIS:

```
Wheelbase ..............................107.4 in
Tread, F,R ..........................55, 53.4 in
Length ..................................181 in
Suspension: F, ind., coil, wishbones; R, rigid
  axle, cantilevered leaf springs, radius rods.
Turns to Full Lock ........................2.2
Tire Size ............................6.40 x 15
Swept Braking Area—disc .............488 sq in
Curb Weight (full tank) .............3300 lbs
Percentage on Driving Wheels ..........44%
Test Weight .........................3600 lbs
```

DRIVE TRAIN:

Gear	Synchro?	Ratio	Step	Overall	Mph per 1000 rpm
Rev	No	2.96		12.73	6.2
1st	No	2.96		12.73	6.2
			59%		
2nd	Yes	1.86		7.01	11.2
			45%		
3rd	Yes	1.28		4.84	16.3
			28%		
4th	Yes	1.00		3.77	20.9
4th OD		0.78	29%	2.93	26.9

Final Drive Ratios: 3.77 with overdrive, 3.54 with automatic transmission.

Top Speed: 125 mph (estimated)

JAGUAR 3.8
Temperature: 65° F.
Wind Velocity: 5 mph
Altitude above sea level: 80 ft.
Curve is average of 2 runs

75

JAGUAR 3.8 Mk. II

One of the Best Saloon Cars in the World

IT was recently a privilege as well as a pleasure to carry out a long road test of a 3.8-litre Mark II Jaguar saloon, after which only one conclusion could be reached, namely, that this is one of the World's best saloon cars. What justification, if any, is needed for this statement? We consider the explanation easy—here is a car capable of 125 m.p.h., of devouring a ¼-mile from a standing start in not much more than 16 seconds, and endowed with a full complement of Dunlop disc brakes which are well able to cope with high-performance of this calibre. Add to this the ease with which this 220 horse-power Jaguar can be driven, the sense of well-being conveyed by its hide upholstery, deep seats and polished veneers, its silent functioning and its very complete equipment, and no one, surely, will dispute our claim. That such a car can be sold for just over £1,800 is a commercial miracle understood only by Sir William Lyons. Fastidious businessmen and keen motorists can save themselves or their businesses something like £3,000 by bearing these facts in mind, and expensive motor cars with double-barrelled names seem somewhat expensive when compared with this Coventry-built twin-cam machine.

Points that Appeal

The big-engined Jaguar is a fascinating car because it has such enormous powers of effortless acceleration, reaching, for instance, 80 m.p.h. from rest in under 15 seconds, 50 m.p.h. as quickly as 6½ seconds, that there is little need to wear oneself out hurling it at corners or playing angry bears in traffic. Like a certain well-known big-twin motorcycle it hunches itself up and streaks away from corners and congestion and, with retardation to match, can afford to behave with dignity in adversity. For this reason alone the 3.8 Jaguar is an effortless motor car in which to cover many miles. If its road-holding is bettered in some sports cars or in Continental G.T. vehicles costing fabulous sums, this is scarcely relevant if the driver is in sympathy with the style of driving this Jaguar encourages.

In fact, the wider rear track of the Mk. II version has cured a certain tendency to rear-end skittishness, and this compact but not claustrophobic saloon will hustle round corners with little roll and the RS5 Dunlops mute. Naturally, if you turn on all the horses on a wet road in a low gear you will have the tail round, and as the steering pays for being unexpectedly light by asking nearly five turns lock-to-lock, there is every reason to employ the driving methods advocated above. Otherwise the latest Jaguar saloon is safe in spite of its very high speed and power. Incidentally, criticism of the low-geared steering, heard in some quarters, must be met by remarking on the usefully taxi-like turning circle, and here it is appropriate to comment that the steering is commendably free from kick-back or shake, has an adjustable column, and vigorous castor return action to ease the driver's task after acute corners. The new wheel is neat, with single spoke and a half-horn ring, and is small and placed sensibly low.

The remote, rigid central gear-lever could hardly be bettered, except that unless a driver has long legs and arms and likes his or her seat right back to gain the fashionable straight-arm stance, its position is apt to be too far aft. The synchromesh can be beaten all too easily but, especially if the optional overdrive is ordered, there can be few grumbles about the spacing of the gear ratios in this best saloon car (certainly at the price) in the World. The engine, in spite of its rather ancient dimensions of 87 × 106 mm., runs up to 6,000 r.p.m., actually peaking at 5,500 r.p.m. Yet it is so docile that it is quite permissible to run down to 20 m.p.h. in o/d top gear and, in fact, this Jaguar has vintage qualities, for the engine speed is then below 1,000 r.p.m., 20 m.p.h. in normal top representing still fractionally under this speed, while even when motoring determinedly one changes up at around 3,500 r.p.m., the full potentialities seldom being required, although it is exceedingly satisfying to have on tap a maximum of 98 m.p.h. in third gear

and as high a pace in the 7-to-1 second gear as many cars stagger up to in third. And there is no finer engine to be behind than the twin-cam, six-cylinder of a Jaguar.

The clutch likes to be fully depressed to effect quiet gear changes. The Dunlop disc brakes are superlative, light pedal pressures producing such fantastic stopping rates that it never becomes necessary to tread really hard, while the action is silent, straight-line and progressive. The handbrake is well located on the right of the driving seat.

The new facia is very attractively laid out and the grouping of switches and auxiliary dials on a central black-finished metal panel sunk into the veneered panel is convenient as well as stylish. This is probably the same panel as that used, polished, on the Sunbeam Alpine but if this, and the Lucas switches, are not special to the Jaguar, such semi-standardisation of proprietary parts was not unknown even in vintage times and is entirely permissible at the price. The six flick switches have their functions neatly lettered below them, from left to right: Interior; Bright—Dim—Panel; Fan—Fast—Slow; Ignition; Cigar; Starter; Map; Wiper—Fast—Slow; Washer,—which should be intelligible to those who find themselves behind the wheel of a Jaguar—if not, they should not attempt to drive such a powerful machine! The wipers and horn function only with h.t. current in circuit. A small crank-handle under the facia extends the radio aerial.

Above this row of switches are four dials recording water temperature, oil pressure, amps and fuel contents. Before the driver, visible through the wheel, are the neat, matching Smiths tachometer and speedometer. The former has an inset clock and reads to 6,000 r.p.m., with a red warning from "five-five"; the speedometer goes to 140 m.p.h. and has total and trip odometers, the latter with decimal readings. Further right is a light which stays on while the handbrake is applied. This may seem unnecessary but it would be easy to burn out this brake with over 200 horse-power beneath the bonnet and in any case the idea is very ingenious, because this light also comes on if the level in the brake fluid reservoir becomes too low, while if the light fails due to a bulb going the owner is likely to notice this when using the handbrake and have a new bulb put in, so that the fluid-level warning isn't impaired. There are, additionally, the usual warning lights, including one for low petrol level (which didn't function, a trek to a garage with a can thus becoming part of the test).

Before the front-seat passenger there is a wooden-lidded lockable cubby-hole, which is supplemented by a very useful map shelf under the centre of the facia, in which a lever controlling the scuttle ventilator is placed. Between this shelf and the transmission tunnel is the H.M.V. radio, and behind it a huge lidded ash-tray which brought joy to Robert Glenton when he reported on the Jaguar in the *Sunday Express*.

Besides the cubby-hole and shelf there are spring-loaded flap-type pockets in the front doors and open pockets in the back doors. When the cubby lid is dropped a big blue bulb burns, for map reading. The back compartment passengers inherit the tables, corner lamps, arm-rests with ash-trays and openable quarter-windows of the Mk. IX, and there is a central folding arm-rest.

The front seats are slightly on the hard side and rather flat. There are quarter-windows with tamper-proof catches, but no rain gutters, the driver's apparently spring-loaded against wind pressure. The front window handles need 2¼ turns, the rear ones 2½ turns, for full ventilation. On the test car the driver's window was very hard to open. Naturally there is a very efficient heater/ventilator, with the merit of a quiet fan.

Forward visibility is good (even if the average driver cannot see the near-side wing) for the broad bonnet slopes away, topped only by the low-set leaping-Jaguar mascot, and the screen pillars are thin. A good point concerns the "tell-tales" on the sidelamps; another is the dimming of the overdrive-indicator lamp when the

The lines of the 3.8 Jaguar suit the demeanour of this excellent high-performance car, the bonnet tapering to a small oval radiator grille, flanked by headlamps, and spot-lamps that can be used in lieu of them. At the rear a pleasing touch is a tiny notice proclaiming that the car has disc brakes.

THE 3.8-LITRE MK. II JAGUAR SALOON

Engine: Six cylinders, 87 × 106 mm. (3,781 c.c.). Twin overhead camshafts. 8.0-to-1 compression-ratio. 220 b.h.p. at 5,500 r.p.m.
Gear ratios: First, 12.73 to 1; second, 7.01 to 1; third 4.84 to 1; top, 3.77 to 1; overdrive top, 2.93 to 1.
Tyres: 6.40 × 15 Dunlop RS5 on bolt-on steel disc wheels.
Weight: Not weighed. (Maker's figure: 29.9 cwt. with 5 gallons of fuel).
Steering ratio: 4.8 turns, lock-to-lock.
Fuel capacity: 12 gallons. (Range approximately 200 miles.)
Wheelbase: 8 ft. 11 in.
Track: Front, 4 ft. 7 in.; rear, 4 ft. 5¾ in.
Dimensions: 15 ft. 0¾ in. × 5 ft. 6¾ in. × 4 ft. 9½ in. (high).
Price: £1,255 (£1,779 0s. 10d. inclusive of p.t.). With extras as tested: £1,890 11s. 0d.).
Makers: Jaguar Cars Ltd., Coventry, England.

---

sidelamps are on. Overdrive is selected by the left-hand stalk below the steering wheel and a matching stalk on the right works the flashers or, moved in and out, flashes a full-beam warning with the headlamps even when the lamps are not switched on—good for Jaguar! The former arrangement of the facia switch selecting sidelamps, then headlamps, then spotlamps, is retained, so that it is not possible to use head and spotlamps together, which too many dazzlers do. Dimming is by a big foot-operated knob, and there is a good, roof-hung central mirror, incorporating a map lamp with facia switch. Less clever are the rigid wood anti-dazzle vizors normally flush in the roof. It is hardly necessary to add that the Jaguar's doors shut "expensively" and have good "keeps," or that the switches work beautifully, or that the two-position facia lighting is nicely arranged.

The lockable boot lid rises automatically, the interior then being illuminated; the spare wheel is hidden under the floor. The bonnet lifts to reveal that splendid twin-cam power-unit and its accessible components; the dip-stick is easier to remove than to replace. Curiously, the ignition key would also start the Editorial Mini-Minor!

Petrol consumption averaged 16¾ m.p.g. and in 1,000 miles three pints of Castrol XL topped up the sump. The fuel gauge was apt to be optimistic, showing half-full with but four gallons in the tank. The twin petrol tanks and separate fuel system of the Mk. IX have unfortunately been abandoned and the tank capacity is rather less than 12 gallons, so that the range is limited to a feeble 200 miles, or less if cruised at 100 or more m.p.h., as this Jaguar can easily be along motor roads or on the Continent. The filler is under a flap, making refuelling from a can difficult.

There are many extras available, such as laminated screen, radio, centre-lock wire wheels, etc., but a Powr-Loc limited-slip differential and vacuum-brake-servo are included in the highly competitive price of £1,779 0s. 10d., which increases to £1,842 15s. 10d. if overdrive is specified, while an automatic transmission model is available for £1,927 15s. 10d. Were purchase tax abolished Sir William Lyons would be able to offer the basic 3.8-litre Mk. II Jaguar at £1,255—incredible, especially when it is remembered that it is only 12 years since the first XK120 was introduced.

Such a car is virtually *sans* rivals and if anyone likes to provide the Editor with one of these excellent, race-bred, and very-English motor cars he will raise no objections!—W. B.

JAGUAR'S GLOSSY NEW CLAWS

CONTINUED FROM PAGE 68

In fact, I was reminded of the driving position in a Cooper racing car. The wheel had about the same degree of rake and about the same arm extension was needed to get a position of firm control. Jag owners will just have to get used to the idea that they are meant to sit above the wheel and not behind it.

I was pleasantly surprised with the automatic transmission. The range selector was mounted on the steering column, not on the dashboard like the standard model and had the indicator panel illuminated in a small nacelle above the centre of the steering wheel. It had a PNDLR range (P for park, N for neutral, D for drive, L for low, R for reverse). The intermediate hold switch was high on the dashboard to the right of the steering wheel and could be operated by one finger without taking the right hand from the wheel.

Low range was an emergency low, but I found after experimenting that its judicious use up to 50 m.p.h. gave the Jag better acceleration. Drive had three ratios and unless the car was driven hard, they changed quickly and the car tended to slug a little with too few revs in the direct drive top ratio. But driven hard the little Borg Warner men under the floor boards behaved with great spirit. Without using the intermediate hold, the car would change ratios at about 4000 r.p.m., which took it to over 70 in intermediate. By using the switch and leaving the transmission in intermediate, it would stay in that ratio until the rev counter needle hovered over the red line at 5500. Speed was then a fraction under 86 m.p.h.

For normal touring, the intermediate hold was a wonderful asset. You would be cruising at say, 60 to 70 m.p.h., and the need to overtake arose. A flick of the switch and without the slightest hesitation, the transmission dropped into intermediate and held there. With ferocious acceleration the Jag would flash past the car ahead and when the time was ripe, the hold was flicked out and, again instantaneously, the transmission dropped into direct drive. It was invaluable around the city, too, for passing, for catching lights and so on. Of course, you could drop into intermediate at any time merely by flooring the accelerator. But with this method there was slight delay, and the transmission would often change back into direct drive sooner than you wanted it to.

The Dunlop disc brakes were flawless. Pedal pressure was light, and stops from high speed were firm and straight. There was no trace whatever of fade.

The only complaint I could find with the car was the cooling arrangement, which works through the heater. It was hot in Melbourne when I tested it and even with the vent wide open, the control on "cold" and the two-speed fan going full blast, the air coming into the car was lukewarm. Earlier model Jags had the same fault and the ventilation design department doesn't seem to have been able to do anything about it.

The only other thing that is possibly a doubtful quality is how the body work will stand up to rough Australian roads. A lot of Jaguars develop uncurable rattles fairly early in their life, but I suspect with the Mark 2 range, the Jaguar body builders have gone to a lot of trouble to make sure their work lasts. The body seemed better than in the past; everything appears to be a perfect fit and all possibly suspect points — doors, windows, dashboard, bonnet, etc. seemed to be solid and long lasting.

I hope they are, because that is the only thing that would prevent the Mark 2 Jaguar 3.4 litre from going down in motoring history as one of the all-time greats. It has beauty, power, performance, road-holding qualities and comfort so far in excess of the run-of-the-mill cars that it just doesn't matter. #

CARS ON TEST

Remarkable value. Not only is the car finish and trim of the highest order but the engine appearance, and performance is enough to delight the heart, and eyes of the most hard-bitten enthusiast. Note the large pancake air-filter for the two S.U. carburetters

Subdued or unleashed

THE JAGUAR 3.8 IS STRICTLY FABULOU

The Mark 2 Jaguar series is distinguished by the increased glass area, the chromium window frames, and improved tail light assembly. It is a variation on a well-known theme yet is as smart as anything on the road

This quality car at a "medium price" offers limousine manners and racing car performance, says Douglas Armstrong

THE Jaguar is a unique car. Currently the range includes no less than eight models yet all are powered by the same basic race-bred engine—made in three different capacities! This fabulous twin-overhead camshaft unit in all its forms has grown up from the original XK120 which was introduced in 1949 with a power-output of 160 b.h.p. Nowadays the unit offers outputs from 120 b.h.p. in the short-stroke 2.4-litre version to 265 b.h.p. in the 3.8-litre XK150s. Racing and development of a very intensive nature have resulted in these impressive figures and unusual adaptability.

Maybe this first paragraph has conjured up a vision of rip-roaring sports cars with fruity exhausts, clattering tappets, "top-end" power, and "in-or-out" clutches. If it has you are off track as far as the modern Jaguar range is concerned. All models have an outstanding performance but it is delivered with dignity as well as alacrity.

When the 3.8-litre Mark IX was introduced I think most motoring journalists sensed (and hoped) that the "enlarged" 3.4 unit would eventually find its way into the "handy-sized" Jaguar *monoque* shell. We had already seen the 2.4 grow up into the 3.4 and, the 3.8 seemed a natural development. It would obviously be a car of exceptional performance, and when it did come along it certainly was.

The 3.8-litre Mark 2 Jaguar saloon as tested by CARS ILLUSTRATED, with four-speed gearbox and Laycock overdrive is remarkable value at £1,842 15s. 10d. with Purchase Tax. If the Purchase Tax is discounted and the manufacturers true price of £1,300 considered it is fantastic value. Not only has the car the performance of a racing machine, it has also the looks and finish of a luxury, quality car (which it is), and the docility, smoothness and quiet running of a product which could easily be twice its price.

When the quiet smoothness of this turbine-like engine is sampled, it is difficult to believe that the same basic unit has won five 24-hour races at Le Mans, and other sports car championship events. It will trickle along in top at 13–14 m.p.h. in the manner of a town carriage, then accelerate strongly to around 120 m.p.h. on the same gear without a pink with standard 8 to 1 pistons. Alternative compression ratios of 7 or 9 : 1 are available but the standard pistons seem to offer an excellent compromise as the car was run on premium fuel throughout its test with every satisfaction.

Engine noise is merely a low, pleasing purr, no valve thrash or rattle being audible. Response from the "organ" type accelerator pedal is immediate with a completely clean pick-up from the two S.U. Type H.D.6 carburetters. The car leaps forward like a Jaguar to record 60 m.p.h. in 8½ seconds—a pretty impressive performance from a four/five-seater luxury saloon weighing some 1½ tons. Part of the acceleration story stems from the 3.8's impressive torque figures. This long-stroke edition of the 3.4 (87 m.m. compared to 83 m.m.) has in fact only a 10 b.h.p. plus on the smaller engine but its torque readings are up to 240 lb. ft. compared with 215 lb. ft.

It is almost steam-like to drive. One of the most amusing ways to use a 3.8 overdrive is to treat it as a "two-speed" car on the open road. It is quite surprising what a large mileage can be put in by merely using the direct top and overdrive gears through the medium of the neat steering column lever. It will crawl in top, accelerate smoothly away from junctions, and then cruise at speed in silence with overdrive engaged.

Full marks too for the adjustable steering column and seat positions. One of my pet hobby horses is the very small number of British manufacturers who provide anything like a good driving position. The Jaguar front seats can be set back far enough for the long-legged to be comfortable, and the steering wheel can be positioned over a range of about 3 inches. The Jaguar is one of the few cars in which I was able to gain the ideal driving stance, legs and arms bent just enough to provide movement with no suggestion of cramp.

The seats themselves (separates at the front, bench at the rear with folding

(ABOVE) *Tail view has no fins, squared-off boot or other "modernities", but the lines are attractive and eager. Bumpers are robust, and rear window offers the maximum visibility*

(LEFT) *The 3.8-litre Jaguar is of excellent aerodynamic form, the accent being on air-penetrating curves. Home models have built-in twin Lucas spotlamps, export do not. The "new" sidelights blend into the wings and have been "handed-down" from the pre-war SS Jaguars*

arm-rest) are made in the finest hide with deep cushions and squabs but the front ones could well do with more "shape" in the squabs for lateral support during fast cornering. The interior finish is in the grand manner. There are thick pile carpets on the floor, and highly polished veneers everywhere. Each of the four doors has an arm-rest and a pocket, and there is a locker on the passenger's side of the dashboard.

The instrument panel is an enthusiast's dream with the large diameter speedometer (with trip and season odometer) and rev.-counter (incorporating an electric clock) set ahead of the driver's eyes. The speedometer also incorporates a main beam warning light, and to the right of this instrument is a warning light which operates when brake fluid level is low, or when the handbrake is left on.

The centre instrument board is finished in a discreet black and contains ammeter, and separate gauges for fuel, oil pressure, and water temperature. The main light switch is in the middle of the board and controls side and tail, head, and fog in that order, the switching on of the twin countersunk Lucas foglights extinguishing both headlamps. Under the smaller instruments is a row of six tumbler switches controlling interior light, panel light, heater fan, maplight, screenwiper, and screen washer (electric). In between the tumbler switches is the ignition switch of the key type, cigar lighter, and starter button. Surprisingly there is a separate starter button, although the very efficient automatic choke is a natural to combine with a "key start". All switches are neatly labelled in small white lettering under a transparent strip which prevents their wearing off over the years, and all instruments have dignified black dials with no "gimmicks". There is a shallow shelf under centre instrument panel which is useful for sunglasses, gloves etc., and just above this is the control for the effective scuttle ventilator.

Below the facia is a handsome leather-trimmed "console" which makes full use of the normally wasted space over the gearbox "hump". In this console is mounted the HMV radio and speaker, vertical-operating heater controls, and king-size ashtray with leather covered lid. Non-smokers will quickly convert the last item to an oddments, or small tools box. Aft of the ashtray is the remote floor gearlever.

The steering wheel itself has two spokes and offers maximum instrument visibility, although it cannot be described as a thing of beauty. The half horn ring, and "crossbar" operates the powerful horns efficiently and quickly, and the ring lock for steering column adjustment is excellent. The overdrive switch is a handsome lever protruding from the right hand side of the column where it can be operated without releasing one's hold of the wheel. The self-cancelling winkers are controlled from an almost twin lever on the left of the column. On overdrive models there is a warning light incorporated in the column-top "quadrant" which is utilised as a gear position indicator on the automatic model. When overdrive is engaged a light glows its warning, and the indicator warning arrows are also incorporated in the "quadrant".

Instrumentation is comprehensive and dignified. Large dial 140 m.p.h. speedometer, and 6,000 r.p.m. tachometer are positioned in front of the driver's eyes, smaller dials are on central facia. Adjustable steering wheel is adjusted to its shortest length where it provides a splendid position for most drivers

This may seem a lengthy description of instruments, controls, and trim, but the writer considers this all part of Jaguar magic. To sit in a car such as the 3.8 with every possible piece of high-class equipment about you, to sense the controls in the right places, to know that at a touch of a switch you can clean your windscreen, engage overdrive, or listen to the Archers imparts a feeling of purposeful motoring. The car seems ready to transport you without a moment's delay to Istanbul, or to Eton Wick—and it would, to either! One can sit in the car and enjoy it without turning a wheel.

A turn of the key, a stab at the starter button and the big engine is purring like a true member of the cat family. A smooth second gear start, then drop straight into top and the car is sliding along at a quiet 60 m.p.h. in seconds. Alternatively a busy take-off using all four speeds and full enthusiasm, not to mention 6,000 r.p.m. and you are doing 100 m.p.h. in 25 seconds. A move into overdrive and this speed can be held with the rev.-counter recording a mere 3,800 r.p.m. on an effective axle ratio of 2.933 : 1. Direct top at 100 m.p.h. represents 4,870 r.p.m. so the overdrive plays a big part in reducing wear, tear, and fuel consumption.

The ride is soft and limousine-like but damping is excellent for "ordinary" motoring. Obviously the production suspension characteristics are aimed at the American market where the Jaguar is a great favourite, and where tight speed limits prevail. British and Continental motorists who would like to fully exploit this magnificent car would undoubtedly go for tougher dampers, and perhaps stiffer springs.

The wider rear track of the Mark 2 models has made a big difference to their general roadability. There is appreciable roll but the car can be cornered fast, particularly if the tyres are run at the prescribed high speed pressures of 34 lb. sq. in. front, and 31 lb. sq. in. rear.

The servo-assisted Dunlop disc brakes on all four wheels are fully in keeping with the car's immense potential. In spite of a laden weight of around 32 cwt. they would bring the speed down in a most impressive way with no fade, snatch, or "pull". Wet

```
            QUICK CHECK
     Maximum: 10 for each category

   PERFORMANCE    ...    ...    10
   ROADHOLDING    ...    ...     9
   GEARBOX        ...    ...     7
   COMFORT        ...    ...     9
   FINISH   ...   ...    ...    10
   BRAKES         ...    ...    10
   VISIBILITY     ...    ...     9
   ECONOMY        ...    ...     9
   STARTING       ...    ...    10
   All assessments comensurate with
        type of car and price tag
```

SPECIFICATION

PERFORMANCE:
Through the gears:

0—30	...	...	3.5 sec.	0—70	... ... 11.5 sec.
0—40	...	...	5.0 sec.	0—80	... ... 14.5 sec.
0—50	...	...	6.5 sec.	0—90	... ... 18.0 sec.
0—60	...	...	8.5 sec.	0—100	... ... 25.0 sec.

Maximum speeds:
Overdrive top, 125 m.p.h.
Top, 120 m.p.h.
Third, 99 m.p.h.
Second, 65 m.p.h.

ENGINE: Jaguar six-cylinder in-line, water-cooled. Twin overhead camshafts. Bore: 87 m.m. Stroke: 106 m.m. Cubic Capacity: 3,781 c.c. Compression ratio (as tested): 8 : 1. Power-output: 220 b.h.p. at 5,500 r.p.m. Twin S.U. HD6 carburetters. Lucas coil and distributor ignition.

TRANSMISSION: Moss type four-speed gearbox in unit with engine. Synchromesh on second, third, and top. Central remote lever. Test car fitted with Laycock overdrive unit operating on top gear only. Non-overdrive, or Borg-Warner three-speed automatic transmission optional. Rear hypoid bevel axle with limited-slip differential. Ratio 3.77 : 1 with overdrive. 3.54 : 1 without overdrive, or with automatic.

SUSPENSION: Independent front by unequal length wishbones, coil springs, telescopic dampers and anti-roll bar. Rear suspension by live axle and cantilever/semi-elliptic springs, telescopic dampers and Panhard rod.

BRAKES: Servo-assisted Dunlop discs all round. Front 11 in. Rear 11⅜ in. Pressed steel, bolt-on wheels. (Centre-lock wire wheels optional extra.)

DIMENSIONS: Wheelbase: 8 ft. 11⅜ in. Track: front, 4 ft. 7 in., rear 4 ft. 5⅝ in. Length 15 ft. 0¾ in. Width: 5 ft. 6¾ in. Turning circle: 33 ft. 6 in. Kerb weight: 30 cwt. Fuel tank capacity: 12 gallons (Imp.). Average fuel consumption (over 500 miles, all types of driving): 18 m.p.g. Premium fuel. Tyres: Dunlop Road Speed RS4 Nylon with tubes. 6.40 × 15.

BRAKING FIGURES: Using Bowmonk Dynometer. From 30 m.p.h. with maximum useable pedal pressures, 96 per cent = 31.4 ft.

PRICE: As tested with manual gearbox, overdrive, and bolt-on wheels, £1,852 15s. 10d. including Purchase Tax. H.M.V. radio £47 15s. 2d. extra including Purchase Tax.

---

weather did not affect them. The handbrake was a sensible stout "pull-up" lever on the right of the driver's seat.

The Jaguar engine has a massive cast iron crankcase which contributes largely to its undoubted rigidity, and consequent smoothness of running. It also adds up to a preponderance of weight on the front wheels, and a high degree of understeer. For fast cornering it was found that the "aiming" needed to be done in advance of requirements otherwise the car would run wide. There was strong caster action which gained in heaviness as speed and lock increased, in spite of a too low steering box ratio of five steering wheel turns from lock to lock. Here again this is an obvious feature for the "ordinary" motorist who will appreciate the low gear for ease of parking etc., but the experienced driver would go for the higher ratio box which is available on request.

The remote gearlever is positioned for driver convenience and the four forward ratios are well chosen. The gear change however is not worthy of such a car. The movement from third to top is excessive, and the synchromesh mechanism (second, third, and top) is weak. The intermediates are reasonably quiet. As already mentioned, engagement of the Laycock overdrive (top only) is quick and easy but for smoothest results the throttle should be eased for upward changes and *vice-versa* for downward changes.

With such tremendous power on tap the driver must exercise restraint with the throttle when coming out of corners, particularly in the wet. It is easy to spin the rear wheels, even in third gear but an experienced driver will find the appropriate action instinctive. The low geared steering is certainly not conducive to easy control with such an eager and responsive power-unit.

It is very noticeable in this day and age when practically all the world's cars follow prescribed fashions that the Jaguar still manages, to retain a unique and handsome appearance. The Mark 2 models have increased window area and plenty of visibility, there are no tailfins, excessive overhang, or "upswept" boot line, yet they look graceful and as modern as the hour. Sir William Lyons has had the knack of producing good-looking cars for about 35 years, and he seems in no danger of losing his flair. Not only is the body handsome but its aerodynamics are good too as witness the performance and fuel consumption figures.

The equipment is too lavish to be mentioned here in full, but to mention a few features such as built-in heating system which not only heats and demists the front compartment but supplies warm ducted air to the rear passengers, two-speed windscreen wipers, folding tables with polished wood surfaces incorporated in the front seat squabs, electric screen washers, cigar lighter, maplight, twin foglamps, and adjustable steering column gives some idea of the value for money incorporated in the 1960 Jaguars. The test 3.8 was equipped with an H.M.V. radio which was built in to the "console", and possessed a first-class performance and tone. It had four press buttons as well as manual tuning, two waves, and a neat telescopic aerial built in to the right front wing with a "wind-up" handle situated under the dash within reach of the driver's hand.

The quarter windows at front and rear of car are very effective as ventilators and are fitted with spring-loaded retainers which will hold them open at 40, 90, or 140 deg., even at high speed. On the first "notch" there is no great wind roar, but when in the closed position they need safety "click" locks. With the slam locks it is possible to lock one's self out of the car, and if the keys are left in the dash the situation is fraught. This happened to the writer at Goodwood, but it was easy to insert a small screwdriver through the quarter window rubbers and unlock the fasteners. Hence the comments on the need for "safety" locks.

Perhaps the greatest noticeable improvement on this particular Mark 2 model is the fitting as standard of the Powr-Lok limited-slip differential. This modification has eliminated the rear axle tramp associated with earlier models under full throttle acceleration. A car for coveting. ★

In this interior view the remote gearlever, radio, speaker, and large ashtray can be seen mounted in the gear-box "console". There is a locker on the passenger's side, and an armrest on each door. Entry and egress is easy due to wide doors

Soft hides and polished veneers abound in the luxury interior. The separate front seats have picnic tables with polished wood top surfaces folded into the squabs. There is an arm-rest and pocket in each door

MOTORWAY—and the 3.8 Jaguar is in its element: a surge of power, and the car is doing an easy 100 m.p.h., then a quick flick to overdrive and it is effortless high-speed cruising for mile after mile.

The Best Value in the World

Mark II 3.8-litre Jaguar's Unique Place in the High-Performance Field

BY GREGOR GRANT

How do they do it? That is the question invariably asked whenever Sir William Lyons announces a new Jaguar. Cost for cost, the products of the Coventry factory offer quite remarkable value in the luxury, high-performance category, and there is little doubt that the latest Mark II 3.8-litre Jaguar stands absolutely supreme in the ever-increasing market for quality cars.

With a maximum speed of over 125 m.p.h. and the ability to accelerate from standstill to 100 m.p.h. in 25 seconds, it would be quite reasonable to assume that the "3.8" is basically a competition machine. This is far from being the case. The car is a perfectly normal saloon, tractable and possessing a degree of comfort which must be tried to be appreciated. True, the Jaguar has been developed from racing, but the designers have managed to endow their product with all the qualities that will appeal to owners who dislike any tendency to fussiness, and to performance which is possible only with a great amount of noise.

The charm of the 3.8-litre Jaguar is its effortless performance allied to the smoothness of a six-cylinder, twin overhead camshaft engine, which must be regarded as one of the major engineering achievements in Great Britain's motor industry today. With 220 b.h.p. under the bonnet, the "3.8" can be regarded as a fast tourer, *par excellence*, which very few machines can match for all-round performance.

For autobahn, autostrada or motorway cruising, this car is ideal. It is possible to maintain 100 m.p.h. for as long as road conditions will allow. With the big engine turning over at around 4,000 r.p.m. in overdrive, M.1 becomes very short indeed, and it is a case of always being on the outside lane, overtaking everything on the road with consummate ease. Because of its deceptive speed, I should say that the "3.8" is a car for experienced drivers, who will not be tempted to put their foot down willy nilly and attempt to do fast cornering without realizing that here is a machine which is built to be handled properly, and which will not readily react satisfactorily to ham-handedness. Indeed, the Jaguar must always be regarded as a vehicle which possesses such tremendous performance potential that it deserves to be treated from the connoisseur's angle.

Naturally one could not possibly use the car properly without an efficient braking system, and I must say that the 12-in. Dunlop disc brakes are both powerful and free from any tricks whatsoever. Judicious choice of pad materials has resulted in brakes which work as well from two miles per minute as they do in traffic crawls. "Fade" is virtually non-existent, and only panic stops require noticeably higher pedal pressures. This, of course, is due to the vacuum servo system, which, like the brakes themselves, was thoroughly tested under racing conditions before being put on production cars. The handbrake was far more efficient than ones tried on earlier Jaguars, but will not hold the car on really steep hills. A good point here is the use of a red tell-tale light, which, in addition to providing a "handbrake on" warning, will also flash on should the level of the hydraulic fluid drop below the normal capacity.

Steering and roadholding have been the subject of considerable thought and research. Without a doubt the modifications to suspension, and the provision of a wider rear track, have given the "3.8" much greater stability than was possessed by some of its predecessors. Progressive development has caused the technicians to arrive at a most happy combination of i.f.s. and rigid rear axle, keeping roll to a minimum and completely eliminating axle patter, even during fierce acceleration from standstill. However, much of the latter must in some measure be due to the adoption of the Powr-Lok, limited-spin differential, producing a judder-free getaway at about 2,000 r.p.m., and a commendable absence of wheelspin even when the clutch is engaged more suddenly than is normally the case. The clutch itself appears well able to cope with the immense torque available, and at no time was there sign of slip or spin.

Naturally it is comparatively easy to

NEW LOOK to the current series of Mark II Jaguars is provided by the thin door pillars and full-width rear window.

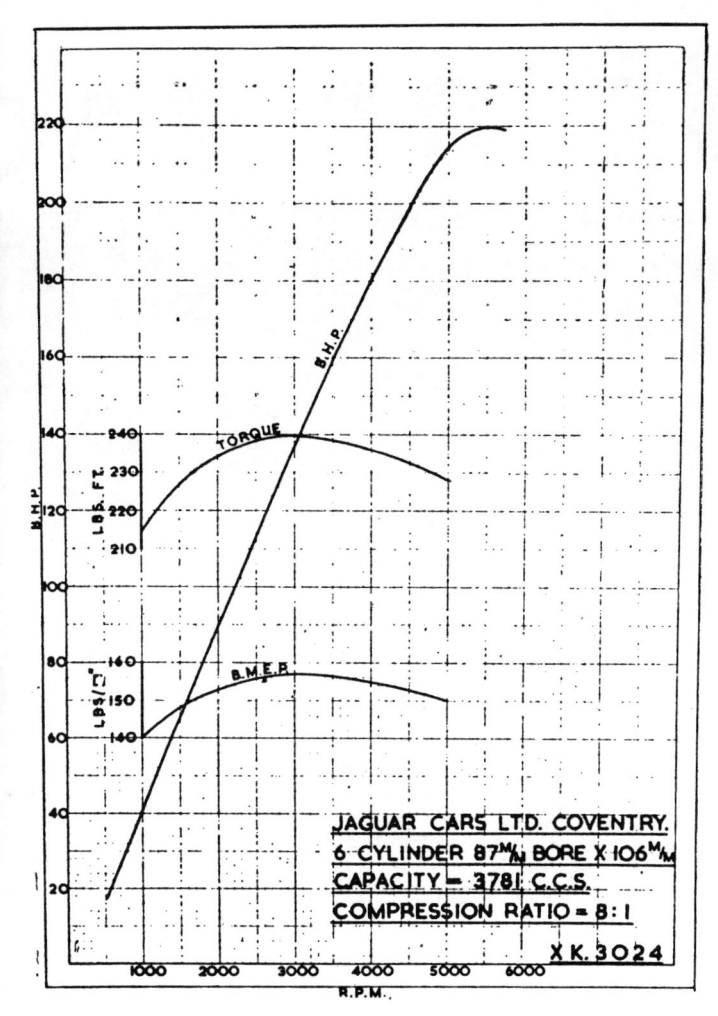

(Above) Picnic tables are recessed behind the front seats.

(Left) B.H.P., torque and b.m.e.p. figures for the 3,781 c.c. engine.

promote wheelspin by injudicious use of the accelerator pedal. Driven, as the French say, *"doucement-doucement"*, full use can be made of the car's accelerative powers—particularly on slippery roads. On ice, the Jaguar handles remarkably well, although I must say that the somewhat low steering ratio makes for slowish slide-correction. Personal preference would be for a slightly higher ratio, even at the expense of increased heaviness in the lower speed ranges. However, I understand that the largest percentage of buyers prefer the standard ratio, so obviously the manufacturers must accede to desires.

Also, I would like to try a Jaguar fitted with an all-synchromesh gearbox of the type found on certain Continental cars. In my opinion, the present box is the least satisfactory aspect of the "3.8"; the synchromesh is decidedly lazy in operation, and the longish gear lever travel is against rapid changes. On the credit side, the ratios are well chosen, and the gears themselves are quiet.

I can find no fault with the general roadholding, possibly because I do like a car with decided understeer characteristics. This will satisfy the majority of people who appreciate the relationship between the competition car and the production vehicle which it inspires. Suspension is on the firm side, but it gives a most comfortable ride and is exceedingly well damped. This was particularly noticeable during a recent run up to Scotland, where repairs on A.1 near Ferrybridge have produced a highly corrugated surface. Whilst the majority of other cars bumped and bucked their way along at greatly reduced speed, the irregularities scarcely troubled the Jaguar at all. Again, no road shocks whatsoever were transmitted to the steering.

At Silverstone I have watched the Jaguars "lean" when being taken through fast bends, but this is entirely due to almost full-throttle cornering, which one could seldom attempt on the public highway. During ordinary fast touring there is no noticeable tendency to roll, and provided tyre pressures are carefully maintained, tyre squeal is virtually non-existent—at any rate with the Dunlop Road Speeds fitted on the car.

I did find that the car is sensitive to tyre pressures, and the 25 p.s.i. front and 22 p.s.i. rear recommended for normal touring produced somewhat "dead" steering, and that increasing this to 30 and 27 respectively made no difference whatsoever to the ride, but made the car more responsive. For M.1 work, when it is intended to go really quickly, I would have no hesitation in recommending 35/32 or even slightly higher pressures.

Returning on a critical note, I would like to see the pedals made adjustable, in view of the large variety of seating positions available due to the generous travel of the seat, and to the use of a telescopic steering column. The positioning of the pedals makes "heeling-and-toeing" virtually impossible, a feature invariably desired by knowledgeable drivers. A larger fuel tank would provide a more practical range for such a high-performance machine (at present 12 gallons, giving approximately 200 miles maximum); the dip-stick could be made easier to re-locate; interior heating could be augmented.

The interior cannot be faulted, and I must compliment the designers on a facia-panel which is not only a model of the way in which instruments should be displayed, but is immensely attractive. The modern flick-switches make one wonder how we put up with the old knobs for so long. Seats are comfortable and tastefully finished in high-grade hide. Visibility all-round has been improved out of all recognition by the adoption of slimmer window pillars and a full-width rear window. The absence of wind-noise at high speeds is almost uncanny, and here at least is one car in which the radio can be operated at over 100 m.p.h. Apart from a satisfying burble, the exhaust note is subdued and unobtrusive.

Sealing is of the highest possible standard, and despite a longish run in torrential rain, not a drop of water reached the interior. The headlamps gave a good spread, and one appreciates such niceties as a flick-on switch for overtaking purposes. I had every opportunity to test the built-in auxiliary lamps, covering about 200 miles in fog varying in density from 100 yards to 10 yards visibility. Candidly they tend to be too bright for fog-driving, and produce back-glare. For winding roads they are ideal, but it should be possible to arrive at a useful compromise.

This article does not contain full road-test performance figures, as compiled by J. V. B., but a few times were taken, and it was possible to record the maximum speeds in gears. These were: 1st, 34 m.p.h.; 2nd, 62 m.p.h.; 3rd, 96 m.p.h.; 4th, 118-120 m.p.h., O.D., 128 m.p.h. Overall fuel consumption, including some fairly fast driving, worked out at 17 m.p.g., using premium grade petrol.

A standing quarter-mile was covered in 16 secs. dead, a remarkable figure for a touring saloon weighing over 32 cwt. (as tested). Using only 1st and 2nd gears, 0-60 m.p.h. can be accomplished in 8.3 secs.

Another excellent feature is a turning circle of 35 ft. 9 ins. better than several cars of much smaller capacity and wheelbase. However, the steering wheel from lock to lock takes exactly five turns.

83

A PRIDE OF JAG

INSIDE, all is comfort and opulence—polished walnut, quality leather and instruments: 3.8 Mark II automatic.

PEOPLE often ask me what I consider the best passenger car, **irrespective of price**, in the world today—and, after much thought, I place the 3.8 automatic Jaguar at the head of the list.

This is bound to displease many who have their favorite cars, and some who still feel that a Jaguar and a "Flash Harry" go together (a hangover from the old "SS" days, particularly in pre-war England, when young men in over-loud checks and over-long hair steered these eye-catching vehicles around Mayfair).

But if you consider performance, safety, comfort, luggage space, appearance and long life, and allot marks for each, the Mark II 3.8 Jaguar will probably come out on top.

I know no other car that will do a genuine 130 and has really fierce acceleration to go with it, offers maximum safety in first-class roadholding **and disc** brakes, comfortably carries four adults and their luggage, looks superb and is quite docile and trouble-free.

I can hear grumbles from Ferrari and Aston Martin fans, but these are GT cars, barely four-seaters (yes, I include the new 2-plus-2 Ferrari and DB4 Aston in this group). The Bentley boys might object, too, but they must admit that their beloved juggernaut, fine as it is, could never hope to run down a 3.4, let alone a 3.8 — Mark II's, of course.

For the Mark II's are as different from the Mark I's as Dr. Porsche's original VW is to the present-day model.

Improving the Breed

Let's go back 5½ years, to when Jaguar first announced the 2.4. Enthusiasts agreed it was a step in the right direction — but they soon decided it was underpowered and would be better with the full C-type engine.

Jaguar gave us just that with an improved cylinder head, plus disc brakes to stop it. This was the 3.4 —the best touring car available in its day.

But there were several things people didn't like about the models. The cow-hocked, tied-in look at the rear wasn't popular and didn't make for easy handling — fair in the dry, but decidedly dicey in the wet.

Jaguars countered complaints by saying the 3.4 handled as well as any car with comparable power and weight. But, even while making this claim, they already had another model on the drawing board—and at Earls Court in 1959 they unveiled the Mark II series, which gave the 2.4 and 3.4 buyers just about all they could desire.

In addition they unleashed the 3.8 Mark II, aimed at enthusiasts who wanted the utmost in performance. The Mark IX engine was used, and a limited-slip differential fitted to help cope with the additional power and torque.

All the new models used the same body and interior styling; all had disc brakes, lighter steering, a modified front end — and the wider, more stable rear track.

Only the power output varied — from 120 b.h.p. for the 2.4 to 210 and 220 for the 3.4 and 3.8 litres.

The 3.4 Mark II was tested in **Modern Motor** last June and all modifications dealt with in detail.

These are common to all the models, so I shan't delve into this subject.

What I want to tell you about is the performance of these new cars as compared to previous models.

Difference Is Startling

The 2.4 Mark I was, in my opinion, a nice little car which weighed too much and consequently didn't do anything in the performance department. Personally, I never managed to get one here for road-testing — perhaps for obvious reasons — so a comparison is difficult.

But the surprising thing about the

JARS

David McKay doffs his cap to the latest Mark II's and compares 'Junior' of family with its 'Old Man,' the 3.8

BODY lines suggest speed, power.

modern MOTOR ROAD TEST

SPECIFICATIONS, 3.8 & 2.4

NOTE: Specifications for the 3.8 model are given first, with different features of the 2.4 listed at the end.

ENGINE: 6-cylinder, twin o.h.c.; bore 87mm., stroke 106mm., capacity 3781 c.c.; compression ratio 8 to 1; maximum b.h.p. 220 at 5500 r.p.m.; twin S.U. HD6 carburettors, electric fuel pump; 12v. ignition.

TRANSMISSION: Borg-Warner three-speed automatic; ratios—Low range, 17.6 to 8.16; Intermediate, 10.95 to 5.08; Top, 3.54 to 1.

SUSPENSION: Front independent, by coil springs, semi-trailing wishbones and anti-roll bar; trailing-link suspension by cantilever semi-elliptic leaf springs and radius rods at rear; telescopic shock-absorbers all round.

STEERING: Recirculating-ball; 4.3 turns lock-to-lock, 33½ft. turning circle.

WHEELS: Pressed-steel discs with Dunlop 6.40 by 15in. tyres.

BRAKES: Dunlop servo-assisted discs all round.

CONSTRUCTION: Unitary.

DIMENSIONS: Wheelbase, 8ft. 11 3-8in.; track, front 4ft. 7in.; rear 4ft. 5 3-8in.; length 15ft. 0¾in.; width 5ft. 6¼in.; height 4ft. 9½in.; ground clearance 7in.

KERB WEIGHT: 28½cwt.

FUEL TANK: 12 gallons.

2.4 MODEL: Identical to 3.8, except for: Bore 83mm., stroke 76.5mm., capacity 2483c.c.; maximum b.h.p. 120 at 5750 r.p.m.; twin Solex down-draught carburettors; manual 4-speed gearbox — ratios, first and reverse, 15.36; 2nd, 8.46; 3rd, 5.84; top, 4.55 to 1; kerb weight 28cwt.

COMPARING THE PERFORMANCE

NOTE: The 3.8 figures are given first, with 2.4 figures shown in brackets.

BEST SPEED: 130 m.p.h. (105).

STANDING quarter: 17.0s. (20.0s.).

MAXIMUM in indirect gears: Low 40 m.p.h.; Intermediate 80 (1st, 25; 2nd, 55; 3rd, 75).

ACCELERATION from rest through gears: 0-30, 3.2s. (3.4); 0-40, 5.2s. (6.2); 0-50, 6.2s. (10.0); 0-60, 9.4s. (14.4); 0-70, 12.4s. (20.0); 0-80, 17.4s. (26.0); 0-90, 22s.

PASSING ACCELERATION: 3.8 model in Intermediate: 20-40, 4.0s.; 30-50, 4.6s.; 40-60, 5.0s.; 2.4 model, in top (with third in brackets): 20-40, 9.0s. (6.8); 30-50, 8.0s. (6.2).

HILLCLIMB: 2min. 18sec. (2min. 30sec.).

MOUNTAIN CIRCUIT: Average 55.4 m.p.h. (55.3).

BRAKING: 33ft. 6in. to stop from 30 m.p.h. in neutral (30ft. 2in.).

FUEL CONSUMPTION: 15 m.p.g. overall (20 m.p.g.).

PRICE with tax: 3.8, £3217; 2.4, £2604

Mark II 2.4 is that it lapped my mountain-circuit course at virtually the same average speed as a Mark I 3.4, despite giving away almost 100 horses and weighing only a shade less!

This amazing performance was made possible by the far better handling of the Mark II car and the use of disc brakes — now standard equipment.

Wet-weather driving is also indicative of the improved suspension. Whereas previously in the wet with a Mark I the driver was often embarrassed by the surplus of power at his command, he can now punt along happily, moving the car as he wishes with the throttle — getting the tail out in the approved attitude — and even wish for an additional horse or two.

This applies also to the 3.4 and 3.8 Mark II models, the limited-slip diff being such an advantage in the 3.8 that I would like to see it incorporated in the 3.4 at least.

Body roll is far less, due to the altered roll centre, and the steering is so effortless that I've had people tell me it's power-assisted (it isn't in Australia as yet, and frankly it doesn't seem necessary).

Despite the number of turns, the lock causes no embarrassment except

CONTINUED ON PAGE 135

The Motor Road Test No. 29/61

Make: Jaguar **Makers:** Jaguar Cars, Ltd., Coventry.
Type: 3.4 litre Mark 2 (automatic transmission and power steering)

Test Data

World copyright reserved: no unauthorized reproduction in whole or in part.

CONDITIONS: Weather: Mild and dry with light breeze. (Temperature 54°-58°F., Barometer 29.4 in. Hg.) Surface: Dry tarred macadam and concrete. Fuel: Premium-grade pump petrol (approx 97 Research Method Octane Rating).

INSTRUMENTS
Speedometer at 30 m.p.h. .. .. 1% fast
Speedometer at 60 m.p.h. .. .. 3% fast
Speedometer at 90 m.p.h. .. .. 3% fast
Speedometer at 120 m.p.h. .. .. 3% fast
Distance recorder .. .. .. 3% slow

WEIGHT
Kerb weight (unladen, but with oil coolant and fuel for approx. 50 miles) .. 30¼ cwt.
Front/rear distribution of kerb weight.. 59/41
Weight laden as tested .. .. 34 cwt.

MAXIMUM SPEEDS
Flying Mile
Mean of six opposite runs .. .. 119.9 m.p.h.
Best one-way time equals .. .. 120.8 m.p.h.
"Maximile" Speed (Timed quarter mile after one mile accelerating from rest)
Mean of four opposite runs .. .. 113.4 m.p.h.
Best one-way time equals .. .. 115.4 m.p.h.
Speed in gears (automatic upward changes)
Max. speed in 2nd gear .. .. 73 m.p.h.
Max. speed in 1st gear .. .. 45 m.p.h.
Speed in gears (at 5,500 r.p.m. using manual "hold" controls)
Max. speed in 2nd gear .. .. 82 m.p.h.
Max. speed in 1st gear .. .. 51 m.p.h.

FUEL CONSUMPTION
26.5 m.p.g. at constant 30 m.p.h. on level.
25.5 m.p.g. at constant 40 m.p.h. on level.
23.5 m.p.g. at constant 50 m.p.h. on level.
22.0 m.p.g. at constant 60 m.p.h. on level.
20.5 m.p.g. at constant 70 m.p.h. on level.
19.0 m.p.g. at constant 80 m.p.h. on level.
17.0 m.p.g. at constant 90 m.p.h. on level.
14.5 m.p.g. at constant 100 m.p.h. on level.

Overall Fuel Consumption for 1,674 miles, 104.6 gallons equals 16.0 m.p.g. (17.7 litres/100 km.)

Touring Fuel Consumption (m.p.g. at steady speed midway between 30 m.p.h. and maximum, less 5% allowance for acceleration) 19 m.p.g.
Fuel tank capacity (maker's figure) 12 gallons.

STEERING
Turning circle between kerbs:
Left .. .. 35 ft.
Right .. .. 36¾ ft.
Turns of steering wheel from lock to lock 4

BRAKES from 30 m.p.h.
0.97 g retardation (equivalent to 31 ft. stopping distance) with 100 lb. pedal pressure
0.83 g retardation (equivalent to 36½ ft. stopping distance) with 75 lb. pedal pressure
0.58 g retardation (equivalent to 52 ft. stopping distance) with 50 lb. pedal pressure
0.32 g retardation (equivalent to 94 ft. stopping distance) with 25 lb. pedal pressure

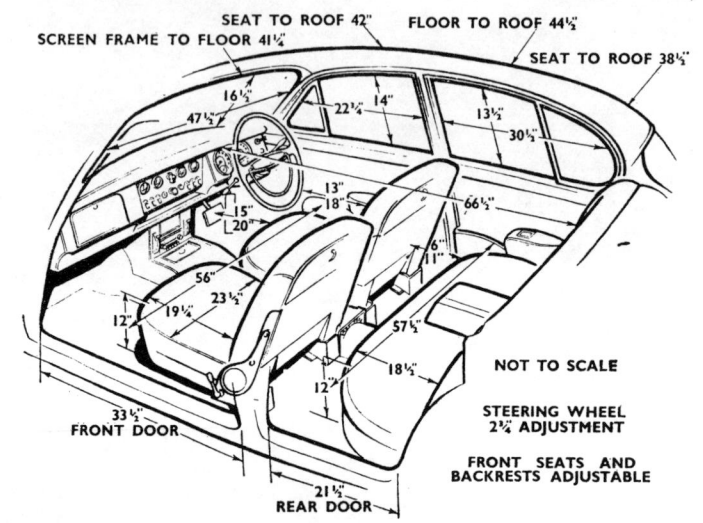

ACCELERATION TIMES from standstill
(Automatic gearchanges)
0-30 m.p.h. .. .. .. 4.5 sec.
0-40 m.p.h. .. .. .. 6.4 sec.
0-50 m.p.h. .. .. .. 9.0 sec.
0-60 m.p.h. .. .. .. 11.9 sec.
0-70 m.p.h. .. .. .. 15.3 sec.
0-80 m.p.h. .. .. .. 20.0 sec.
0-90 m.p.h. .. .. .. 26.0 sec.
0-100 m.p.h. .. .. .. 33.3 sec.
0-110 m.p.h. .. .. .. 44.5 sec.
Standing quarter mile .. .. 19.1 sec.
Standing quarter mile (using 5,500 r.p.m. in lower gears with manual "hold" controls) .. .. .. 18.4 sec.

ACCELERATION TIMES from rolling start
"Kick down"
10-30 m.p.h. .. .. .. 3.4 sec.
20-40 m.p.h. .. .. .. 3.8 sec.
30-50 m.p.h. .. .. .. 4.6 sec.
40-60 m.p.h. .. .. .. 5.6 sec.
50-70 m.p.h. .. .. .. 6.3 sec.
60-80 m.p.h. .. .. .. 8.1 sec.
70-90 m.p.h. .. .. .. 10.7 sec.
80-100 m.p.h. .. .. .. 13.3 sec.
90-110 m.p.h. .. .. .. 18.5 sec.

HILL CLIMBING at sustained steady speeds
Max. gradient on top gear .. .. 1 in 8.1 (Tapley 275 lb./ton)

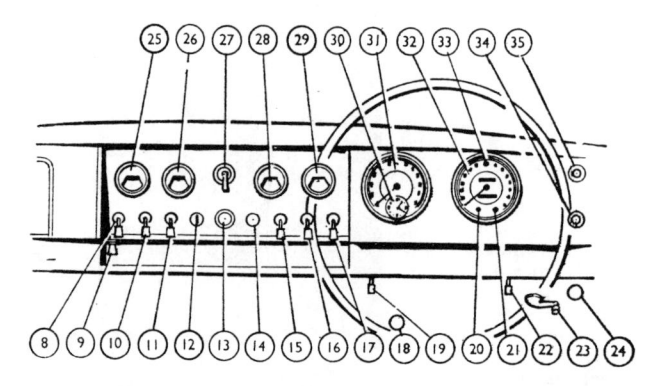

1, Heater temperature control. 2, Radio. 3, Heater air control. 4, Direction indicators switch and headlamp flasher. 5, Horn ring. 6, Automatic transmission selector. 7, Handbrake. 8, Interior lights switch. 9, Scuttle vent lever. 10, Bright/dim panel lights. 11, Two-speed fan switch. 12, Ignition switch. 13, Cigar lighter. 14, Starter button. 15, Map-light switch. 16, Two-speed windscreen wipers control. 17, Screenwasher control. 18, Headlamp dipswitch. 19, Clock resetting knob. 20, Dynamo charge warning light. 21, Fuel level warning light. 22, Trip resetting knob. 23, Aerial winder. 24, Bonnet catch release. 25, Ammeter. 26, Fuel gauge. 27, Lights switch. 28, Oil pressure gauge. 29, Water thermometer. 30, Clock. 31, Rev. counter. 32, Speedometer and mileage recorder. 33, Main beam indicator light. 34, Brake fluid and handbrake warning light. 35, Intermediate gear hold switch.

The JAGUAR 3.4-litre Mark 2

Power Steering and Automatic Transmission on a car of Brilliant Versatility

CUSTOMARILY those of our staff who draft Road Test Reports set out to indicate what sort of tastes and requirements a particular car can satisfy. The eight who drove this 3.4 litre Jaguar while it was on test, with others who also were invited to comment upon it, represent widely varied sizes, shapes and sorts of motorist, but the unanimity of their praise was virtually unprecedented. This model is only designed to carry four people or an occasional five, and it is only seen to best advantage on reasonably well surfaced roads, but within this frame of reference it offers an outstanding combination of speed, refinement and true driving ease. When price also is considered, it is easy to see why Jaguar competition has been driving one make after another out of existence.

From an extensive range, our test model might reasonably be described as a very good average Jaguar. The twin-overhead-camshaft XK-series engine is available also in larger-bore 3.8 litre form, or, with shorter stroke, as a 2.4-litre. Smaller two-seat cars are listed, and there is the roomier Mark IX Jaguar with longer wheelbase and wider track. Optional items on the 3.4 saloon tested included the Borg Warner fully automatic transmission, power-assisted steering and variable-rake front seat backrests.

Our first impressions were almost completely favourable. For a driver there were minor surprises, in that the starter button has not been supplanted by key-starting, and the gear selector lever is on the right of the steering wheel with the turn signal control on the left instead of vice-versa. The driving seat is extremely comfortable, however, and has an adjustment range adequate for any save quite exceptionally long legs, driving vision is good between slim windscreen pillars, and it seems easy to find a natural position for the telescopically-mounted steering wheel. Although it is hard to see the nearside front wing, the bonnet drops well below normal sight lines, and even if the waistline of the body is higher than is fashionable, one cannot complain about the size of the thinly-framed side and rear windows. After

COCKPIT of the Mark 2 Jaguar shows the walnut facia and neat instrument layout. The big pedal controls servo disc brakes, steering is power assisted and the transmission is automatic.

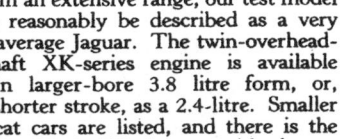

In Brief
Price (including automatic transmission and power steering as tested) £1,337 plus purchase tax £614 0s. 7d. equals £1,951 0s. 7d. Price with synchromesh gearbox and manual steering (including purchase tax), £1,717 13s. 11d.
Capacity ... ... 3,442 c.c.
Unladen kerb weight ... 30¼ cwt.
Acceleration:
  20-40 m.p.h. in drive range 3.8 sec.
  0-50 m.p.h. through gears ... ... 9.0 sec.
Maximum top gear gradient 1 in 8.1
Maximum speed ... ... 119.9 m.p.h.
"Maximile" speed ... 113.4 m.p.h.
Touring fuel consumption 19.0 m.p.g.
Gearing: 21.4 m.p.h. in top gear at 1,000 r.p.m.; 30.8 m.p.h. at 1,000 ft./min. piston speed.

dark, the headlamps gave long but not very wide beams.

Interior decoration is excellent, with heavily grained polished walnut on the facia and below the windows, leather upholstery over a layer of Dunlopillo, and pile carpets underlaid with soft felt. Large m.p.h. and r.p.m. dials face the driver, the latter incorporating the clock, and other instruments are central on the facia. The neat line of lift-up switches, which might otherwise be confusing, has identifications for each switch illuminated when the dim-or-bright instrument lighting is on.

Essentially this is a close-coupled four-seater body, but most people will find sufficient knee-room and headroom in the back of these Mark 2 cars, with plenty of foot-room under the front seats. The car is wide enough to seat three abreast, but as the seat cushion is very thin over the central transmission line it is better to fold down the central armrest and regard this as a car for four adults. Rear-seat travellers are most certainly not given "poor relation" treatment; an air supply from the interior heater is ducted to them between the front seats, and the latter have folding picnic tables behind them.

A luggage locker of quite generous size has a flat carpeted floor and heavy trunks are easily loaded into it. The spare wheel and tools are carried beneath a trap door in the boot. Inside the car, adequate

87

INTERIOR TRIM uses leather, pile carpets and polished woodwork to produce a quietly pleasing effect. Rear-seat legroom is adequate rather than generous, but special provision for back-seat comfort includes hot-air ducts, folding picnic tables and three armrests.

rather than generous stowage for oddments is provided by a lockable glove box, but it would be better if the lid had a catch so that it would stay closed without use of a key. A map shelf lies below the instruments, and there are pockets in the four doors. Four interior lights, with courtesy switches on all doors, plus a map-reading lamp and a light inside the glove box, are details typical of this car's comprehensive equipment.

It may seem odd to place emphasis on the equipment when describing a model so renowned for its top speed of 120 m.p.h. but the Mark 2 3.4 is an outstanding car irrespective of price, not merely because it is faster than most others with which it might be compared, but because this performance is provided so smoothly and effortlessly by such a well-furnished car.

Most journeys start in a town, and this Jaguar is entirely at ease in heavy traffic. Modest pressure on the accelerator pedal inches the car gently away from rest, or progressively firmer pressures give graduated response until it really surges away very fast indeed. An unusual but very convenient device holds some pressure in the hydraulic braking system after the car has been stopped, until the ignition is switched off or the accelerator pedal is touched lightly, so that the car stays put on the level or on moderate slopes without any creep due to tick-over drag in the automatic transmission. The stop lights do not continue to shine whilst this device is holding the car still.

There being a more powerful engine available, it is appropriate that this 3.4-litre Jaguar should be silenced with extreme effectiveness, and its automatic transmission tuned to give very high performance (as distinct from the maximum possible) without fuss. Simply by depressing the accelerator fully, one moves away from rest without wheelspin; the car reaches 80 m.p.h. in 20 seconds and is doing 110 m.p.h. within less than 45 seconds of starting. This happens with upward changes of gear occurring automatically at engine speeds below 5,000 r.p.m., and if on some special occasion for haste the manual controls are used to delay upward changes until the maximum-power speed of 5,500 r.p.m., even better acceleration can be recorded. This is not the smoothest automatic transmission that we have sampled; changes of gear are quite evident and alternation between 2nd and top gears sometimes occurs rather easily with small changes of speed or throttle opening, but it is very satisfactory. There is a convenient switch for holding 2nd gear in use when top gear might otherwise engage automatically, but the keener driver might prefer this switch to be combined with the selector quadrant, one position on which allows 1st gear to be engaged by the driver —primarily for braking down very steep hills.

Nobody will buy this as an economy car, but our overall fuel consumption of 16 m.p.g. (recorded in driving which was usually quite fast or in very heavy traffic) suggests that a fuel tank holding more than 12 gallons would be advantageous. The test car with 8/1 compression ratio was entirely happy to burn any Premium-grade petrol, a 9/1 ratio being available to exploit 100-octane fuel, and a 7/1 ratio for countries where fuel quality is very low. With about 12,000 miles of running-in behind it, the test model used more oil than most modern engines.

One expects the combination of a 210 b.h.p. engine with a fully automatic transmission, power-assisted disc brakes and power-assisted steering to take most of the physical effort out of driving, and this they certainly do. Unlike other manufacturers who offer similar features, Jaguars have a vast fund of competition experience which includes five outright victories in Le Mans 24-hour sports car races, and this car goes far towards taking the mental strain, as well as the physical effort, out of fast travel. Nothing eliminates the need for alertness at the speeds which it can attain, but the driver of this car has no worries about its response to the accelerator, brakes or steering.

Hydraulic power assistance does almost all the work of steering, although servo action has been limited so that, whilst the car steers easily at a fraction of an m.p.h., it is not easy to strain the steering by turning it with the car at rest on a hard surface. For all its extreme lightness, the power steering has definite self-centering action, and although it never kicks back, it can react gently to road irregularities, letting a driver know just what he is doing. Most people felt that its gearing (four turns from lock to lock) was not as quick as they would

TAIL DETAILS include a flat-floored luggage locker which, like the rear window, is wider than on early 3.4-litre Jaguars. Rolling back the carpet discloses a cover over the spare wheel and tools.

OVERHEAD CAMSHAFTS give the 6-cylinder Jaguar engine its distinctive appearance, and in 3.4-litre form 120 m.p.h. speed potential is combined with notable quietness and smooth response for traffic driving. Accessible details include a brake fluid reservoir with low-level warning lamp, petrol filter, battery, starter solenoid, heater and screen-washer reservoir.

have liked, but with such light control very prompt response to any emergency was possible. At speed, there was no trace of instability.

Fast cornering showed no sway and only a limited degree of body roll, with a safe and consistent but not exaggerated degree of understeer. At the pressures which a fast driver should use, the tyres only squealed under severe provocation. If the front seat backrests were curved to give more lateral support, the driver would be happier and his passenger would be incomparably more comfortable during fast travel along winding roads.

It was difficult to criticize a set of brakes that did their job magnificently. No fade or squeal was encountered, no snatch at town speeds; progressive response to pleasantly moderate pedal pressures, and a tailing-off of servo assistance under extreme braking conditions, effectively prevented unintentional wheel locking. In sharp and welcome contrast with some previous cars with disc brakes on all four wheels, this model's self-adjusting handbrake would hold it on a 1 in 3 gradient in either direction—and, in spite of nose-heavy weight distribution, a smooth re-start from rest was possible on this test hill in reverse, as well as forwards, a trial which most cars fail with wheelspin.

Qualified praise was earned for riding qualities. Small road irregularities such as marking studs were smoothed out and silenced to an exceptional extent, this quiet and smooth "secondary ride" over small bumps giving a fine impression of luxury. Larger bumps were at times less well absorbed, initial spring flexibility seeming to give place to much greater stiffness for larger deflections, so that whilst big bumps encountered at speed did not produce any of the alarming effects that over-soft springing could induce, riding comfort at medium speeds is not this car's best feature. Even the high tyre pressures advised for very fast motorway cruising induced only a bare trace of body shake on cobbled town streets.

An automatic auxiliary carburetter gave prompt engine starting on cool mornings in summer, but two-position mixture strength and idling speed control did not always provide a reliable tick-over with a half-warm engine. As a safety interlock makes it necessary to select "neutral" with the transmission control before the starter button is "alive," a stalled engine in traffic proved an occasional nuisance. Once warm, the engine was a model of quiet smoothness, wonderfully inconspicuous at most times in town, but always ready with a gentle roar of real power if the throttles were slammed open. At a cruising speed of 100 m.p.h. this car has notably little noise from power unit, road or wind to spoil enjoyment of the car radio.

This Jaguar is not a perfect car; the team which is headed by Sir William Lyons and has William Heynes in charge of engineering will no doubt eventually produce something even better. In the present state of the automobile engineering art, however, what they are already making in the summer of 1961 rates as one of the best all-round cars for motoring on civilized roads yet seen anywhere in the world.

The World Copyright of this article and photographs is strictly reserved © Temple Press Limited, 1961

Specification

Engine
- Cylinders ... 6
- Bore ... 83 mm.
- Stroke ... 106 mm.
- Cubic capacity ... 3,442 c.c.
- Piston area ... 50.4 sq. in.
- Valves ... Inclined o.h.v. (2 o.h. camshafts)
- Compression ratio ... 8/1 (optional 7/1 or 9/1)
- Carburetter ... Two horizontal S.U. type HD 6
- Fuel pump ... S.U. electrical
- Ignition timing control ... Centrifugal and vacuum
- Oil filter ... Tecalemit full-flow
- Max. power (gross) ... 210 b.h.p.
- at ... 5,500 r.p.m.
- Piston speed at max. b.h.p.: 3,820 ft./min.

Transmission (Borg Warner automatic)
- Clutch: Hydraulic torque converter, maximum multiplication 2.15/1, working with 1st, 2nd and reverse gears only.
- Top gear ... 3.54
- 2nd gear ... 5.08
- 1st gear ... 8.16
- Reverse ... 7.11
- Propeller shaft ... Hardy Spicer
- Final drive ... Hypoid bevel
- Top gear m.p.h. at 1,000 r.p.m. ... 21.4
- Top gear m.p.h. at 1,000 ft./min. piston speed ... 30.8

Chassis
- Brakes: Dunlop disc type (all wheels) with vacuum servo.
- Brake disc diameters ... Front 11 in. rear 11⅜ in.
- Friction area: 31.8 sq. in. of pad area working on 495 sq. in. rubbed area of discs.
- Suspension:
  - Front: Independent by transverse wishbones, coil springs and anti-roll torsion bar.
  - Rear: Rigid axle, cantilever leaf springs, trailing radius arms and Panhard rod.
- Shock absorbers ... Girling telescopic
- Steering gear: Burman recirculating ball type, power assisted (optional) on test car.
- Tyres: Dunlop Road Speed (tubed), 6.40—15

Coachwork and Equipment

- Starting handle ... None
- Battery mounting ... On scuttle behind engine
- Jack ... Bipod pillar type with ratchet handle
- Jacking points: 2 external sockets under each side of body.
- Standard tool kit: Jack and handle, wheel brace, set of open-jaw spanners, set of box spanners, adjustable spanner, pliers, screwdriver, feeler gauges, grease gun, brake bleeder tube.
- Exterior lights: 2 headlamps, 2 foglamps, 2 sidelamps, 2 stop/tail lamps, reversing lamp.
- Number of electrical fuses ... 2
- Direction indicators: Self-cancelling amber flashers
- Windscreen wipers: Electrical two-speed twin-blade, self parking.
- Windscreen washers ... Electrical pump type
- Sun visors ... 2, universally pivoted
- Instruments: Speedometer with total and decimal trip distance recorder, r.p.m. indicator, clock, fuel contents gauge, coolant thermometer, oil pressure gauge, ammeter.
- Warning lights: Dynamo charge, headlamp main beam, low fuel level, turn indicators, brakes (handbrake on or low fluid level).
- Locks:
  - With ignition key: Ignition switch, petrol filler cover and either front door.
  - With other key: Glove box and luggage locker.
- Glove lockers: One on facia, with locking lid
- Map pockets: Map shelf under facia panel and pockets in all four doors.
- Parcel shelves ... One below rear window
- Ashtrays ... One on facia, two in rear armrests
- Cigar lighters ... One on facia
- Interior lights: 2 on centre pillars and 2 in rear quarters (with courtesy switches on all doors), also glove-box and map-reading lamps.
- Interior heater: Fresh air heater fitted as standard, with screen de-misters and warm air ducts to rear compartment.
- Car radio: Optional extra, Smiths Radiomobile
- Extras available: Automatic transmission or overdrive, power steering, PowrLok differential, radio, centre-lock wire wheels, reclining seats, etc.
- Upholstery material ... Vaumol leather over Dunlopillo
- Floor covering ... Pile carpets with felt underlay
- Exterior colours standardized: 10 (others to order at extra cost).
- Alternative body styles: None (similar bodywork available with 2.4-litre or 3.8-litre engine).

Maintenance

- Sump: 11 pints plus 2 pints in filter, S.A.E. 20 winter, S.A.E. 30 summer, S.A.E. 40 tropical.
- Gearbox (automatic): 15 pints, automatic transmission fluid type "A".
- Rear axle: 3½ pints, S.A.E. 90 hypoid gear oil
- Steering gear lubricant: S.A.E. 140 gear oil in manual steering, or Automatic Transmission Fluid in power steering reservoir.
- Cooling system capacity 22 pints (2 drain taps)
- Chassis lubrication: By grease gun every 2,500 miles to 8 points, and by oil gun every 2,500 miles to 4 points.
- Ignition timing ... 7° before t.d.c. static (with 8/1 compression)
- Contact-breaker gap ... 0.014-0.016 in.
- Sparking plug type ... Champion N5
- Sparking plug gap ... 0.025 in.
- Valve timing: Inlet opens 15° before t.d.c. and closes 57° after b.d.c.; exhaust opens 57° before b.d.c. and closes 15° after t.d.c.
- Tappet clearances (cold) ... Inlet 0.004 in. Exhaust 0.006 in.
- Front wheel toe-in ... Parallel to 1/16 in.
- Camber angle ... ½°-1° positive
- Castor angle ... −¼ to +¼°
- Steering swivel pin inclination ... 3½°
- Tyre pressures: Front 28 lb.; Rear 24 lb. (increase front and rear pressures by 5 lb. for very fast driving).
- Brake fluid ... S.A.E. Spec. 70.R.3
- Battery type and capacity: Lucas BV 11 A, 12 volt, 60 amp. hr.
- Miscellaneous: Check fluid levels in automatic transmission and in power steering reservoir every 1,250 miles.

Road Test
JAGUAR 3.8

MT's exclusive test of one of the most respected imported luxury cars

TOP **AUTOMOTIVE IDEAS** from both sides of the Atlantic are combined in the Jaguar 3.8 sedan. A compact package only 180.8 inches long, it contains a potent six-cylinder engine basically similar to that used in Jaguar sports cars, together with a superbly roadable chassis highlighted by such features as disc brakes.

To these exotic components from European racing machines, the 3.8 adds American concepts of convenience in the form of an optional automatic transmission and power assists for both the steering and brakes.

In displacement, Jaguar's 3.8-liter, six-cylinder engine is roughly equivalent to Detroit Sixes. Refinements, however, have made it far more powerful.

At first glance, the specifications of the 3.8 engine read very much like those of a big U.S. Six. Displacement is 230.6 cubic inches, or, to explain the car's designation, 3.8 liters, while the compression ratio is a modest 8-to-1 and the two carburetors are of the single-barrel type.

Yet the 3.8 has 225 hp at 5500 rpm, nearly one full hp per cubic inch, and 240 lbs.-ft. of torque at 3,000 rpm.

The secret of its efficiency is its race-bred design. Similar to the XK powerplant, it is a classic example of an engine seasoned by competition experience. Typifying its overall ruggedness, seven main bearings support the crankshaft. The stroke is relatively long, the aluminum pistons thrusting 4.17 inches through cylinders measuring only 3.42 inches in bore. The block is cast iron but the head, which incorporates hemispherical combustion chambers developed from the Jaguar "C" and "D" racing units, is made of aluminum. Finally, topping the whole works are dual overhead camshafts.

The result of such a sophisticated approach is an engine which delivers its high output with a minimum of fuss. The 3.8 performs beautifully with the smoothness and flexibility one expects in a luxury vehicle. It is a relatively quiet powerplant, too, normally operating with a gentle purr. It will raise its voice under full throttle, but just enough to please the enthusiastic sort of driver apt to choose such a car in the first place.

Jaguar's automatic transmission is the Borg-Warner unit consisting of a torque converter and three-speed gearbox and offered in this country by Ford, American Motors and Studebaker-Packard. A four-speed manual with overdrive is actually standard in the 3.8 but is fitted to very few of the cars shipped across the Atlantic. The rear axle, which includes a limited-slip differential as regular equipment, has ratios of 3.54 with the automatic and 3.77 with the manual.

As adapted for Jaguar, the B-W automatic has a special anti-creep device actuated by the brake pedal and a manual control to engage second gear.

The normal B-W unit can be downshifted to second by pulling from "D" range to "L" above first-gear road speed. If the car is slowed, however, the transmission will drop to first and not return to second without being shifted back through "D."

But in the 3.8, it can be dropped into second or held there during acceleration with, literally, a flick of the finger. The control is a simple lever at the left side of the dash, replacing the overdrive switch used in the manually-geared Jaguar. And when the feature is used, the transmission not only shifts down to first but back up to second as the speed of the car demands.

continued

Sleek Jaguar sedan embodies many of the best elements of a modern European sports car in a package that has adopted many American concepts of convenience.

Instrument cluster is neat, comprehensive and has a practical layout which makes driver controls easy to see and use. Although car is an automatic, tach is standard.

One area in which the sedan is disappointing is in the trunk, where luggage space is considerably limited.

An excellent feature is the manner in which the instrument cluster quickly disengages and drops down to permit repairs or adjustments to the electrical system.

JAGUAR 3.8 ROAD TEST

JAGUAR SEDAN HANDLES LIKE THE SPORTS CAR FOR WHICH ITS NAME IS FAMOUS, YET IS COMFORTABLE AND EASY TO DRIVE.

In effect, it works like the "2" range in Chrysler's Torqueflite.

Performance of the automatic 3.8 compares favorably with that of our better V-8's. The test car's acceleration times from zero to 30, 45 and 60 mph were 4.2, 7.2 and 10.8 seconds, respectively.

Judging from experience with other makes using the B-W transmission, that 0-to-60 figure is between 1.5 and two seconds slower than could be expected with the manual gearbox. The four-speed Jaguar should be able to hold its own against all but the hottest Detroit machines.

Fuel consumption is quite moderate for a car of such performance. The test car operated in a 15-to-19-mpg range and, with the manual shift, it would be reasonable to hope for averages one or two mpg higher.

The 3.8 is capable of safe travel well in excess of most legal speed limits. Stable suspension, consisting of independent coils in front and quarter-elliptics in the rear, provides the ideal complement for the car's performance. The spring rates are relatively firm by U.S. standards, and Jaguar mechanics place an unusually strong emphasis on varying tire pressure for different kinds of driving.

The primary goal is a steady ride at high speeds and it is achieved admirably. Extremely sure-footed, the 3.8 imparts a remarkable sense of security to both driver and passengers. It squats right on the highway with no floating tendencies whatever. Even at city speeds, the feeling of stability is apparent.

The road is felt by the car's occupants, though seldom with any harshness except on very hard, irregular surfaces. Severe shocks are noticed just once, however; rebound is cancelled immediately.

Side windows and vents seal the passenger compartment from external noises and drafts exceptionally well. Notice the rear wind vents plus a roll-down window.

Considering its short overall length, the Jaguar sedan has an excellent back seat with comfort that exceeds most American-made cars with the same dimensions.

Cornering is firm, flat and accurate. Body sway is slight for a sedan, and a skilled driver soon finds the proper amount of throttle pressure to keep the car in the groove on tight turns. Quick correction is a slight problem, though, until the feel of the power steering is mastered. While more sensitive to the road than the systems used in most domestic makes, it seems unusually light to drivers coming to the 3.8 from other high-powered imports.

The steering requires 4.3 turns lock-to-lock but is quicker than such a figure might indicate because it turns the car around in less than 36 feet.

Drivers of all sizes and shapes have little trouble arranging themselves comfortably in the cockpit. The seat moves not only fore and aft, but also up and down. In addition, a telescoping steering column allows the wheel to be repositioned.

Matching five-inch dials house a speedometer and tachometer right in front of the driver. The tach may seem dilettantish in a car with an automatic shift, but Jaguar's modification of the B-W transmission allows enough manual control to justify the instrument. Besides, a Jaguar would not seem right without a tach!

Secondary instruments and toggle switches for minor controls are at the center of the dash, all clearly marked and well illuminated. The panel is easily removed for access to the wiring. Immediately below on the transmission hump is a leather-covered console that includes radio and heater controls and an ash tray.

The front seats are not fully contoured but the foam rubber padding in their pleated centers is softer than that around the edges. Occupants sink into them slightly and get a bucket effect.

With the windows closed, anything happening outside the car can barely be heard by the passengers, yet plenty of fresh air is delivered through the cowl ventilator. The front vent wings are carefully designed to cock at 45 or 90 degrees. In the lesser position, wind noise is astonishingly low. There is almost none of the usual howl and pitch from the air rushing past.

Considering the 3.8's potential as a highway car, it is disappointing to find it has a rather small, 13.5-cubic-foot trunk.

The sloping deck restricts cargo space considerably. The compartment is deep, though, extending to the back of the rear seat. The floor is level, except for a slight kick-up over the axle, and is covered with a detachable piece of leatherette.

The spare tire and tools are carried in a space of their own underneath the trunk. The tire rests horizontally, carrying in its middle a neat case for the tools. In addition to the usual pliers and wrenches, the kit includes such items as a tire gauge, grease gun and spark plug and tappet feeler gauges. /MT

MOTOR TREND TEST DATA

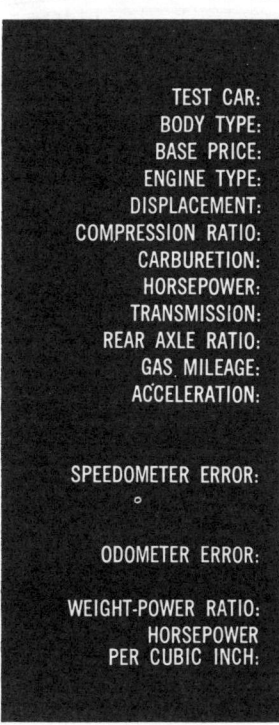

TEST CAR:	Jaguar 3.8
BODY TYPE:	Four-door sedan
BASE PRICE:	$5195
ENGINE TYPE:	Dohc 6
DISPLACEMENT:	230.6 cubic inches
COMPRESSION RATIO:	8-to-1
CARBURETION:	Two single-barrel
HORSEPOWER:	225 @ 5500 rpm
TRANSMISSION:	Three-speed automatic
REAR AXLE RATIO:	3.54
GAS MILEAGE:	15 to 19 miles per gallon
ACCELERATION:	0-30 mph in 4.2 seconds, 0-45 mph in 7.2 seconds and 0-60 mph in 10.8 seconds
SPEEDOMETER ERROR:	Indicated 30, 45 and 60 mph are actual 30, 45 and 60 mph, respectively
ODOMETER ERROR:	Indicated 100 miles is actual 104 miles
WEIGHT-POWER RATIO:	13.9 lbs. per horsepower
HORSEPOWER PER CUBIC INCH:	.98

THE 2·4-LITRE IN DETAIL

The 2·4-litre motor which, with a capacity of 2,483 c.c. develops 120 b.h.p. at 5,750 r.p.m. The cylinder head is of high tensile aluminium alloy with hemispherical combustion chambers.

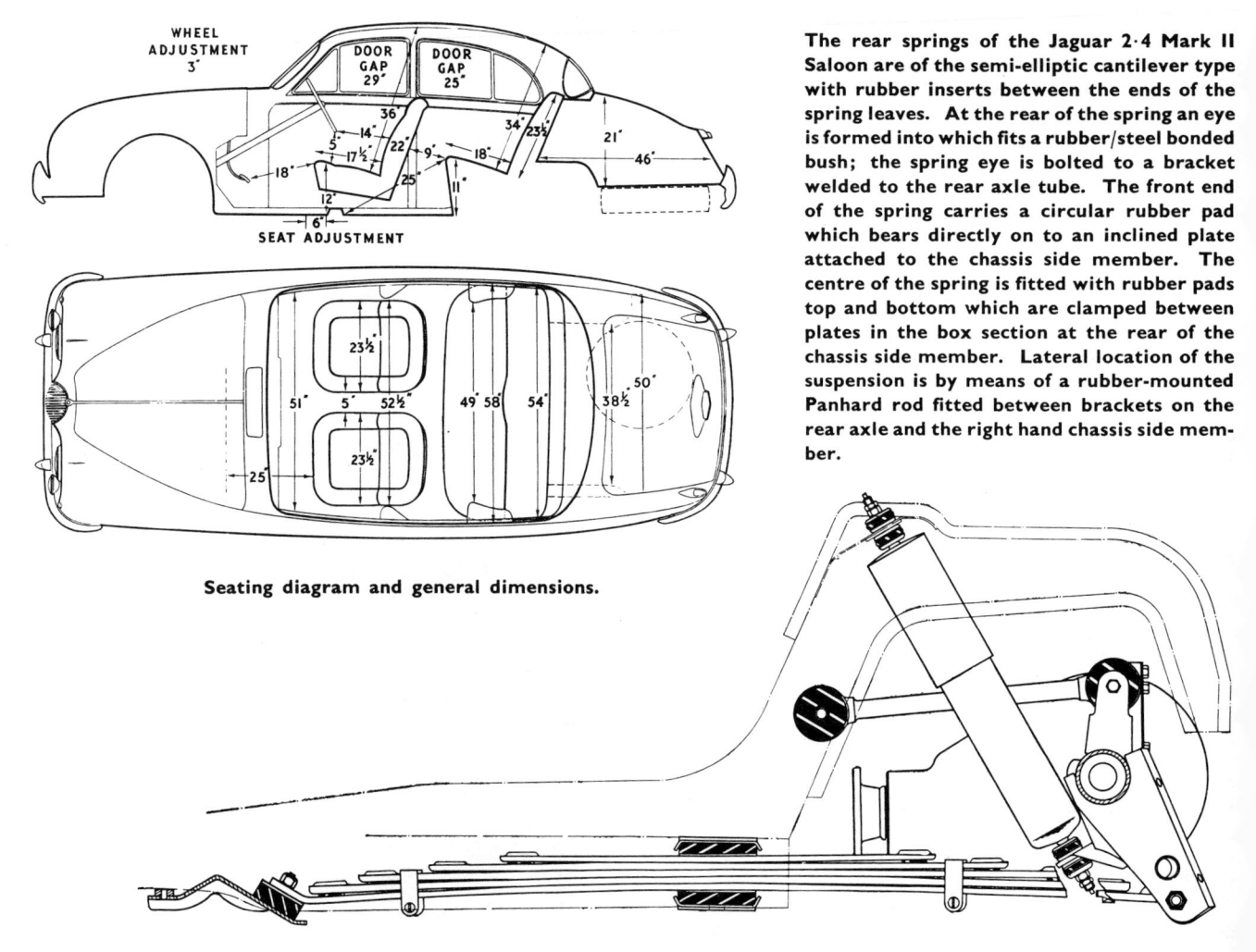

Seating diagram and general dimensions.

The rear springs of the Jaguar 2·4 Mark II Saloon are of the semi-elliptic cantilever type with rubber inserts between the ends of the spring leaves. At the rear of the spring an eye is formed into which fits a rubber/steel bonded bush; the spring eye is bolted to a bracket welded to the rear axle tube. The front end of the spring carries a circular rubber pad which bears directly on to an inclined plate attached to the chassis side member. The centre of the spring is fitted with rubber pads top and bottom which are clamped between plates in the box section at the rear of the chassis side member. Lateral location of the suspension is by means of a rubber-mounted Panhard rod fitted between brackets on the rear axle and the right hand chassis side member.

A CAR ROAD TEST

The 2·4-litre Jaguar Mark II saloon

ONE of the fastest and most expensive pieces of machinery to be road-tested by CAR in 1959 was a Mark II Jaguar 2.4-litre saloon, models of which are now available at the South African coast at a cost of nearly £1,600. Little or no introduction to this famous marque is needed; let it merely be said by way of preface that this latest Jaguar provides a superb combination of the comfort one looks for in an elegant town carriage with the performance one expects from a factory which has five outright wins in the Le Mans 24-hour Endurance Race to its credit.

The Mark II version of the 2.4 Jaguar is, outwardly, but little changed from its predecessors. While retaining the same shape as former 2.4 models, modifications to the bodywork include the fitting of a larger windscreen and larger rear windows — the latter, in particular, giving rear-seat passengers increased outward visibility. Mechanically, the track of the rear wheels has been slightly increased in the interests of better road-holding and Dunlop disc brakes are now standard equipment on all four wheels.

Interior changes include the fitting of a completely re-designed instrument panel. In older models, the rev. counter and speedometer were mounted centrally on the dashboard; in the Mark II, however, these two dials are now directly ahead of the driver, the array of smaller dials, tumbler switches, starter button, cigar lighter, etc., having been moved to the centre of the facia. Also centrally located, and below the instrument panel, is a loudspeaker grille for a radio (optional equipment) and to the left and right of this grille the controls to operate the heating and ventilating equipment which is standard.

Of special interest on the extreme right of the dashboard is the red warning light which glows when the pull-up handbrake (on the right of the driver's seat) is on and which also glows when the level of the brake fluid drops.

The all-steel four-door body is of integral body-chassis construction, the front suspension being independent by means of semi-trailing wishbones and coil springs, while the rear is by trailing link and cantilever semi-elliptic springs.

Interior appointments and finish are all that one could expect: there are four courtesy lights — two above the central door pillars and two mounted in the roof above the rear seats. Picnic tables fold out from the backs of the front seats and a separate heating duct is incorporated for the rear compartment.

We can fault these interior appointments in three

Good looks combine with racy lines in this three-quarter frontal view of the Mark II Jaguar.

minor aspects: the sun visors do not swivel but fold out in one plane from the roof; the lid of the "cubby-hole" can only be opened by being unlocked with the ignition key (which cannot be removed if the motor is running) and the appearance of the rear-seat courtesy lights did not strike us as being quite in keeping with the conservative nature of the rest of the interior. In fact, they excited the comment that they looked more like lights inside a refrigerator!

Once seated in the driver's seat, one is immediately conscious of the quality of the Jaguar and of the care which has obviously gone into its interior design. The seat is extremely comfortable and holds the driver securely even under hard cornering; he can adjust his driving position to his exact requirements not only by moving the seat itself but also by making use of the telescopic adjustment to the steering column. The Jaguar started from cold on two successive mornings without use of the choke — this operation being accompanied audibly only by the whirr of the starter motor; the engine itself was absolutely silent at ticking-over speeds.

On the open road, it at once becomes apparent that silence is one of the Jaguar's most laudable qualities. At all speeds — whether in traffic or in the 80's and 90's — there is absolutely nothing audible in the car's behaviour to disturb the passengers. From the engine itself comes a satisfying hum; there is very little tyre howl when the car is cornered. In addition to this silence, the car conveys a marked air of steadiness and smoothness: it literally sits on the road and even at maximum — 102 m.p.h. — there was no vibration of any kind. We know of few other cars available in South Africa to-day better suited to the type of long-distance motoring which our national roads afford. With the aid of the excellent disc brakes we estimate that averages of up to 10 m.p.h. faster can easily be attained by the Jaguar over comparable cars which are fitted with conventional drum brakes.

In city and suburban traffic, the merest touch on the brake pedal is enough to slow the car; violent applications in the 90's brought it to a smooth and rapid halt

The purposeful lines of the Mark II version of the Jaguar 2·4-litre are shown to advantage in this side view. Unlike former 2·4 models, it has no spats covering the rear wheels.

The dignified driving compartment. In this picture can be seen the comprehensive instrumentation — the brake-fluid warning light is on the extreme right of the facia.

— with little dip of the bonnet — and with no swerving to one side or the other. Only the clutch seemed out of keeping with the rest of the car: during our acceleration tests, we detected some slipping. The makers, very wisely in our view, have seen fit to incorporate a small triangular sign on the rear bumper with the warning "Disc brakes"!

Synchromesh between first and second gears could be beaten when making rapid changes. (We are told by the distributors that this is a trait common to all Jaguars). For the rest, however, gear-changing was a delight — the lever having a short, positive movement. On rough country roads, the suspension, so ideally suited, as we have said, to high speeds on macadam, did not show up quite as well; shocks were transmitted to the interior of the car and the rear axle, under heavy acceleration, did, once or twice, hop about somewhat. The steering — although needing five turns from lock-to-lock — was light at all speeds; although the Jaguar gives the impression of being a "heavy" car there is nothing but lightness in its immediate response to steering and, in fact, to all other controls.

The chief attribute of the Jaguar, to our way of thinking, is the excellence of the motor. This is a six-cylinder XK unit fitted with the "B" type cylinder head which has twin overhead cams acting on valves set at 70° in hemispherical combustion chambers and which delivers its power with silky smoothness throughout its revolution range. It is extremely flexible; the car draws away smoothly in top from speeds as low as 15 m.p.h. The crankshaft is carried in seven bearings of $2\frac{3}{4}$ ins. diameter to ensure absence of vibration and "whip". (According to the makers, the most careful inspection is applied throughout every stage of manufacture and assembly of this motor — each engine being bench-tested for four hours and thereafter checked on the road.) The motor is a short-stroke version, in fact, of the world-famous 3.4 litre unit.

When turning at more than 5,000 r.p.m. we could detect no evidence that the unit was working "over time" — so smooth is it. We do think, however, that the fitting of overdrive — regrettably not available in models imported into the Union — to be advisable in the interests of even greater longevity. (Overdrive would reduce the top gear ratio of 4.27 to 1 to 3.54 to 1.)

We have already mentioned that the Jaguar 2.4 is an expensive car and, as is to be expected, it is somewhat expensive to run. In our hands — admittedly being driven hard and fast most of the time — it returned a fuel consumption figure of 18.9 m.p.g. Only one free service at 500 miles is provided — lubricants used, being chargeable. In addition, the cost of licensing it amounts to £12 a year and the basic insurance premium is a further £52. On the credit side, the car's mechanical

Lifting the bonnet reveals "a lot of engine" — neatly installed and easily accessible.

components are guaranteed for a period of six months — regardless of the mileage covered in this time — but electrical equipment is guaranteed for six months or 6,000 miles — whichever occurs first.

Nation-wide service facilities are available and a full range of spare parts is carried. The car can be serviced by an authorized Jaguar distributor or dealer, regardless of where it was bought.

In summary, the Jaguar 2.4 is among the foremost *gran turismo* vehicles available in South Africa to-day. It will find ready acceptance not only among those who regard the ownership of a Jaguar to be "tres chic" but also among those members of the sporting fraternity who have got beyond the stage of sports-car ownership and who seek in a car the exhilaration of high-speed motoring in comfort as opposed to high-speed motoring in the open air.

The boot of the Mark II Jaguar is 13½ cu. ft. in capacity — the spare wheel being mounted in a separate compartment below.

SPECIFICATION AND PERFORMANCE

BRIEF SPECIFICATION

Make ... ... JAGUAR
Model ... ... 2·4 Litre, Mark II
Style of Engine Straight 6-cylinder, water-cooled, twin o.h.c., twin Solex carburetters.
Bore ... ... 83 mm.
Stroke ... ... 76·5 mm.
Cubic Capacity ... 2,483 c.c.
Maximum Horsepower 120 b.h.p. at 5,750 r.p.m.
Brakes ... Dunlop bride-type disc.
Steering Burman recirculating ball type, with adjustable two-spoke wheel. 5 turns from lock-to-lock.
Front Suspension Independent, semi-trailing wishbones and coil springs.
Rear Suspension Trailing link by cantilever semi-elliptic springs.
Transmission System Four-speed single helical synchromesh manually-operated. Synchromesh on 2nd, 3rd and top. S.d.p. clutch.
Gear Ratios (overall) 1st 14·12 to 1
2nd 7·94 to 1
3rd 5·48 to 1
Top 4·27 to 1
Rev. 14·12 to 1
Final Drive Ratio ... 4·27 to 1
Overall Length ... 15 ft. 0¾ ins.
Overall Width ... 5 ft. 6¾ ins.
Overall Height ... 4 ft. 9½ ins.
Ground Clearance ... ... 7 ins.
Dry Weight ... ... 3,100 lbs.
Turning Circle ... ... 33½ ft.
Price ... £1,598 at the S.A. Coast

PERFORMANCE

Acceleration 0-30 m.p.h. 4·7 secs.
0-40 m.p.h. 8·2 secs.
0-50 m.p.h. 11·4 secs.
0-60 m.p.h. 16·5 secs.
0-70 m.p.h. 21·5 secs.
0-80 m.p.h. 28·9 secs.
In top gear from a steady 20 m.p.h. to 40 m.p.h. ... ... 8·6 secs.
In top gear from a steady 30 m.p.h. to 50 m.p.h. ... ... 8·9 secs.
In top gear from a steady 40 m.p.h. to 60 m.p.h. ... ... 9·5 secs.
In top gear from a steady 50 m.p.h. to 70 m.p.h. ... ... 10·1 secs.
In top gear from a steady 60 m.p.h. to 80 m.p.h. ... ... 12·5 secs.
Maximum Speed ... 102 m.p.h.
Reasonable Maximum Speed in:
1st gear ... ... 30 m.p.h.
2nd gear ... ... 55 m.p.h.
3rd gear ... ... 78 m.p.h.
Fuel Consumption ... 18·9 m.p.g.
Test Conditions Sea level. Fine and mild. Wet road (during acceleration tests). 90-octane fuel.

3·8 Sir William's Big-Bore Cats

The front of the Jaguar has a large presentation of lights, all of which are built-in. Dark colored lamps are for the winking turn indicators.

Sir William Lyon's biggest engined Jaguar saloons are a combination of docility when required or fiery power when floorboarded by the driver.

WHEELS FULL ROAD TEST

By PETER HALL

WHEN the sensational new E-type Jaguar sports car was released in Britain earlier this year, a controversy as old as the motor car was revived.

The outcry was against speed, just as it had been when the first car to exceed 10 mph was built.

Those who condemned the E-type said that appalling risks were entailed in selling a car to the general public that had a top speed of 150 mph.

Some said that special "experts" licences should be issued before drivers could pilot these machines on the public highway. Others said the police should vet every intending purchaser of the new model.

Some went as far as to say that the sale of this car to anyone who did not intend to use it exclusively on the racing circuit should be prevented by law.

The Americans ordered 20 million dollars worth of E-type as soon as they saw it and the motoring critics in England lavished enthusiastic praise on it, at the same time defending its power and speed.

It will be a long while before those in Australia who are down on speed will be able to view the E-type with alarm. Even then the market for a car costing nearly £4000, as its price is anticipated to be, will be small enough and specialised enough to escape some of the anti-speed criticism it got in car-conscious Great Britain.

So it will be a long time, too, before we get the chance to assess this car with its fantastic performance and to see whether it is or is not a true road car.

About the nearest thing we can get is the car which played a big

Large, pan-cake type air cleaner filters air for the two SU carburettors. Most components are packed in, but are reasonably accessible.

The other difference between the 3.4 and 3.8 litre models is one of the latter's greatest features — a limited slip differential.

I found on testing both the automatic and standard transmission models that I could do things with the car that if it was fitted with a conventional diff, would have been positively dangerous.

For example, there was no feeling of insecurity in putting the nearside wheels into dirt at the side of a bitumen road at 100 mph. The car sped along untroubled, the limited slip diff making all the necessary corrections between the different tractive quality of the two surfaces, automatically, with no sign to the driver apart from the occasional suspension thump when a wheel hit a deep rut or hole, that it was doing its job.

On wet surfaces, or when a wheel found an oil slick on a corner at high speed, the back axle made its own power correction and the driver just sat and revelled in its remarkable security.

One advantage attributed to the limited slip differential was not put to the test. It is that the car is almost impossible to bog, if only one wheel is in trouble. The diff merely transfers power from the useless, spinning wheel to the one sitting on firmer ground. I have no doubt it works but was unable to find a suitable testing area for such an unorthodox experiment. Owners of this

part in the E-type's development — the Mark 2, 3.8 litre Jaguar saloon.

Not that we mean in any way to depreciate the 3.8. Indeed, that would be almost impossible for one who appreciates engineering ingenuity, performance and impeccable design.

The test had an added interest to the fact that the 3.8 is the newest and best performer of the Jaguar range of saloons.

It was, in fact, a test of two 3.8 Jaguars, one fully automatic, the other the four-speed gearbox with electric overdrive model. The automatic was Bryson Industries' demonstration car. The standard model belonged to Bill Clemen's one of Melbourne's better-known young trials and racing drivers, a suburban Jaguar agent and proprietor of a rapidly expanding tuning and garage business.

The 3.8 has now been selling in Australia for more than a year but only in the last few months has it become readily available.

It was introduced to the Jaguar range when the Mark 2 range of 2.4 and 3.4 models were released. The bodywork and general chassis and mechanical specifications are no different from the 3.4 litre Mark 2.

There are indeed only two main differences. The 3.8 engine is basically the smaller engine with the bore increased from 83 to 87 mm and the overall capacity from 3442 cc to 3781 cc. Developed horsepower is up from 210 to 220 and torque in the medium and lower rev ranges is improved.

The engine is, of course the now justly famous twin overhead camshaft six-cylinder unit that has been de-

veloped from the XK120 unit that broke record after record in the early 1950's. On both the 3.4 and 3.8 models, the Jaguar B-type head is fitted as are twin SU type HD6 carburettors.

The 3.8 litre engine is fitted to the new E-type sports car, but a different head, carburetion and manifolding produce 265 bhp, as in the 3.8 litre XK150S sports coupe.

Automatic's interior is basically the same as the manual shift model, except that there is no gear lever poking out of the transmission tunnel.

The 3.8's styling is conservative, but large glass area permits good visibility in all directions — a shortcoming in earlier Mark One models.

model assure me that this claim is thoroughly justified.

Car lovers have become used to the Jaguar reputation for extraordinary performance in the speed and acceleration departments.

But even the most ardent Jaguar fan whose experience has not gone beyond a good 3.4 could not stop his heart pounding with excitement as a passenger or driver in the 3.8.

The acceleration of the standard transmission model was as good as any other production car available in Australia, better than anything else in its price range. And the automatic was little behind it.

Top speed of the standard model was a timed 120 mph and the automatic was a bit more than 115. Both cars could have been coaxed to higher speeds, given the room, the road and the right weather, but the maximums as tested make them very fast cars anyway.

But brilliant acceleration and the ability to cover two miles every minute is only half the 3.8 story.

Much more important, and much more exciting from the car lover's point of view is the silence, comfort and remarkable safety that accompany this performance.

The original Mark I Jaguar 3.4 models were occasionally criticised for being tricky handling cars at high speeds. Among other things their steering characteristic would change from understeer to oversteer unexpectedly and without warning.

That criticism no longer holds true as far as the 3.4 Jag goes, nor does it apply to the 3.8. A wider track between the back wheels and better weight distribution are the main factors behind the improvement.

At any rate, 3.8 models I tested handled with all the sureness and stability of the Queen Mary sailing down a queen-size mill pond.

You do not need to be a seasoned grand prix driver to handle one of these queens of the road safely at high speeds. By the same token, you need to be competent before you start burling along in the eighties, nineties and over 100 mph.

The point is that this Jaguar has no vices in the handling department. Steering is almost exactly neutral, with a slight understeering tendency in the middle speed ranges (50 to 80). It scoots around corners at speeds that in lesser cars would mean instant destruction without giving the driver or passengers even the sign of a bad moment. Roll is almost imperceptible — it seems like a slight lean to the outside wheels to give them added traction to hurl the car around the corner in a clean, unwavering line.

The brakes could not be faulted. Dunlop disc brakes are fitted on all wheels and they are servo-assisted to allow even the lightest foot to operate them with maximum power. A red warning light is fitted to the dashboard which shines brightly to let the driver know when he has not taken the handbrake off. It also flashes on if the fluid in the brake reservoir drops below a certain level. Fortunately, I did not witness it gleaming for that reason.

The Jaguar company has been one of the pioneers of disc brakes and they are accepted as so much part and parcel of every Jaguar product now that the latest Jag catalogues merely mention them in passing as part of the accepted equipment, like headlights or a fuel tank.

I gave the brakes on the automatic test car a really severe thrashing. They failed to show any sign of fade and there was no sign of an uneven pull to the side even after several murderous stops from high speed one after the other.

The silence of the 3.8 was remarkable, considering the power and performance it produced.

Tail badges are the clue to the transmission fitted in Jaguars. Manual job is unmarked, but automatic is clearly marked. Note twin exhausts.

CONTINUED ON PAGE 104

wheels ROAD TEST

TECHNICAL DETAILS OF THE JAGUAR 3.8

PERFORMANCE

TOP SPEED:

	Overdrive Model	Automatic Model
Fastest run	120 mph	115.4 mph

MAXIMUM SPEED IN GEARS:

First (low)	40 mph	52.0 mph
Second (intermed)	70 mph	81.0 mph
Third (drive)	103 mph	115.4 mph
Top	120 mph	—
Overdrive	120 mph	—

ACCELERATION:

Standing quarter mile:

Fastest run	16.1 sec	16.90 sec
Average of all runs	—	16.95 sec
Through gears:		
0-30	2.80 sec	3.20 sec
0-40	4.15 sec	4.80 sec
0-50	6.30 sec	6.60 sec
0-60	8.25 sec	8.65 sec
0-70	11.00 sec	11.10 sec
0-80	15.00 sec	—
0-100	24.80 sec	24.90 sec
20-40 (top/drive)	5.80 sec	3.00 sec
40-60 (top/drive)	6.00 sec	4.05 sec

SPEEDO ERROR:

Overdrive Model:
- Indicated 30 mph .... Actual 29.1 mph
- Indicated 50 mph .... Actual 48.0 mph
- Indicated 70 mph .... Actual 68.0 mph

Automatic:
- Indicated 30 mph .... Actual 29.3 mph
- Indicated 50 mph .... Actual 47.8 mph
- Indicated 70 mph .... Actual 67.6 mph

GO-TO-WHOA:

0 to 60 to 0 — 12.55 sec

FUEL CONSUMPTION:

Overall for test .... 15.83 mpg   15.7 mpg
(38 miles in overdrive, 232 miles in auto)

TEST CARS FROM:

Automatic from Bryson Industries Ltd, Exhibition St, Melbourne. Overdrive from Clemens Sporting Car Services, Murrumbeena, Victoria.

SPECIFICATIONS

ENGINE:

Cylinders	Six, in-line, water-cooled
Bore and stroke	87 x 106 mm
Cubic capacity	3781 cc
Comp. ratio	8 to 1
Valves	Overhead by twin overhead camshafts
Carburettors	Two SUs type HD6
Power at rpm	220 bhp at 5500 rpm

TRANSMISSION:

(Overdrive Model) Type: four-speed gearbox with Laycock de Normanville overdrive; ratios, first and reverse 12.73, second 7.01, third 4.84, top 3.77, overdrive 2.93.
(Automatic model) Type: Borg-Warner three-speed with intermediate gear hold. Ratios: Low 17.6 to 8.16; intermediate 10.95 to 5.08; drive 3.54. Rear axle type: limited slip diff on both models.

SUSPENSION:

Front	Independent by coils
Rear	Cantilever semi-elliptics
Shockers	Telescopic

STEERING:

Type	Burman recirculating Ball
Turns 1 to 1	4.3
Circle	33 ft 6 in

BRAKES:

Type .... Disc, Servo-assisted

DIMENSIONS:

Wheelbase	8 ft 11⅝ in
Track, front	4 ft 7 in
Rear	4 ft 5⅝ in
Length	15 ft 0¾ in
Width	5 ft 6¾ in
Height	4 ft 9½ in

TYRES:

Size .... 6.40 x 15 in

WEIGHT:

Dry .... 29 cwt (approx)

PRICE:

Overdrive	£3128
Automatic	£3217

The suspension system of the Jaguar 3.8 has been lowered and much modified. Wire wheels are white.

THE DADDY OF ALL JAGUARS

Incredibly fast and amazingly smooth, this is the sedan to make high performance enthusiasts drool.

By IAN FRASER

ALTHOUGH each particular branch of motor racing has its devotees, I firmly believe all fans have been truly impressed by the monumental battles which have raged between the Jaguars in sedan car events.

Actually these battles have only had three main participants, but since this trio has raced on circuits in various parts of Eastern Australia, many thousands of people—most of them awestruck—have witnessed the spectacle.

Two of these Jaguars have their homes in NSW, while the other comes from Brisbane. There are, of course, other Jaguar sedans which have appeared in competition, but real success has eluded them.

The three real battlers are Queenslander Bill Pitt and New South Welshmen Ian Geoghegan and Ron Hodgson. The former two pilot 3.4s and the latter a 3.8 Mk2.

Between them, they have completely dominated sedan car racing for about the last 18 months. No other driver has really been able to get near them whilst they themselves almost invariably streak arcoss the finish line within fractions of second of each other.

The Geoghegan 3.4 came down the line firstly from David McKay then Ron Hodgson who sold it off to its present owners. They changed the color scheme from grey to black.

Both this car and Pitt's came to Australia with the factory prepared competition engine installed.

After Ron Hodgson sold the 3.4 he bought a 3.8 off the showroom floor in Sydney and set about preparing it for competition use. This, however, was not as easy as it sounds because the Mk2 models weigh considerably more than the earlier cars—a fact which is not fully counteracted by the extra 400 cc of engine capacity.

Before we go any further, there is one thing which must first be explained: these competition Jaguars bear almost no resemblance to the models you see driving about the streets. Certainly the normal Jag has plenty of steam, but I venture to say that if the owner of a normal 3.8 got behind the wheel of one of these competition cars and flattened it, his hair would turn grey inside 15 minutes.

Now comes the sad part. For business reasons Ron Hodgson has retired from motor sport for the time being and has consequently sold his car to a man who intends to use it purely for road transport.

Before de-tuning it and passing it over to its new owner, Ron invited me to take a run in the car just to see how it really does go in racing tune and trim.

The first impression one gets of the 3.8 is that its brightly polished red paintwork is consistent with a car which has been carefully used to 3000 miles. It differs in that the wire wheels are painted white and there is a white surround to the radiator grille.

One other very noticeable difference is that the car sits much lower to the ground than the standard Jaguar. Dropping the suspension is one of the modifications which were done to the 3.8 to make it raceable. Just how much lower it sits and how it was done are secrets. Ron claims it cost him hundreds of pounds to find just the right combination which, although simple and relatively cheap to do, consumed a great deal of time in meticulous research.

Armstrong adjustable shockers are used on the rear and normal Armstrongs on the front.

Because the regulations governing sedan car racing permit only limited modifications, the interior of the car is completely standard and, of course, has not been lightened. A Vane electric tachometer has been fitted in place of the standard Jaguar instrument. Apart from that, the interior was quite normal.

Probably the biggest surpise is that the car is quite flexible to drive. I spent the best part of an hour in thick Sydney traffic trying to get out to our test strip. The car idled for 10 minutes at one stage but the temperature gauge needle did not rise more than five degrees over the normal running figure.

Apart from an exhaust system which is inclined to be shatteringly loud when the throttle is opened wide, there is no real reason why the car cannot be driven comfortably on suburban streets. The

This is how the 3.8 takes a rather sharp corner at a speed of around 85 mph. Note moderate body roll, characteristic understeer.

motor willingly purrs with only 1000 rpm on the clock in overdrive top (a good healthy engine speed which keeps the noise level down).

Clear of the restrictions, I gave the Jaguar full throttle in third at 40 mph. The triple, double choke, horizontal Weber carburettors (worth around £165 *each*) gave a mighty splutter and the engine died, only to recover its breath a moment later to literally hurl the car forward in the most breathtaking burst of acceleration I have ever encountered in a sedan.

An electric tacho has been fitted to the Jaguar, but the interior is standard in other ways. Seat is loose covered with cloth.

Three huge Weber carburettors feed the engine through a D-type cylinder head. Both camshaft and crankshaft are D-type, also. Overhead trunk supplies fresh air to carbies.

DADDY OF ALL JAGUARS

At 105 mph I pulled it back into top and tramped again. The accelerative urge did not diminish and it seemed only fractionally less in overdrive top.

At 100 mph there is such a lot of acceleration still left that it makes the "ton" quite a normal figure to see on the speedo. The Jag will simply rip from 100 to 125 mph after which the rate falls off slightly.

With Ron Hodgson at the wheel we saw 140 mph on the speedo and the car was still accelerating. I think the true maximum would be more than 140 mph but a long, long straight would be necessary to achieve it.

Well, that it is the kind of performance this car has to offer. Because the vehicle was about to be prepared for its new owner we did not record any figures.

The engine of the car has had extensive modifications, as the performance would indicate. The three Webers bolt onto a D-type cylinder head which differs from both the E-type and XK150S heads. The camshafts and crankshaft are both D-type, also, the former accounting to why there is so much upper range acceleration. Good slow speed running is a characteristic. Maximum power is 265 at 5800 rpm. Compression is 9 to 1.

The gearbox is the Jaguar optional close ratio unit designed specifically for competition purposes. The overdrive is manually controlled by a small arm extending from the right hand side of the steering column. Changing into overdrive was smooth and simple, but when re-engaging direct top a better result was obtained by using the clutch.

Firm pressure was needed to disengage the clutch, but the take-up was gentle enough not to cause any alarming moments in thick traffic.

To give the car more acceleration, the optional low ratio rear axle has been fitted. It is 4.27 to 1 instead of the standard 3.77 to 1. A limited slip differential is standard equipment, by the way.

Modifications have been made to the brakes—disc, of course—to improve their efficiency from high speeds. We came to a crash stop from 125 mph to test their effectiveness. Result: very strong stopping power with no apparent vices.

Jaguar offer an alternative high geared steering ratio which reduces the number turns lock to lock to 3¼ from 4¼. The turning circle is unaltered at 33 ft.

Additionally, Ron Hodgson has carried out various other modifications to the steering which make it light and extremely accurate. Even at very high speeds the steering feels good and suffers no feeling of lost motion which sometimes occurs when production cars are taken past the speeds for which they are designed.

The cornering capabilities of the car are outstandingly good. Push the Jaguar into a long bend at 100 mph-plus and it becomes very obvious that the money spent of modifying the suspension and steering was well spent. It is purely a matter of selecting the right amount of steering lock, applying it and sitting back to more or less leave the car to thunder around. Very stable but with some body roll, I consider this Jaguar to be one of the safest handling cars I have ever driven. It is very faithful, very true to the controls and sticks to the bitumen as if held by some subterranean magnet.

The cost of getting this 3.8 up to its current specification has been very high indeed. Not that parts have cost much, but the enormous amount of time spent in experimenting to get everything just right has cost a great deal of money.

For instance, the steering was rearranged (only Jag parts are used) but it took hundreds of hours to find out what had to be done. For this reason Ron is reluctant to discuss the details, but he points out with a grin that he is willing to undertake the work quite cheaply for Jaguar owners who want their cars to handle and go better than standard. #

3.8 SIR WILLIAM'S BIG BORE CATS

CONTINUED FROM PAGE 100

It did not matter whether you were crawling around the city at legal speeds, accelerating full bore on the test track or cruising at near maximum speeds, the only noise that sneaked inside the cabin was a pleasant purr from the engine and an occasional murmur from the air as it was flung aside with great force.

Open the driver's window, however, under full acceleration and a magnificent growl rent the air from the big twin exhaust pipes protruding beneath the back bumper bar.

Although most enthusiasts would prefer the standard gearbox model with its four forward speeds and overdrive operated by a flick switch on the dashboard near the steering wheel, they would have little reason to treat the automatic with scorn.

A Borg-Warner unit is teamed up well with the big six-cylinder engine and seemed to consume little of its power.

An intermediate gear hold is fitted with its switch on the dashboard in a position corresponding with that of the overdrive switch on the overdrive model.

The intermediate hold was useful, especially for cornering in the medium speed ranges and for ensuring around the city that maximum power was always at the driver's command. But for most driving, I had little need to use it. When extra power was wanted the response of the automatic gear changing mechanism to accelerator pressure was instantaneous. Anyway, the average driver is rarely likely to want more power than the slap-in-the-back variety that is available right up to maximum speed in drive range.

The gear lever of the overdrive model was well-placed on the transmission hump just ahead of the division between the front bucket seats.

The movement was positive but, as in other Jaguars, a little long between first and second. It was possible, also, to beat the synchromesh on second gear until you had had ample practice.

The overdrive operated on top gear and would undoubtedly prove a great fuel condenser. On the test car, it did not improve top speed at all, 120 mph being attained in both top and overdrive gears.

The performance figures were little different between the two cars. But it is fair to point out that the two tests were not exactly similar. All the acceleration figures for the overdrive model, except the standing quarter mile, were taken with two people aboard. All the figures for the automatic model were taken with only the driver in the car. The overdrive model, therefore, suffered a weight penalty of about 160 pounds.

In both forms, the 3.8 litre Jaguar proved itself to be a rare breed of car. It combined race-circuit performance with the safety demanded of road driving and the utter luxury we have come to expect from British makers of fine cars. #

USED CARS on the Road

No. 169   1958 JAGUAR 3.4

PRICE: Secondhand £995;   New—basic £1,114, with tax £1,776

Petrol consumption 20-23 m.p.g.	Date first registered 11 February 1958
Oil consumption 1,600 m.p.g.	Mileometer reading 27,520

HAND-IN-HAND with high initial cost goes heavy depreciation, and this is one of the factors which makes the larger and more costly cars expensive to own. In the case of this three-year-old Jaguar 3·4 the initial price, including over £100 extra for disc brakes and overdrive, has dropped some £780—an average of more than £250 per year. A comparably rapid decline in the value of the car may be expected to continue, and repairs and replacements when needed may be relatively costly; but this Jaguar should not prove prohibitively expensive for the used car buyer able to afford £1,000, particularly when the good fuel consumption is taken into account.

In comparison with the original Road Test car, this virtually identical used example—bearing in mind its impressive acceleration and high-speed cruising capabilities—proved unexpectedly economical. The lowest figure returned, 20 m.p.g., was measured on a fast main road run in which 100 m.p.h. was reached on several occasions, and in which a high average speed was returned in spite of delay among dense traffic on some parts of the route. Part of this improvement may be attributed to thorough loosening-up of all the mechanical components as a result of the mileage covered, while the use of a more economical setting of the twin S.U. carburettors is suggested by the slightly inferior acceleration figures returned.

Starting is immediate, and the automatic choke is working satisfactorily; the engine warms up quickly after a cold start, but there is a tendency for it to stall before normal running temperature is reached. The remarkable combination of extreme smoothness and low-speed tractability with almost startling acceleration and response to the throttle is still the same outstanding characteristic of the car that was appreciated in the Road Test of the new model, and the engine is also virtually inaudible throughout the normally-used speed range. Only at very high revs and wide throttle is the angry, purposeful snarl of the Jaguar heard.

Laycock-de Normanville overdrive raises the effective gearing in top from 20·6 m.p.h. to 26·4 m.p.h. per 1,000 r.p.m., and in this high cruising ratio the Jaguar 3·4 will pull strongly from speeds even below 40 m.p.h., and cruise at almost any figure which traffic and road conditions allow.

Weak point of the car is the gearbox, and it is this component alone which suggests that a somewhat higher mileage may have been covered than the total indicated in the data. All the indirect ratios are sufficiently noisy to discourage use of high revs, and the synchromesh is particularly weak. Gear changes have to be made fairly slowly and the first and third gear positions involve the driver in a considerable reach forward.

Clutch take-up is smooth, and the pedal movement for release is less long and heavy than is normal on the model. Clutch spin occurred only in full-throttle acceleration when changing up from first to second gears. When cruising in top gear a mild degree of rear axle whine is audible, but this is more noticeable than it would be on many comparable cars as a result of the commendable silence of the Jaguar; for the same reason, the ticking of the rear-located fuel pump is annoying.

No deterioration is noticed in the suspension, which still gives an admirably comfortable ride, free from pitch or any feeling of over-firmness. A tremor from the road wheels as the car passes over minor surface irregularities is a characteristic of the suspension, and gives a false impression that the tyres are too hard.

On dry, well-surfaced roads, the directional stability of the Mark I series Jaguar 3·4 is good, but in the wet, or if there is ice about, the car slides unduly readily, and control is not easy. The steering, although free from lost movement in the straight-ahead position, lacks the precision of control desirable for such a fast car. The brakes, on the other hand, are fully equal to their task. Little more than a touch on the pedal brings instantaneous response from the four servo-operated Dunlop discs, and the maximum braking power available at high speeds is such as to give confidence to the driver. The handbrake is only just adequate for normal requirements.

In appearance, the condition of the Jaguar is good even for a life of only three years. The chromium is only slightly scratched and is free from rust, and the paintwork, in letter-box red, is unmarked. Careful examination revealed that an unusually good respray has been carried out, though there is no associated evidence of repaired accident damage to explain why this should have been necessary. Inside the car, the grey leather seats are in practically new condition, and the matching door trim and floor carpets show equally little sign of wear. The cloth roof linings have darkened, but are still sound, and deterioration of the polished wood trim is confined to the lower edge of the side window surrounds. This excellent interior condition is spoilt by the unsightly trim panel below the facia, which appears to have been damaged in the process of removing the radio.

A standard equipment fresh-air heater warms the car quickly after a cold start, but requires the fan all the time it is in use. The Trico windscreen washer is not working, nor are the interior lights and the cigarette lighter. The Jaguar is shod with Dunlop Road Speed RS4 tyres, all of which are approximately two-thirds worn. The toolkit, nestling in its tray on the spare wheel, is practically complete, and there is a handbook.

As an example of a used car of extremely high performance, this Jaguar 3·4 is reassuring. Mechanically it is quite fit enough for full use to be made of its 120 m.p.h. capabilities.

There are three interior and two exterior mirrors on the Jaguar. The knob is missing from the vent control. Behind the front bucket seats were seen mounting points to which safety belts, now removed, were fitted

PERFORMANCE CHECK
(Figures in brackets are those of the new car Road Test—13 June 1958)

0 to 30 m.p.h... 3·2 sec (3·1)	0 to 90 m.p.h............24·5 sec (20·5)
0 to 50 m.p.h... 8·3 sec (7·0)	20 to 40 m.p.h. (top gear).. 7·4 sec (7·1)
0 to 60 m.p.h...10·3 sec (9·1)	30 to 50 m.p.h. (top gear).. 6·5 sec (6·7)
0 to 70 m.p.h...14·6 sec (12·4)	
0 to 80 m.p.h...17·8 sec (16·0)	Standing quarter-mile....17·7 sec (17·2)

Provided for test by Twickenham Cars Ltd., 55-57, London Road, Twickenham, Middlesex. Telephone: POPesgrove 5128.

Aptly fitting the expression, "Corners as if it's on rails," the 3.8 Jaguar is an extremely roadable sedan. At extreme right, it brakes down smooth, fast.

Below, the compact four-door stands out as an excellent example of functional, quality design. We found it to be one of most "make-sense" cars we've tested.

ROAD TEST/20-62
JAGUAR 3.8 SEDAN MK II

PHOTOS: PAT BROLLIER

UNAVAILABILITY OF TEST CARS has led to an absence of Jaguar road tests out of SCG's main office in California, but a wide-awake dealership, Holiday Motors of Sherman Oaks, recently came to our rescue with a car we've been anxious to evaluate for some months now — the 3.8 Jag sedan. It's a rare bird in that it's one of the few performance/luxury four-door sedans on the market. Based on our test evaluation, it seems far above any of its competition in the degree that it fulfills the requirements of its design. Offered our choice at Holiday, we picked a metallic grey 3.8 equipped with four-speed gearbox and overdrive; a combination that would produce the best performance and economy figures. Readers interested in the optional automatic, a two-speed Borg-Warner unit, can rationalize that this choice will net a 10-15% reduction in these entities under most circumstances.

Aside from trim and other superficial alterations, the styling of the 2.4, then the 3.4, now the 3.8 series has remained unchanged since its inception; the nose basically that of the XK series, the remainder having some resemblance to the Mark series. Far from any hodge-podge, it's a well-balanced design that is both functional and pleasing. Frontal area is quite low in comparison to the amount of interior room, and visibility is exceptional. The bodywork and finish are what one would expect of an above-$5000 car — top quality and virtually flawless. There has been minute attention to detail, but this is something relatively easy to accomplish when not faced with the hassle of yearly drastic restyling.

Likewise beautifully finished is the all-leather interior. The front seats are individual, with at least vestigal bucket configuration. The rear seat is a bench type but molded into two segments. Two passengers are very comfortable back there, with a surprising amount of footroom, and three can be carried without excessive crowding or discomfort. Our one point of criticism was with the actual shape and angles of the front seat. The backrest seems too vertical and a person of moderate weight sits strictly on top of the seat with very little lateral support. Fore and aft adjustment is adequate and the tracks can easily be placed back about an inch to fit an over-six-footer. We had hoped that shims under the front would tilt the seat a bit to give the backrest some angle, but this would immediately force the front cushion up into the backs of the knees. This can be even more annoying, so shorter folk will have to grin and bear the minor discomfort or have an upholsterer re-shape the seat a bit. Even the tallest driver will find that headroom is ample and, thanks to the adjustable steering column, the distance between wheel and chest can be set to the ideal for everyone. With the wheel shoved forward it's almost completely out of the driver's range of vision through the windshield, and this we found to be very pleasant.

The instrumentation and other controls are *par excellent* in both design and location. The Smith instruments, with five-inch tach and speedo directly behind the wheel, and other gauges for pressure and temperature located in a single row across the center panel, are wonderful; steady, accurate, and extremely legible by virture of no-nonsense design and large numbering. Directly below the gauges is a row of togggle switches, each identified by white-on-black lettering. The backing for this panel is a flat-black vinyl, but the major part of the dash remains of polished walnut in the Jaguar tradition.

Located on the low transmission tunnel is the short, sturdy gearshift handle. Shifting, at least in a low-mileage car, requires some effort. Though it is relatively easy to get into Reverse when trying for First, the remainder of detents are wide-spaced and positive. First is non-synchro and Overdrive is locked out to operate only in top gear. A small lever on the column actuates the latter and, while we found engagement postive, it was also a bit harsh if power wasn't being applied. This is another item, however, that may improve with mileage.

Power-assist for both steering and brakes were included on our test car, the former optional, the latter standard. Both contributed largely to the general flexibility of the car as well as ease of control. Since the steering wheel is small and the assist pressure moderate, the "feel" is very satisfying, especially for a sedan. The high ratio of 4.3 turns from lock to lock is not annoying under normal circumstances but means some fast hand work if you get in trouble. On the other hand, parking in tight places is a real cinch; a more logical factor to have on the plus side in this type of car. The turning circle is very small.

From a ride and handling standpoint, the 3.8 Jag is not the best in the world but is certainly well above average for anything in its weight class, including several sports cars. The thank-you-ma'm type of bumps are felt in the passenger compartment, but every other type and variation of road surface was absorbed, pillow-smooth, by the suspension. With

107

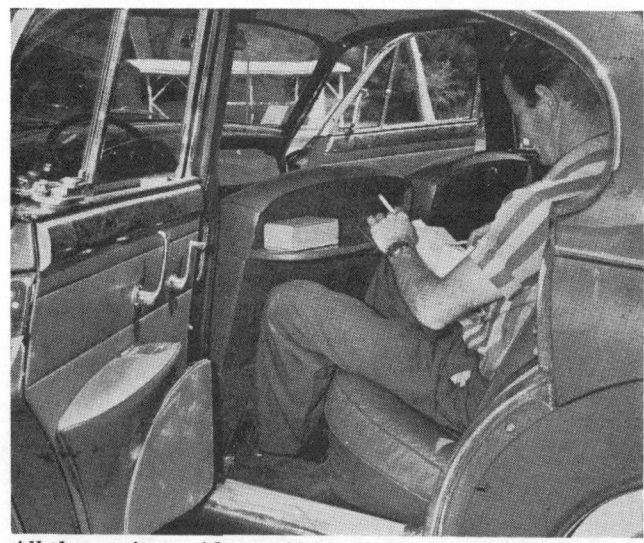

All the comforts of home. Although legroom is just a bit close for rear passengers, they have the conveniences of small trays in the backrests, pockets and ashtrays in each door panel.

At left, the 3.8 is very much at home in the back country as well as in high-speed touring on the highway, as demonstrated here. The limited-slip differential saved us a long walk....

1. Design and placement of instrumentation and controls are first-rate. Steering column is adjustable. Heater knobs are located in the console atop the square transmission tunnel.

2. The twin-cam engine develops 225 horsepower and lots of torque. Its long-stroke crankshaft is carried in the block by seven main bearings. Exhaust manifolding is porcelainized.

3. Spare tire and tool kit are carried in the floor of the ample trunk. Lid is carried by counter-balanced hinges. Note how the rugged bumper is sealed flush against the body metal.

a slight tendency to be slippery on standard Dunlops, the car handles very well and almost neutral . . . as long as power is being applied. Lifting your right foot in a corner approach will produce a marked-but-controllable understeer. It is an effortless car to drive *very* fast under all conditions except the wet. We got carried away while whomping through a little-used section of paved mountain road at 70-per and were chopping the apexes when a forestry truck appeared SUDDENLY around a left-hander. A slight decrease of throttle pressure just increased the slip angles enough for us to drift out and past their left fender before the driver had time for any evasive action. Since there was no reason to move our hands at all, he was convinced it was going to be a head-on and is probably still frozen to the wheel awaiting impact. We do not recommend this stunt, good handling car or not, but thank you, Sir William, for the 3.8's stability.

The big disc brakes do a very creditable job of pulling the car down from high speed. Even with the power assist, pedal pressure is fairly high and it's just above impossible to lock the wheels above 50 mph. With more squat than nosedive, the car reacts to fast braking beautifully and is completely stable. At lower speeds, the booster seems more effective and less pressure is required; perhaps on purpose.

We were very happy with the combination of four-speed and overdrive as applied to the torque-healthy Jag engine. Our acceleration figures were impressive and fuel economy, ranging between 21 and 26 mpg, was almost unbelievable. In First and Second there's little point in taking the engine over 5000, and we garnered the best standing-start times by shifting at 4200 in First and 5000 in Second. From 1500 rpm on up, you've got ample suds in all gears including Overdrive Fourth. The engine is, under all circumstances, extremely quiet and the chassis compliments this by being one of the few cars in which normal-volume conversation can be carried on at high speeds.

From a draft standpoint, the passenger package isn't too keen if you like to drive with the windows rolled down. The fairly flat windshield creates a vacuum that swirls air into the car just behind the driver's head. The little vent windows, front and rear, can be set to provide more than adequate ventilation in the car but emit quite a wind-roar at normal cruising speed. A scoop-type cowl vent can be opened to provide a moderate circulation to the front floor. The climate controls — wipers, heater, defrosters — are fine, well-designed and located. Despite the really crammed-full engine compartment, the floor doesn't get uncomfortably hot. We were also impressed that the engine never exceeded 180°F in coolant temperature as we ranged from sea-level traffic jams to second-gear climbing at 6000 feet.

A limited-slip differential is standard equipment on the 3.8 and occasionally can be heard engaging, but this makes the car a pleasure for off-road usage and high-speed cornering. We plunked into a sandbar hub-deep, in crossing a dried up river bed, let some air out of the rear tires, and then backed out easily from a situation that would have buried most sedans due to wheelspin.

As can be seen from the above, we gave the 3.8 a really diversified test — with the exception of cold-weather operation — and it came through with flying colors in virtually every aspect; a most enjoyable car for every application. It is by far our favorite in the Jaguar line and seems well worth the asking price as compared to any other practical automobile. It remained tight and rattle-less throughout our test, gave every indication of being a strong and durable unit. It's been called "a most improved car" over its predecessors by those who have worked on and driven them.

Jaguar can justly be proud of this one. — *Jerry Titus*

ROAD TEST 20/62
TEST DATA

VEHICLE .... Jaguar   MODEL .... 3.8 Sedan Mk II
PRICE (as tested) .. $4810.00 POE, L.A. OPTIONS .... Overdrive

ENGINE:
Type: .... 6 cyl. in-line XK type
Head: .... Alloy, removable
Valves: .... DOHC direct, chain cam drive
Max. bhp. .... 225 @ 5500 rpm
Max. Torque .... 242 lbs. ft. @ 3000 rpm
Bore .... 3.42 in. 86.9 mm.
Stroke .... 4.17 in. 106 mm.
Displacement .... 230.6 cu. in. 3781 cc.
Compression Ratio .... 9.0 to 1
Induction System: .... Two SU HD. 6's with elect. auto choke
Exhaust System: .... Cast headers, dual exhausts (tandem)
Electrical System: .... Lucas 12V distrib. ign.

CLUTCH: ... Borg & Beck single disc   **DIFFERENTIAL:** .... Salisbury with
Diameter: .... 10 in.   limited slip
Actuation: .... Hydraulic   Ratio: .... 3.77 to 1
TRANSMISSION: .... four speed,   Drive Axles (type): .... Enclosed
    synchro top three, with Overdrive   semi floating
Ratios: 1st .... 3.38 to 1   **STEERING:** .... Burman Recirc.
        2nd .... 1.86 to 1   ball, adj. column
        3rd .... 1.28 to 1   Turns Lock to Lock: 4.3
        4th .... 1.0 to 1
        O.D. .... 0.778 to 1   Turn Circle: .... 33.6 ft.
        BRAKES: .... Dunlop caliper/discs
Drum or Disc Diameter .... 11.25 in.   Swept Area .... 498 sq. in.

CHASSIS:
Frame and Body: .... Unitized steel
Front Suspension: .... Unequal arms, coil springs, tubular shocks
Rear Suspension: .... Semi-elliptic leaf springs, live axle, trailing arms
Tire Size & Type: .... 6.40 x 15 Dunlop

WEIGHTS AND MEASURES:
Wheelbase: .... 107.4 in.   Ground Clearance .... 7 in.
Front Track: .... 55 in.   Curb Weight .... 3210 lbs.
Rear Track: .... 53.4 in.   Test Weight .... 3450 lbs.
Overall Height .... 57.5 in.   Crankcase .... 12.5 qts.
Overall Width .... 66.75 in.   Cooling System .... N.A.
Overall Length .... 180.75 in.   Gas Tank .... 14.5 gals.

PERFORMANCE
0-30 .... 3.5 sec.   0-70 .... 13.4 sec.
0-40 .... 5.6 sec.   0-80 .... 17.8 sec.
0-50 .... 7.4 sec.   0-90 .... 24.2 sec.
0-60 .... 9.8 sec.   0-100 .... 31.5 sec.
Standing ¼ mile 17.5 sec. @ 76 mph   Top Speed (av. two-way run) 119 mph

FUEL CONSUMPTION
Test: .... 21 mpg   RPM Red-line .... 5500 rpm
Average: .... 23 mpg   Speed Ranges in gears:
Recommended Shift Points   1st .... 0 to 30 mph
Max. 1st .... 24 mph   2nd .... 15 to 60 mph
Max. 2nd .... 55 mph   3rd .... 25 to 90 mph
Max. 3rd .... 90 mph   4th .... 32 to top mph
Brake Test: 69 Average % G, over 10 stops.
Fade encountered on no stops.

REFERENCE FACTORS:
Bhp. per Cubic Inch 0.976
Lbs. per bhp. 14.3
Piston Speed @ Peak rpm .... 3723 ft./min.
Swept Brake area pr Lb. .... 0.1551 sq. in.

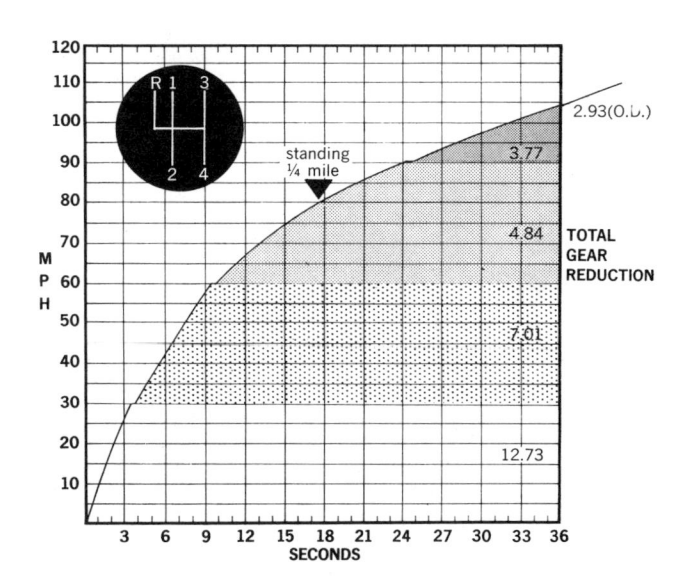

Total Gear Reduction: 2.93 (O.D.), 3.77, 4.84, 7.01, 12.73

CHARMING JAGUAR CUB

By PETER HALL

Baby of the Jaguar family, the 2.4 has a charm all of its own.

From the front the 2.4 is the same as any of the other Mk 2 series Jaguars. Basic difference is under the bonnet.

THESE days the trend in the motor industry is for the number of companies making cars to grow less and less and their size to grow even larger.

The last 10 years have seen a greater spate of mergers than any since the formation of General Motors in the few years before 1920.

Some of the biggest mergers have been between smaller companies who saw in combining their great chance to maintain or improve their competitive positions.

Thus Austin of England and the Nuffield group merged to become the British Motor Corporation, Studebaker and Packard became Studebaker-Packard Nash and Hudson became American Motors. On a smaller scale in Europe, DKW merged with Daimler-Benz.

Many knowledgable observers of the motor scene predict that the present decade will see an extension of this trend. The intensification of competition which showed itself on every one of the world markets in the 50's will tend to increase rather than diminish in the next 10 years.

So, it is almost a parodox to find a small number of comparatively small and apparently vulnerable car companies doing better than ever as each financial year comes to a close. The family owned Peugeot company in France is one such company.

The British Jaguar company is another.

There are several reasons for Jaguar's tremendous success in the years since the second world war ended.

Some say it is all due to the genius of Jaguar's

Travel stained, but not weary (Jaguars rarely get tired) the 2.4 still looks handsome in spite of a little dirt.

founder and still very active head, Sir William Lyons. Others hold to the theory that Jaguar has never fallen into the expensive trap of making annual model changes.

A recent shrewd observer put much of the company success down to the fact that it has a very low — probably the lowest in the industry — ratio of non-productive management to production line personnel.

I, and just about any other road tester who has spent any time behind the wheels of the company's products, lean more to the theory that Jaguar's success lies in the cars it makes.

Recently I had the rare pleasure to test the latest model "little" Jaguar, the Mark 2 model of the 2.4 litre sedan. It is little only because it is the cheapest —or, should we say, least expensive—of the current Jaguar range, and is powered by the smallest, least powerful engine.

The Mark 2 model of the 2.4 can be had for as "low" as £2603 including tax, for the standard four speed gearbox model. Overdrive is extra and the fully automatic model runs out to £2759, tax paid.

The 2.4, which was first released to the world in 1956 in the Mark I version, has been seldom if ever tested in this country, mainly because the Australian agents have rarely had one in their demonstration fleets.

However, Bryson Motors, the Melbourne, Sydney and Adelaide distributors recently put one on their Melbourne fleet (they were expanding their agent network in the country and also boosting their activities in Adelaide). They needed an extra vehicle to

Tail end view of the Jaguar shows the usual Mark 2 features of wide back window, new-styled tail lamps.

let their men get rapidly around the country and registered the 2.4 which I tested after it had done 4000 miles.

To my slight regret the car they provided was the automatic model. The reason, however, was quickly apparent.

Despite the appeal of the Jaguar range to keen enthusiasts, the type of drivers who, above all others, are gearbox men to the last, by far the great majority of Jag buyers specify automatic transmission.

Brysons estimate that nearly 99 percent of buyers of the big models (Mark 9 and 10) buy automatic, between 80 and 90 percent of 3.4 litre buyers and well over 70 per cent of the 2.4 men.

As the automatics would have a few pounds more profit, there would be little reason indeed for the distributors to attempt to swing buyers' taste across to the standard gearbox models.

The fact of the test car being automatic was disappointing on two grounds. First, I frankly prefer a good gearbox to a robot over whom I have little or no control. Secondly, the 2.4 Jag stands out in my opinion as a car that would perform much better and be much more exciting to driving with the benefit of a conventional gearbox.

However, forced to accept the dictates of public taste, I was only too happy to analyse it for traces of sanity.

WHEELS FULL ROAD TEST

The interior is beautifully appointed. The carpet is thick and seat richly upholstered in good quality hide. Note immense brake pedal.

And there were more than traces there. More than £600 cheaper than the 3.8 model, top liner of the Mark 2 Jaguar range, the 2.4 seemed only to concede price and racing circuit style performance to the bigger car.

In some ways, indeed, I felt the 2.4 to be a better car.

Its engine is based on the classic Jaguar XK design —six cylinders in line with twin overhead camshafts and twin carburettors to help it deliver its power economically and smoothly. Unlike other current Jag engines which use SU carbies, the little bloke under test was fitted with two Solex downdraught models.

It develops 120 brake horsepower from its 2483cc at a fairly high 5750 rpm—that is 100 horsepower less than the 3.8, and it is hard to see any other difference between the cars, especially matters of weight or size.

Performance is noticeably more gentle than the almost brutal 3.8 and 3.4 models. But take it right away from comparisons with other Jaguars, and it appears more in its right context.

For indeed, the 2.4, even hampered by an automatic transmission, performs with remarkable agility.

The fastest run in the test car was just under 98 mph, a speed which even in 1962 a very small percentage of motor cars available off the showroom floor would have the faintest hope of attaining.

The Jag reached it smoothly and held it without any sign of engine strain and not the slightest bad moment for the driver.

Acceleration was not startling, but was adequate enough to beat most modern cars. This is one aspect of the baby Jag's performance which would improve noticeably with the standard four-speed gearbox and its intelligent use.

The most impressive part of this car's performance was the silkly smoothness with which it delivered. The engine ticked over throughout its rev range with all the silence and evenness of a straight-eight Buick of early postwar vintage.

There was no impression, to the driver or passengers, that this Jag revved hard—or there would have been none had it not been for the large and easy-to-read rev counter sitting up in front of the driver.

From the comfort and styling points of view, there was little to distinguish this cheapest of the Jaguar sedan range from the dearer, more powerful models.

It had the same, low compact body with the same modifications that distinguish the Mark 2 range from the original 2.4s and 3.4s—the wider rear track, bigger windows and redesigned interior.

This Jaguar was superbly comfortable for the driver, with one annoying black mark to mar its design copybook. The black mark was the placement of the steering wheel relatively to the driver's chest and the footbrake pedal—at least as far as my fairly average size was concerned.

The trouble was this; with the steering wheel in the correct position—far enough away to allow my arms reasonable movement and the seat far enough back to allow a fair stretch of the legs, I found it impossible to move either leg onto the broad brake pedal without jamming my thigh against the steering wheel rim. Smaller people (I am a fraction under 5ft 11in) would probably be better off, as would the strange characters who like to caress the steering wheel with their hairy chests.

But a permanent solution to suit all drivers, remembering that both the steering wheel and seat in the Jag are adjustable, would require either a smaller steering wheel or a re-designed brake pedal. The latter would probably be the most satisfactory solution all round, since the present pedal, a pendant affair, is slung high off the floor and requires a substantial lift of the foot and probably is the main reason for the thigh trouble.

But apart from this one grumble, all was well in the 2.4's driving compartment. The wheel was nicely raked and vision in all directions almost flawless.

The main instruments—speedo and rev counter, sat in a tilted panel right in front of the driver. The intermediate ratio hold switch (or the overdrive switch on non-automatic cars) was high on the right-hand side of the speedo where it could be flicked in and out by the finger-tip without lifting the hand from the steering wheel rim.

Below this was an invaluable red light which shone when the handbrake was inadvertently left on — or, and this did not happen during the test—when the brake fluid supply dropped below a certain level.

The other minor instruments and a long row of toggle switches were built into a matt black panel in the centre of the polished wooden dashboard. The automatic selector lever and winker turn indicator switch were on either side of the steering column.

The rest of the interior was luxuriously fitted. Rich leather covered the seats, thick carpets caressed the floor and a magnificent heater included a built-in duct to direct warmed air to the back seat.

Back seat passengers admired the soft seating and its restful, steep rake. But some complained the seat was so far tilted that they could not easily see through the windscreen. Others complained of restricted knee room.

These were drawbacks almost unavoidable in a car the Jag's size.

But the size had other advantages. Being a compact size (just over 15ft long) it was easily parked, and its moderate wheelbase was one of the reasons for its outstanding handling properties.

After nearly 300 miles behind the wheel of the 2.4 I am prepared to say it handled better than any of the current range of Jaguar sedans (excepting, of course, the Mark 10, which I had not had the opportunity of sampling at time of writing).

The major factor here, I think, is the smaller, lighter engine. Not only does it lighten the load on the steering wheel—an important factor in such a high-geared steering system—but it allows the car's total weight to be distributed more evenly over front and back wheels.

The result, anyway, was a car with impeccable road manners with neutral steeering characteristics to all normal intents and purposes and just a trace of easily managed oversteer when pressed to the limit.

It was a joy to drive this little Jaguar, a joy that was enhanced by the knowledge that by Jag standards, it cost comparatively little to buy and run.

But I still believe standard transmission would give it even more appeal, even though my view is clearly a minority one.

\#

wheels ROAD TEST

TECHNICAL DETAILS OF THE JAGUAR 2.4

PERFORMANCE

TOP SPEED:
Fastest run .... 97.8 mph
average of all runs .... 95.3 mph

MAXIMUM SPEED IN GEARS
low .... 44 mph
intermediate .... 71 mph
drive .... 97.8 mph

ACCELERATION
Standing quarter mile:
fastest run .... 20.9 sec
average of runs .... 21.05 sec
Through gears:
0-30 mph .... 5.6 sec
0-40 mph .... 7.8 sec
0-50 mph .... 12.15 sec
0-60 mph .... 16.25 sec
0-70 mph .... 22.75 sec
0-80 mph .... 31.0 sec
20-40 mph .... 5.2 sec
30-50 mph .... 6.0 sec
40-60 mph .... 8.1 sec

SPEEDOMETER ERROR
indicated   actual
30 mph .... 29.0 mph
50 mph .... 48.1 mph
70 mph .... 67.2 mph

GO-TO-WHOA
0-60-0 mph .... 20.45 sec

FUEL CONSUMPTION .... 18.4 mpg

PRICE: £2759 inc. Tax (automatic).

TEST CAR FROM
Bryson Industries, Exhibition-st. Melbourne.

SPECIFICATIONS

ENGINE:
Cylinders .... six, in-line, water-cooled
Bore and stroke .... 83 x 76.5cc
Cubic capacity 2483cc
Compression ratio .... 8 to 1
Valves .... overhead by twin overhead camshafts.
Carburettors .... A twin Solex downdraught.
Power at rpm .... 120 at 5750.

TRANSMISSION
Type .... Borg Warner three-speed automatic.
Ratios .... low 21.2-9.86; intermediate 13.2-6.14; drive 4.27 (overall) rear axle hypoid.

SUSPENSION:
Front .... independent with coil springs.
back .... cantilever semi-elliptics.
shockers .... telescopic.

STEERING:
Type .... Burman re-circulating ball, turns 1 to 1, 4.3.
Circle .... 33½ ft.

BRAKES:
Type .... discs, servo-assisted.

DIMENSIONS:
wheelbase .... 8 ft 11¾ in
track front .... 4 ft 7 in
track rear .... 4 ft 5⅜ in
length .... 15 ft 0¾ in
width .... 5 ft 6¾ in
height .... 4 ft 9½ in

TYRES:
Size .... 6.40 by 15

WEIGHT:
Dry .... 28¼ cwt

106 m.p.h. for 10,000 miles

A standard 3.8 Jaguar saloon breaks International class records at Monza

AT 10 p.m. last Wednesday, a standard Jaguar 3.8 Mark 2 saloon finished four hectic days and nights of cruising round the Monza banked circuit at 114 m.p.h. During those four days, the car and its team of five drivers broke four international Class C records and achieved their objective of covering 10,000 miles in four days. The run was organized by Castrol Ltd., with Castrol competition manager Jimmy Hill in charge, as a research project into the performance of lubricants under sustained high speed. Save for additional tankage, the car was run in standard form without the high performance extras employed for saloon car racing, but was meticulously prepared for the run by Jaguar Cars Ltd.

After one false start, the run began at 10 p.m. local time on Saturday, March 9, with Peter Lumsden at the wheel. His co-drivers were Geoff Duke, one-time motor cycling world champion, Peter Sargent, Andrew Hedges and John Bekaert, and each drove for three hours at a time. Each pit stop for a driver change and the addition of 30 gallons of fuel took 50 seconds, the time stationary increasing to 80 seconds when the two rear tyres were changed. The Jaguar ran on Dunlop RS5s and during the

Drivers, from left to right, are Peter Sargent, John Bekaert, Geoff Duke, Peter Lumsden, and Andrew Hedges.

Monza's high banking (left) and uninterrupted straights allowed continuous cruising at 114 m.p.h.

Pit stops for fuel, oil and water and a new driver took about 50 seconds, every 3 hours—longer when the tyres were changed too.

Lofty England of Jaguar lending a hand.

four days four front wheels and 14 rear wheels were changed.

The run began in rain and mist which did not cease until Sunday morning. Thereafter the circuit dried for a time but on Sunday evening drifting fog patches made the drivers' task more difficult. Just over 2,500 miles were covered in the first 24 hours and the car was running with impressive steadiness, it being a point of honour among the drivers not to let the lap time vary by more than a second. The team had been looking forward to basking in the Italian sun but it poured all day Monday, all Monday night and a thin drizzle was still falling on Tuesday morning. Nevertheless, by 10 p.m. on Monday 5,117 miles had been covered in the 48 hours since the start. Dense patches of fog on Tuesday night made it difficult to place the car on the bankings and to avoid the bumps produced by the frost and snow. These bumps, especially a fierce one on the exit from the second banking, were probably responsible for the ever-increasing smell of petrol inside the car that brought Geoff Duke into the pits on Wednesday for an unscheduled stop at 12.17 p.m. It was found that one of the three fuel tanks had split, and in order to isolate it, a second tank also had to be cut out of the system, the pit stop for draining the damaged tank and modifying the plumbing taking just under 14 minutes. Thereafter, the car ran on a single 11-gallon tank which meant that a refuelling stop was now necessary every hour. Nevertheless, the set average speed was maintained and at ten minutes to eight on Wednesday evening the Jaguar completed 10,000 miles, but the run was continued until 10 p.m. to allow for any miscalculations.

During the four days, the car had used 21 pints of Castrol XL, 28 pints of water and had averaged around 14 m.p.g.

International Class C Records (3,000-5,000 c.c.)
3 days   107·023 m.p.h.
15,000 kilometres
                    106·615 m.p.h.
10,000 miles
                    106·58 m.p.h.
4 days   106·622 m.p.h.

Jaguar 3·8 Mark 2 Automatic 3,781 c.c.

ONE of the most impressive sights today is the rapid and purposeful progress of a Mark 2 Jaguar on a motorway, eating up the miles in the fast lane. Like the nose of a bullet, the rounded frontal shape looks right for high speed, and the sheer velocity attained is usually exhilarating. This is the outside view, familiar to many; but what is it like to be within? Over three years have passed since our previous Jaguar 3·8 Road Test (26 February 1960), and it is time to reassess.

In the interval no significant exterior changes have been made, and the now-almost-universal sealed headlamps are the only difference in appearance for the keen schoolboy "car spotter" to identify; but many detail improvements have contributed to even higher standards of quality. Also, as a further contrast with our previous test of the manual gearbox model with overdrive, we have taken the Borg-Warner automatic transmission version for this one.

Behind the wheel for the first time, one is conscious immediately of the Jaguar's unrestricted visibility, thanks to the low waistline, slender screen pillars and the well-tailored driving position. Few cars today permit one to stretch out as well as does the Jaguar, whose adequate seat adjustment allows even the long-legged ample distance from the pedals. By releasing the twist-lock on the column, the steering wheel can be set to give the arms-extended driving position which many drivers consider ideal, while others who disagree can move the wheel nearer to them. The wheel can be set so that it clears the driver's legs, and yet the top of the rim still does not protrude into the line of vision. Part of the curved bonnet, the lithe back of the Jaguar mascot, and the right sidelamp tell-tale are in view; but only those who are long in the body can see the left wing from the driving seat. The orderly layout of the instruments and the clearly marked controls soon become familiar.

An auxiliary starting carburettor cuts in automatically and stays in action until cylinder head temperature reaches 35 deg. C. It is adjustable for richness and on the test car the setting was a little over-generous, accounting for a tendency to stall when the transmission selector was moved

PRICES	£	s	d
Four-door saloon (Four-speed gearbox)	1,288	0	0
Purchase Tax	268	17	11
Total (in G.B.)	1,556	17	11
Extras (including P.T.):			
Automatic transmission	126	18	6
Overdrive (with manual gearbox)	54	7	6
Power-assisted steering	66	9	2
Reclining seats (pair)	16	6	3
H.M.V. 620T radio	44	12	8

Autocar road test · No. 1917

Make · JAGUAR Type · 3·8 Mark 2 Automatic

Manufacturers: Jaguar Cars Ltd., Browns Lane, Allesley, Coventry, Warwickshire

Test Conditions
Weather...... Dry, overcast with 10–20 m.p.h. wind
Temperature... 10 deg. C. (50 deg. F.). Barometer 29·5in Hg.
Dry concrete and tarmac surfaces.

Weight
Kerb weight (with oil, water and half-full fuel tank) 30·75cwt (3,444lb–1,562kg).
Front-rear distribution, per cent F, 57·2; R, 42·8
Laden as tested ...... 33·75 cwt (3,780lb–1,713kg.)

Turning Circles
Between kerbs .... L, 36ft 9in; R, 37ft. 10in
Between walls .... L, 38ft. 8in; R, 39ft. 9in
Turns of steering wheel lock to lock .... 4·3

Performance Data
Top gear m.p.h. per 1,000 r.p.m................. 21·4
Mean piston speed at max. power ... 3,820 ft/min.
Engine revs. at mean max. speed ...... 5,600 r.p.m.
B.h.p. per ton laden (gross) 130

MAXIMUM SPEED AND ACCELERATION (mean) TIMES

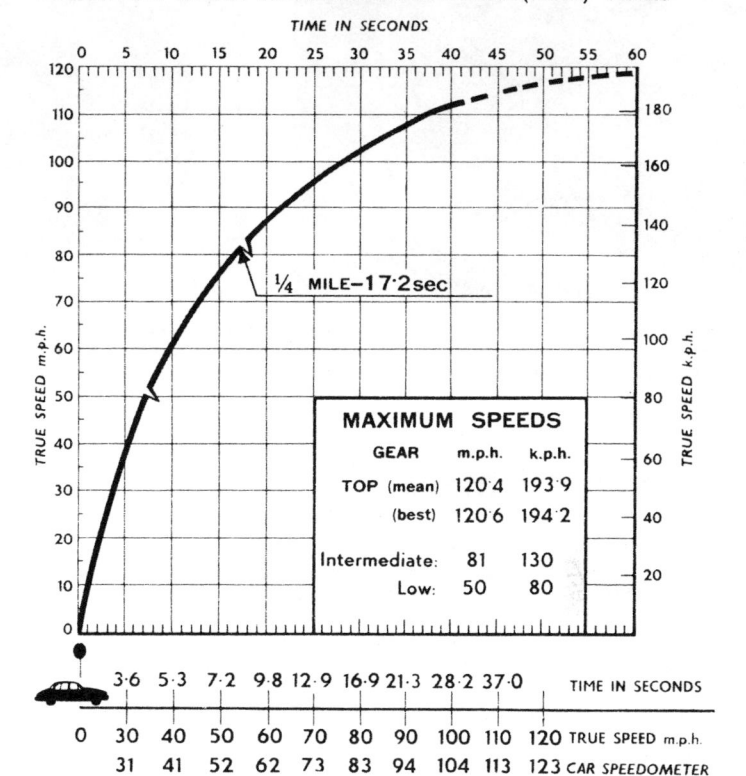

¼ MILE—17·2sec

MAXIMUM SPEEDS
GEAR	m.p.h.	k.p.h.
TOP (mean)	120·4	193·9
(best)	120·6	194·2
Intermediate:	81	130
Low:	50	80

TIME IN SECONDS	3·6	5·3	7·2	9·8	12·9	16·9	21·3	28·2	37·0	
TRUE SPEED m.p.h.	30	40	50	60	70	80	90	100	110	120
CAR SPEEDOMETER	31	41	52	62	73	83	94	104	113	123

Speed range and time in seconds
m.p.h.	Top	Inter	Low
10—30	—	—	2·8
20—40	—	4·3	3·3
30—50	7·4	4·7	3·7
40—60	7·1	5·0	—
50—70	7·6	5·6	—
60—80	8·4	7·1	—
70—90	9·6	—	—
80—100	11·3	—	—
90—110	15·7	—	—

FUEL AND OIL CONSUMPTION

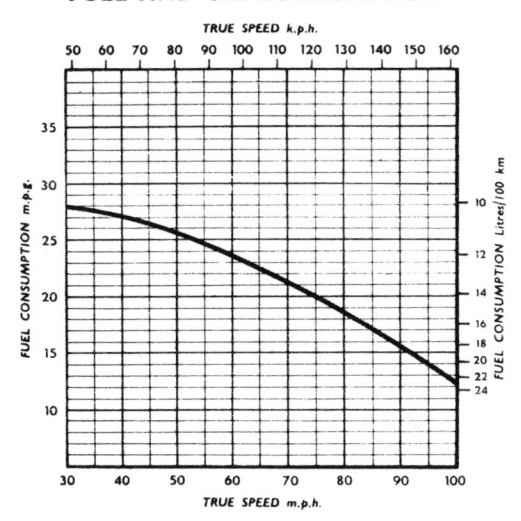

FUEL................Premium Grade (97 octane RM)
Test distance........................1,634 miles
Overall Consumption............17·3 m.p.g. (16·3 litres/100 km.)
Normal Range.....................16–20 m.p.g. (17·7-14·1 litres/100 km.)
OIL: SAE 10–30......Consumption 3,800 m.p.g.

BRAKES
(from 30 m.p.h. in neutral)

Pedal load	Retardation	Equiv. distance
25lb	0·16g	189ft
50lb	0·42g	72ft
75lb	0·64g	47ft
100lb	0·90g	33·6ft
Hand brake	0·35g	86ft

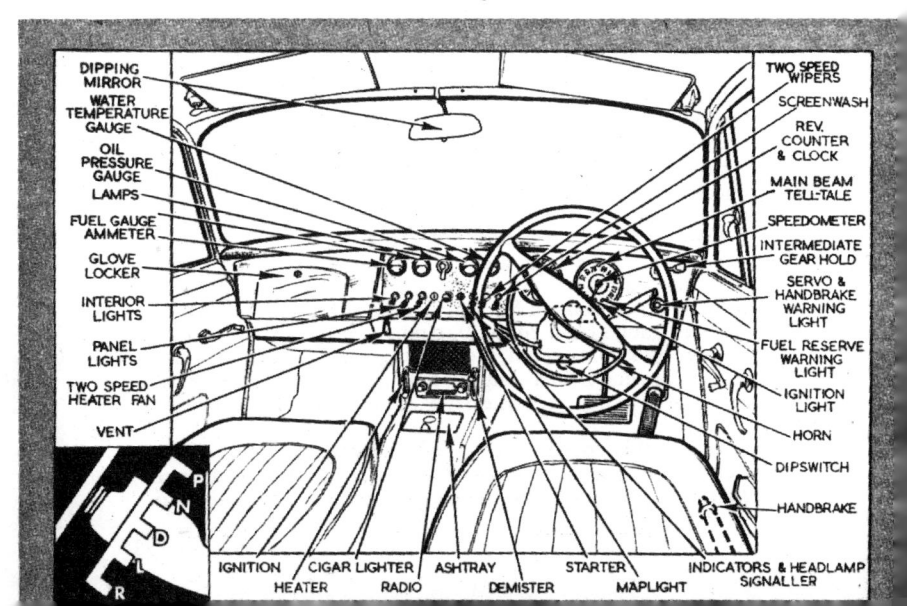

HILL CLIMBING AT STEADY SPEEDS
Inter: 1 in 4·4
Top: 1 in 7·4

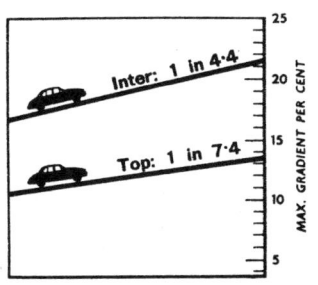

GEAR	Top	Inter
PULL (lb per ton)	300	470
Speed range (m.p.h.)	55–62	45–55

Dashboard labels: DIPPING MIRROR, WATER TEMPERATURE GAUGE, OIL PRESSURE GAUGE, LAMPS, FUEL GAUGE, AMMETER, GLOVE LOCKER, INTERIOR LIGHTS, PANEL LIGHTS, TWO SPEED HEATER FAN, VENT, IGNITION, HEATER, CIGAR LIGHTER, RADIO, ASHTRAY, DEMISTER, STARTER, MAPLIGHT, INDICATORS & HEADLAMP SIGNALLER, HANDBRAKE, DIPSWITCH, HORN, IGNITION LIGHT, FUEL RESERVE WARNING LIGHT, SERVO & HANDBRAKE WARNING LIGHT, INTERMEDIATE GEAR HOLD, SPEEDOMETER, MAIN BEAM TELL-TALE, REV. COUNTER & CLOCK, SCREENWASH, TWO SPEED WIPERS

Left: Space for maps and small packages is provided below the central instrument panel. The Radiomobile all-transistor radio, an extra on the test car, has a retractable aerial with wind-down handle under the facia. Right: Latest seats for the Mark 2 have indented backrests to give rear passengers more knee room. Fold-down picnic trays are still provided

Jaguar 3·8 Mark 2 . . .

to "Drive" after the first start of the day; but the well-known trick of flicking off the ignition switch for an instant can be used to cut out the starting carburettor prematurely. After a night of severe frost, the engine never failed to fire at the first touch on the starter button. When cold, timing chain rattle is audible, but quickly diminishes as the oil pressure rises.

With the automatic gearbox, the Jaguar is something of a top-gear car. Full-throttle kickdown allows the Intermediate range to be held up to the governed point at 5,000 r.p.m. (74 m.p.h.) when direct drive re-engages. Intermediate hold control, within fingertip reach of the steering wheel, enables the change-down to be made on part-throttle. With Intermediate hold switched in there is limited engine braking on the overrun. In either the Intermediate hold or the Low position on the selector, the driver can over-ride the normal full-throttle change-up points. At the maximum recommended engine speed of 5,500 r.p.m., 50 m.p.h. is reached in Low—this ratio gives overrun braking—and 80 m.p.h. in Intermediate. In full-throttle take-off from rest with Drive selected, Intermediate comes in at 43 m.p.h.

General Quietness

Throughout almost the whole top-gear range the engine is scarcely heard, but at high revs through Low and Intermediate, a distinctly audible combination of induction and exhaust noise accentuates the impression of vivid acceleration. With all windows closed there is unusual freedom from wind noise at speed. This commendable quietness is not spoilt when the rear quarter vents are opened, though the front vents do provoke slipstream shriek.

Because of the smoothness and good low-speed torque of its six-cylinder engine, the Jaguar lends itself admirably to automatic transmission. Hustling along secondary and minor roads, the driver feels that he is more often holding the car back than he is urging it forward; so quickly and unobtrusively does it gather speed that it seems almost to bound away after each hold-up or corner. Similarly, the ability to overtake safely in short distances means that the Jaguar is never delayed for long behind slower traffic. Over the speed range from 30 to 70 m.p.h., it takes less than 8sec for any 20 m.p.h. speed increment even in top; and the spectacular under-half-minute time for acceleration from rest to 100 m.p.h. is a measure of the lusty performance available.

Even at this speed, there is still vigorous acceleration in reserve, and in a brisk cross-wind a top speed of just over 120 m.p.h. was obtained in each direction. A following wind might have helped to an even higher one-way maximum, but the average for runs in opposite directions is a representative mean figure for the car, only 5 m.p.h. slower than the manual gearbox model in overdrive. Naturally this top performance calls for 1,000 r.p.m. higher engine speed than with the overdrive car, and takes the rev counter needle round to the red segment, which starts at 5,500 r.p.m.

Fuel consumption at a constant 90 m.p.h. was the same (15·5 m.p.g.) as the previous Jaguar achieved in overdrive at 100 m.p.h. Yet the overall consumption for the full test distance was 17·3 m.p.g.—a creditable figure when related to the car's size and performance, and considerably better than the 15·7 m.p.g. obtained with the manual model. The tank holds only 12 gallons, so that, within about 170 miles of filling to the brim, the winking "evil eye" of the fuel warning light reminds the driver that it is time to stop and take out his wallet again. Gentle driving is rather out of character with the car, but does offer a consumption improvement to about 20 m.p.g. fairly readily; however, it would be unrealistic to hope for better than this.

Experience of this engine in previous models prepares one for a rather heavy oil consumption, so it was a surprise to find that only three pints were used during the 1,600 miles of the test—equivalent to nearly 4,000 m.p.g. from a demonstrator which had already covered 52,000 miles. Clearly, the Brico Maxiflex oil control rings now fitted have helped to solve this problem.

At total extra cost of £66, power-assisted steering is an option which had been added to the test car. It certainly reduces the effort needed to hold the car to its line through

A pancake-type paper element air filter is now fitted, and an extra-long dipstick near the top water hose allows the oil-level for the automatic transmission to be checked easily

Separate amber indicator lamps are fitted at front and rear. Twin exhaust pipes are standard, and a tiny disc brake badge on the rear bumper warns of the Jaguar's stopping power

a fast bend, against the rather pronounced understeer. Low-speed manœuvring is made much easier, and coupled with the turning circle of less than 38ft between kerbs—which is good in relation to the wheelbase—it makes parking and turning in confined spaces easy.

It is, however, a pity that Jaguar have not taken the opportunity to raise the steering gear ratio to go with power assistance; at 4.3 turns from lock to lock the steering is unusually low geared for so fast a car, and there is little response to small movements of the wheel around the straight-ahead position. The nose-heaviness contributes to good directional stability, but in cross winds at high speeds the driver has to do a lot of sawing at the wheel.

Ample power is available to help the tail round, neutralizing the understeer in fast cornering, but if this technique is used on wet or icy roads, quick reaction is needed to correct a rear-end skid. When they do lose adhesion, the back wheels start to slide somewhat abruptly, and have to be checked at once. A higher steering gear ratio would help here, too.

Excellent Suspension

Girling gas cell dampers now used in the suspension give improved control on rough roads. The springing is certainly an excellent compromise in providing the stability for such high performance, coupled with the comfort and insulation from bad surfaces expected from a car of this quality. It is especially good over humps or dips causing large spring deflections, and the recoil is damped out effectively. *Pavé* was traversed with unusually little bucketing and bouncing, while a corrugated section, which produces violent shake and vibration in most cars, was taken relatively smoothly.

On normal road surfaces, small suspension tremors are felt, as in a car whose tyres are too hard, and the Jaguar feels firm on its springs. There is little road noise, and only the occasional squeak is heard from the coachwork. The recommended normal tyre pressures to include motorway use up to 110 m.p.h. are 28 p.s.i. front, 24 rear. They may be reduced by 3 p.s.i. for low-speed work and bad surfaces, which improves ride comfort even further, but not unreasonably results in some tyre squeal in spirited cornering.

As spectacular as the acceleration of the Jaguar is the performance of its brakes, the all-disc system by Dunlop being most reassuring at any speed. The brake pedal is wide to suit those who like left foot braking with automatic transmission. In fact, there is even room for both feet together on the pedal but once 100lb effort is exceeded at 30 m.p.h., the wheels tend to lock, and 0.9g was the maximum efficiency obtained. The magnificent response at high speeds, complete freedom from fade, and the ability to stop quickly without sliding on wet or slippery roads, are the braking characteristics best appreciated. On braking to rest, a small residual pressure is trapped in the hydraulic lines which is released on initial opening of the throttle, and this eliminates the annoying creep often associated with automatic transmissions.

An automatic transmission asset is the positive locking pawl when Park is engaged, but the latest car also has considerably improved handbrake efficiency. The sturdy and well-placed lever to the right of the driving seat now provides 0.35g braking effort at 30 m.p.h., and holds the car securely—provided the lever is pulled on really hard—on a 1-in-3 gradient, although the previous one failed on 1-in-4.

A handbrake warning lamp on the facia is sufficiently sensitive to light up, with the ignition on, if the handbrake is on one notch, and it serves a second duty in warning of any serious drop in brake fluid level.

Both improved main beam illumination and a longer throw when dipped are provided by the sealed headlamps now used on the Mark 2. Their freedom from scatter on dipped beam also made them more effective in dense mist than were the twin fog lamps provided as standard equipment. Also appreciated for night driving are a bright map-reading light at the top of the facia, shaded from the driver's eyes; an automatic reversing lamp; well-diffused interior lighting by four lamps switched on at the dash or by opening any door; and automatic illumination of both the facia glove box and the luggage compartment, when the respective lid is open with the sidelamps in use. The lever switch beneath the steering wheel for the indicators is also the headlamps flasher.

Jack and wheelbrace are clipped to the back of the boot. The boot key also locks the lid of the facia glove box. Below the floor are spare wheel, jack handle and tool kit

Jaguar 3·8 Mark 2 . . .

Jaguar owners have long had cause to complain of inadequate heating in their cars; but detail improvements have been made to the fresh-air system provided as standard with the Mark 2. The heater matrix is always at engine temperature, and incoming air from a flap vent above the scuttle either goes through it or by-passes it, as selected with the heat control. Windscreen de-icing is effective, and variable settings of the controls are possible except for the main air inlet: the flap is either closed or open, without intermediate positions. Trunking along the top of the propeller shaft tunnel carries hot or cold air to the rear compartment.

Revised front seats were introduced on Mark 2 Jaguars while the car was on test, and the demonstrator was returned to the factory for the new ones to be fitted. Their backrests are scalloped away to a depth of about 1½in. on the reverse side, to give the rear passengers more knee room. The cushions are comfortably upholstered and reach well under the thighs; but, especially when new, the squabs offer little sideways support, although they extend around the shoulders. Reclining seats, as fitted to the test car, cost £16 extra including tax; and the modified backrest is fitted whichever pattern of seat is chosen. Lack of rear compartment knee room has been a frequent complaint of some Jaguar owners, and this useful improvement is a valuable gain for those who carry the full complement of passengers.

Some who drove the car complained of the bad angle of attack to the organ-type throttle pedal, its stiffness of movement and its position well forward of the brake pedal. While improvements have been made since earlier models, these faults are not yet eliminated.

Aids to Jaguar enjoyment include an electric windscreen washer, with sensibly large reservoir under the bonnet; two-speed wipers which self-park at the base of the windscreen; a dipping interior mirror mounted on a stalk which is adjustable for length; map pockets and armrests on all doors; a large ashtray in the front console, and smaller ones built into the rear door armrests; and a cigarette lighter. Of course, the comprehensive array of instruments includes a rev counter, thermometer, oil pressure gauge and an uncalibrated ammeter; and there is a clock set in the rev counter. On the test car, it kept perfect time.

At intervals of 2,500 miles the engine oil should be changed and attention to seven chassis grease points (eight on models without power steering) is recommended. For access to the spare wheel some luggage may have to be removed to allow the trap door in the boot floor to be raised. A useful set of tools housed in a fitted tray nestles within the spare wheel.

There are competitors, both British and foreign, which can match the superb silence and five-seater comfort of a Jaguar 3·8, and no doubt some of them can also rival the vivid acceleration and 120 m.p.h. performance; but no car in the world can offer all this and still compete with Jaguar on price.

Specification

Scale: 0·3in. to 1ft.

Cushions uncompressed.

ENGINE
- Cylinders ... 6 in-line
- Bore ... 87mm (3·43in.)
- Stroke ... 106mm (4·17in.)
- Displacement ... 3,781 c.c. (230·6 cu. in.)
- Valve gear ... Twin overhead camshafts
- Compression ratio 8·0 to 1 (7·0 or 9·0 to 1 optional)
- Carburettors ... Two S.U. HD6 with automatic cold starting mixture control
- Fuel pump ... One S.U. electric
- Oil filter ... Tecalemit full-flow, replaceable element
- Max. power ... 220 b.h.p. (gross) at 5,500 r.p.m.
- Max. torque ... 240 lb. ft. at 3,000 r.p.m.

TRANSMISSION
- Gearbox ... Borg-Warner three speed automatic with torque converter
- Overall ratios ... Top 3·54; Inter. 10·95-5·08; Low 17·6-8·16; Reverse 13·36-6·21
- Final drive ... Salisbury hypoid bevel 3·54 to 1 with Powr-Lok limited-slip differential

CHASSIS
- Construction ... Integral with steel body

SUSPENSION
- Front ... Semi-trailing wishbones and coil springs, with Girling gas cell telescopic dampers. Anti-roll bar
- Rear ... Live axle on cantilever leaf springs with radius arms and Panhard rod. Girling gas cell telescopic dampers
- Steering ... Burman recirculating ball, with optional power assistance. Wheel dia., 17in.

BRAKES
- Type ... Dunlop discs, vacuum servo assisted
- Disc dia. ... F. 11in. R. 11·75in.
- Swept area ... F. 242 sq. in.; R. 253 sq. in. Total 495 sq. in. (294 sq. in. per ton laden)

WHEELS
- Type ... Dunlop pressed steel disc with 5 bolts; rim width 5·0in. Centre-lock wire wheels optional extra
- Tyres ... 6·40—15in. Dunlop RS5 with tubes

EQUIPMENT
- Battery ... 12-volt 60-amp. hr.
- Headlamps ... Two sealed units 60-45 watt
- Reversing lamp ... One automatic
- Electric fuses ... 2
- Screen wipers ... Two-speed self-parking
- Screen washers ... Lucas electric
- Interior heater ... Fresh air with two speed electric booster
- Safety belts ... Built-in anchorages provided
- Interior trim ... Leather seats, cloth roof lining
- Floor covering ... Pile carpet
- Starting handle ... No provision
- Jack ... Screw pillar with ratchet handle
- Jacking points ... Two each side near wheels
- Other bodies ... None

MAINTENANCE
- Fuel tank ... 12 Imp. gallons (no reserve)
- Cooling system ... 26 pints (including heater)
- Engine sump ... 11 pints SAE 20W-30. Change oil every 2,500 miles. Change filter element every 5,000 miles
- Automatic transmission ... 15 pints Automatic Transmission Fluid, Type "A." Change oil every 10,000 miles
- Final drive ... 2·75 pints Hypoid 90. Change oil every 10,000 miles
- Grease ... 7 points every 2,500 miles
- Tyre pressures ... F. 25; R. 21 p.s.i. (town and low-speed driving); F. 28; R. 24 p.s.i. (normal driving up to 110 m.p.h.); F. 33; R. 29 p.s.i. (fast driving); F. 33; R. 33 p.s.i. (fast, full load)

Cars ON TEST

JAGUAR 3·8 MARK 2

WITH AUTOMATIC TRANSMISSION

THERE IS NO DOUBT about it: Jaguar motoring is among the finest in the world. Smooth, effortless performance; luxurious comfort; a car which is designed with the driver in mind, yet one which convinces the passengers that it has been produced for them alone—and that just about sums up the Jaguar 3.8 Mark 2, which CARS ILLUSTRATED has recently tested in its automatic transmission form.

This is a car which must be one of the very few which provides an answer to almost every motoring situation. One can idle along in a traffic jam, only the movement of the car informing the passengers that the engine is, in fact, running. When the road clears, depression of the accelerator pedal increases the speed with astonishing rapidity to the three-figure mark until more leisurely progress is called for: powerful, servo-assisted brakes on all four wheels then take hold of the great car and restrain it firmly and without fuss. The Jaguar is one of those cars which will, if called upon to do so, accelerate from a standstill to 100 m.p.h. and back again to rest in well under a minute: similarly, it will idle along for mile after mile in thick traffic while the occupants relax in deep, leather seats and listen to the radio. How they make such a car at such a modest price must be one of the major mysteries of the motor industry.

Bodily and mechanically, the Jaguar saloon has undergone very few changes over the years, yet it still remains one of the really great cars: proof, if proof were needed, of the soundness of the original concept. It is a strong, smooth and silent car: the engine is extremely smooth, and at tick-over speeds it is necessary to glance repeatedly at the rev. counter when one first drives the car to make sure that the unit is, in fact, running. At town speeds this smooth silence is retained, while even cruising at three-figure speeds on the open road does not produce any excessive mechanical noise.

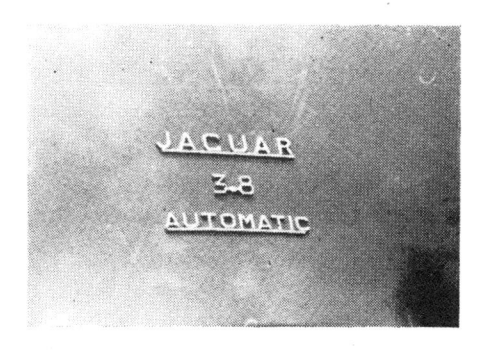

The power unit is, of course, extremely

well-known. The massive six-cylinder, twin overhead camshaft unit, of 3,781 c.c., is a refined version of that which so successfully powered the famous sports-racing Jaguars of a few years ago. On the test car it was always an easy starter, whether hot or cold, and immediately pulled well and evenly. The immense punch of 220 b.h.p. is delivered smoothly openings will result in a change from low to intermediate when the road speed reaches about 11 m.p.h., and from intermediate to direct at 23 m.p.h. Decelerating with a closed throttle will bring about a change from direct to intermediate at 16 m.p.h., and from thence to low at 4 m.p.h. Full throttle acceleration will cause upward changes to take place at rear suspension, and provide excellent control on all types of surface. The suspension is, in fact, a fine compromise between the type required for stability at speeds of the order of 120 m.p.h. and the type giving maximum comfort. The ride is fairly firm at low speeds, but perfect comfort is retained: at high speed there is a gently undulating motion, and bumps

and progressively, and even moderate throttle openings will result in acceleration of an order which will put the car at the head of most traffic streams. At high speed, it is relatively lazy, with an indicated 100 m.p.h. on the speedometer at only 4,500 r.p.m., a brisk enough cruising speed even allowing for the optimism of the speedometer.

The transmission on the test car was the Borg-Warner three-speed automatic unit with torque converter, with provision for "kick-down" changes to a lower ratio and a manually-controlled "hold" switch on the intermediate speed. Under normal conditions, the steering column selector lever will be placed in the position marked "D", or "drive", on the quadrant, when changes from low to intermediate and intermediate to direct, as well as downward changes, will take place automatically according to a combination of road speed and throttle opening. Small throttle 40 and 64 m.p.h. respectively, while use of the "kick-down" over-riding arrangement will engage intermediate from direct at up to 68 m.p.h., the transmission returning to direct drive at 78 m.p.h.

Low gear can also be selected manually, and when this is done no automatic change can occur. Maximum speed in this ratio is 45 m.p.h., in excess of which speed the engine will be over-revved just like an "ordinary" car with a manual gearbox.

The automatic changes go through extremely smoothly: on small throttle openings the change from low to intermediate is extremely difficult to detect, while at worst the changes are more "sensed" than actually felt.

The suspension of the Jaguar 3.8 is of conventional type, being independent at the front with semi-trailing wishbones and coil-springs, with a live axle on cantilever leaf springs at the rear. Girling gas cell dampers are employed on both front and are smoothly ironed out. The car runs straight and true, and remains perfectly controllable at all speeds, right up to its maximum. Heavy braking causes the nose to dip, and there is some roll on corners but in neither case is the movement either excessive or uncomfortable.

The driving position approaches the ideal almost endless variations of adjustment to seat, steering wheel and squab provide position to suit any driver. The seats themselves are extremely comfortable, the deep leather upholstery being firm enough to resist any kind of "sinking" feeling. Power-assisted steering was fitted to the test car, and this provided light, positive steering at any speed. A good "feel" of the road was retained under all conditions and a slight "nervousness" on the part of the driver during a wet spell was quickly allayed.

It goes almost without saying that the Jaguar's instrumentation is complete

Similarly, one need scarcely stress the perfection of layout employed. Immediately before the driver's eyes are matching, large diameter speedometer (incorporating trip and tenths recorder) and rev.-counter, reading to 6,000 r.p.m. In a neat central panel are water temperature, oil pressure and fuel contents gauges, and an ammeter, with the lighting switch centrally mounted among them. Beneath these instruments is a long row of tumbler switches controlling electric screen washers, two-speed, self-parking windscreen wipers, map-reading light, starter, cigar lighter, ignition, heater fan, panel light and interior light: all these switches are clearly labelled and are easily reached from the driver's seat. The flashing direction indicator switch is mounted on one side of the steering column, with the gear selector lever and quadrant on the opposite side. On the extreme right of the facia is a warning light which gives indication of a low brake fluid level, and also indicates that the handbrake is on. Warning lights for headlamp main beam and ignition are incorporated in the speedometer dial, and the rev.-counter dial includes an electric clock. A further warning light in the speedometer dial lights up when the fuel level is low.

Passenger comfort is well considered. Full seat adjustment of both cushion and squab over a wide range ensures that a first-rate seating position is quickly found. The floor of both front and rear compartments is thickly carpeted, and in fact the whole of the interior of the car is trimmed in a manner which suggests a far higher price than the modest price tag which is in fact attached to this model. Visibility is good all round the car, and all hand and foot controls are light to operate and conveniently located. On the automatic transmission model, the brake pedal is large enough to be operated easily by the left foot if the driver feels so inclined.

Almost the only criticism heard from passengers who rode in the car during the test period came from those in the back. Legroom for those occupying the rear seats is, in fact, a little too restricted, even with the front seats well forward.

The main appeal of a Jaguar, however, lies in its performance. And this, without doubt, is tremendous. Full-throttle acceleration brings into play a delightfully smooth surge of power, and the car simply rushes up to around 110 m.p.h. In fact, only just over half-a-minute is required to reach 100 m.p.h., while the standing quarter-mile is covered in a little over seventeen seconds: both most impressive figures for a luxuriously-equipped, four-seater saloon car. Equally impressive is the almost gentle delivery of this sort of power: there is nothing harsh or savage about the way the Jaguar gets off the mark unless the driver deliberately attempts to achieve the absolute maximum acceleration of which the car is capable. A Powr-Lok limited slip differential ensures proper transmission of the power to the road and, under normal circumstances, the getaway is entirely without fuss. It would, in fact, be entirely possible to own one of these cars without ever realising its true potential: the Jaguar is an equally delightful car to drive slowly!

When one does make use of the available performance, however, one does so with complete confidence. The power-assisted steering provides a light, delicate control of the car, coupled with roadholding of a high order. For after-dark motoring, the headlights, on both main and dipped beams, provide illumination of a sufficiently high order to make possible very fast journeys: one trip of some ninety miles, which included the crossing of London, was completed in less than 1½ hours in the early hours of the morning without any tiring effect or even the consciousness of having travelled fast.

Vivid acceleration, a 120 m.p.h. maximum speed, luxurious accommodation for four —and great comfort for five—all this can be found elsewhere: but not at the Jaguar price. Apart from legroom in the rear of the car, there remains but one other major criticism: the capacity of the full tank is only 12 gallons, which, on a hard-driven consumption of, say, 15 m.p.g. provides a too-limited range.

SPECIFICATION AND PERFORMANCE DATA

Jaguar 3.8 Mark 2 with automatic transmission and power-assisted steering. Price: £1,393 plus £290 15s. 5d. P.T.—£1,683 15s. 5d.

Engine: six cylinder, 87 mm. × 106 mm. (3,781 c.c.). Compression ratio 8 : 1, 220 b.h.p. at 5,500 r.p.m. Twin HD6 S.U. carburettors with automatic cold-starting control. Twin overhead camshafts.

Transmission: Borg-Warner three-speed automatic transmission with torque converter. Powr-Lok limited slip differential

Suspension: Front, independent with semi-trailing wishbones and coil springs. Rear, live axle on cantilever leaf springs, with radius arms and Panhard rod. Girling gas cell dampers front and rear. Tyres: 6.40 × 15.

Brakes: Dunlop disc brakes all round with vacuum servo assistance; front, 11-in. discs, rear, 11¾-in. discs.

Equipment: 12-volt lighting and starting. Self - parking, two - speed windscreen wipers. Speedometer, rev.-counter, water temperature, oil pressure and fuel contents gauges. Flashing direction indicators. Cigar lighter. Headlamp flasher unit. Radio. Electric screen washers. Fog and spot lamps. Built-in reversing lamp. Heating, ventilating and demisting equipment.

Dimensions: 15 ft. 0¾ in. overall length; 5 ft. 6¾ in. overall width; 4 ft. 9½ in. overall height. Ground clearance: 7 in. Dry weight: 27½ cwt. Turning circle: 33 ft. 6 in.

Performance: Maximum speed: 120 m.p.h. Speeds in gears: see text. Acceleration: 0–30, 3.8 secs.; 0–40, 5.7 secs.; 0–50, 7.8 secs.; 0–60, 10.8 secs.; 0–70, 14 secs.; 0–80, 18 secs.; 0–100, 32.5 secs. Standing quarter-mile: 17.2 secs. Fuel consumption: 14–17 m.p.g.

Basic styling is ageless. Foglamps are a standard fitting.

MK 2 saloon 3·4 litre
JAGUAR

THERE is a special kind of magic associated with the name "Jaguar". These are cars in a class all by themselves — beautiful in the classical tradition, at home on road or track and ranked with the best in both fields, yet priced so that even the humble salaried man can — and does — aspire to own one.

The Jaguar (pronounce it "Jag-wahr") is not a perfect car, nor does it claim to be.

But it offers a quite extraordinary combination of luxury, performance and sheer lovable character that has made it one of the world's most-sought-after cars.

We are speaking of the Mk II saloons — basically the same car — which come in three engine sizes and are assembled at the C.D.A. plant in East London. These are the bread-and-butter models as far as the Republic is concerned, though the luxury car import scheme is now giving us a sprinkling of E-Types and Mk. X's as well.

The smallest engine is the 2·4-litre, which is now no longer available in South Africa, and which is the XK engine in its lowliest form.

Next one up is the 3·4-litre — the increase in capacity achieved by a longer stroke — with overdrive, which is the subject of this test.

The biggest engine is of 3·8 litres, which uses the long stroke accompanied by a slight increase in bore diameter, and this model is available here with overdrive or automatic transmission.

Jaguar styling is among the world's best, and the basic shape of the Mk. 2 saloons is now 12 years old without showing the slightest sign of becoming dated. In this respect, it is significant that the newest model — the improved S-Type saloon — has virtually the same overall lines.

But there are a few interior features which have distinctly vintage associations: the concentration of the controls at the centre of the dash panel, for instance, and particularly the old-fashioned rotary-type light switch. Yet this switch works perfectly well, and without doubt it contributes to the ageless charm of the interior.

Instrumentation is a Jaguar speciality, and out of curiosity we made a rough count of the individual features. Our tally is 18 hand switches and controls, including hooter and electric overdrive selector lever; 7 gauges of various kinds including rev-counter and clock; and seven warning lights.

The total is 32, without counting the foot controls, all of which have to be used at some time or another during driving. We may even have missed a few!

Instrumentation is profuse. Lockring for steering column adjustment is visible behind wheel boss.

124

Instrumentation of this kind puts the driver on his mettle. It warns him that this is no ordinary car, but one in which driving becomes a challenge, and invitation to teamwork between man and machine.

In essence, the Jaguar is the archetype of the sports saloon. The driver sits in a low, snug-fitting bucket seat which makes him feel part of the car, and quite truthfully gives promise of great comfort.

He grips a slender and sensitive steering wheel, and can select the arm reach he prefers by adjusting the steering column length. (A large twist-ring behind the wheel locks and unlocks it to allow an adjustment range of about 2¼ in.)

In front of him stretches the walnut-veneer fascia with its array of instruments and controls, and at left is the high transmission console with the short gear lever in just the right place.

On the steering wheel boss a jaguar stares at him with its fangs bared in a perpetual snarl; the bonnet curves gently down into the distance, and at its edge the streamlined leaping Central American cat speaks of the capabilities waiting to be unleashed.

The effect on the motorist is to make him want to drive this car: to feel it answer the commands of his hands and feet.

There can be few cars which so clearly invite to be driven, as Jaguar owners can testify. Perhaps by accident — but more likely by subtle design — the Jaguar, any Jaguar from the early SS to the Mk. X, has a compelling individual character, a restlessness which only seems to ease once it is on the road.

It can go very fast, and in the right hands there is no reason why it should not go very fast. In this sense it is a thoroughbred, with sports-car steering, stability and braking, so that its safety factor is of the highest order.

One of motoring's electrifying experiences is to be

A CAR ROAD TEST

overtaken on one of South Africa's many stretches of endless straight national road by a Jag. purring along at 115 m.p.h. in overdrive.

In the same breath, we should mention that we have never seen a Jag. being driven carelessly or foolishly. It might happen, but it is unheard-of in our experience.

The car itself seems to impress the driver with the responsibility for handling extremely potent machinery in adult fashion. It would almost seem that a generous measure of road courtesy comes with the car when it is taken from the showroom floor.

Handling the Jaguar's power itself calls for a polished driving technique.

Acceleration away from a standing start, for instance, very easily provokes that bouncing reaction to the torque force which is commonly called "axle tramp". The right rear wheel will bounce violently when the clutch is engaged at high revs, resulting in almost total loss of traction.

This effect only arises in performance testing, and makes it necessary to practise a bit to get the right balance for an optimum take-off from rest.

In ordinary driving it is never necessary to provoke the rear end like this. In racing cars, where every ounce of acceleration counts, the torque reaction is overcome by independent rear suspension and limited-slip differentials, and this latter feature is used in the Jaguar 3·8 models, where even more power is available.

Once it is moving, the Jag gets to its maximum speed very quickly. In 10 seconds it is doing 60 m.p.h., in 20 sec. 85 has come up, and just on half a minute is needed to reach 100 m.p.h.

Third gear is usable up to 90 m.p.h., and once 110 comes up the driver flicks the overdrive stalk switch on the steering column to take the maximum speed to near 120 m.p.h.

The Laycock de Normanville electric overdrive is a delightful feature of this car, and is the full equivalent of a fifth ratio. At 100 m.p.h. in overdrive the engine is turning over at a mere 3,500 r.p.m.

In top gear at high speeds the car is noisy on the mechanical side, but the switch over to O/D has an almost magical effect on the interior noise level, and normal speech can be carried on inside the car at 100 m.p.h.

Speech interference (over 85 decibels) only comes in above 110 m.p.h. in overdrive, and is caused mainly by wind whistle at the tops of the closed windows.

A noticeable feature in acceleration tests is that the synchromesh can be beaten quite easily, particularly on

A bit of body roll, but handling is unusually good.

second gear. This and the fairly heavy movement of the short gear lever are well-known characteristics of these saloons which are evident in performance tests, but which do not cause much worry in everyday driving when the pace is more leisurely. But the gearbox itself is a thoroughly sound and reliable unit.

Once it is moving the 3·4 seldom needs gearshifts except when stopping and taking off again, and it takes off well in 2nd gear under normal conditions. The overdrive switch takes care of all extra power requirements such as climbing and overtaking.

We can say emphatically that the Jag. saloons are not intended for use on really bad roads. The suspension is sports-type and becomes distinctly harsh on a bad surface, though minor irregularities such as light corrugations are taken in the car's long stride with the greatest of ease.

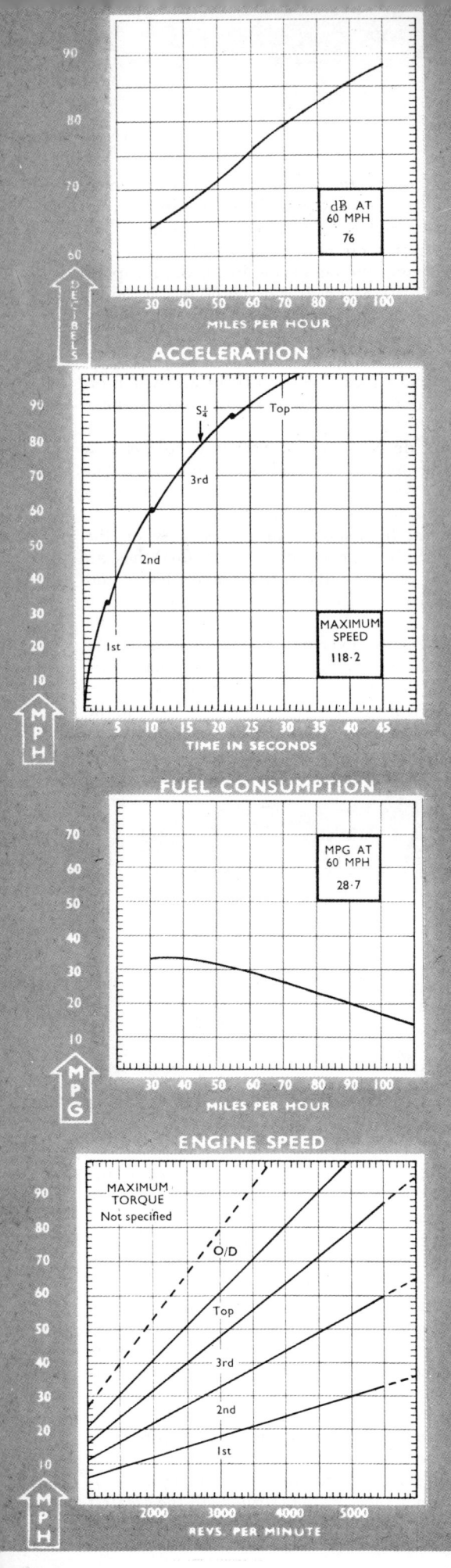

JAGUAR 3·4-litre Saloon

SPECIFICATION

MAKE AND MODEL: Jaguar 3·4-litre saloon, Mk. 2, O/D.
ENGINE: 6-cylinder XK, in-line, water-cooled, twin o.h.c., twin S.U. carburettors.
BORE AND STROKE: 83 x 106 mm.
CUBIC CAPACITY: 3,442 c.c. (210 cu. in.).
COMPRESSION RATIO: 8 to 1.
MAXIMUM HORSE-POWER: 210 b.h.p. at 5,500 r.p.m.
MAXIMUM TORQUE: Not specified.
ROAD SPEED IN TOP GEAR AT 1,000 R.P.M.: 20·25 m.p.h.
ROAD SPEED IN O/D RATIO AT 1,000 R.P.M.: 26·2 m.p.h.
PISTON SPEED AT MAXIMUM HORSE-POWER: 3,838 ft./min.
BRAKES: Dunlop discs, vacuum-assisted.
SUSPENSION: (Front) Independent, semi-trailing wishbones and coil springs. (Rear) Trailing links and radius arms, leaf springs.
TRANSMISSION: 4 forward speeds and electric O/D, synchromesh on upper ratios.
OVERALL GEAR RATIOS:
  1st: 12·73   Top: 3·77
  2nd: 7·01   O/D: 2·93
  3rd: 4·84   Reverse: 12·73
FINAL DRIVE RATIO: 3·77 to 1. **TYRE SIZE:** 6·40 x 15.
LENGTH: 181 in. **WIDTH:** 66·8 in. **HEIGHT:** 57·5 in.
WHEELBASE: 107·4 in. **TRACK:** 55 in. front, 53·4 in. rear.
GROUND CLEARANCE: 7 in. **STEERING:** Re-circulating ball, adjustable reach; 4·7 turns lock to lock. Turning circle: 33·5 ft.
FUEL TANK CAPACITY: 12 gal. **BOOT CAPACITY:** 13·5 cu. ft.
LICENSING WEIGHT: 3,070 lb. **WEIGHT AS TESTED:** 3,436 lb.
ANNUAL LICENCE: R24. **PRICE AT COAST:** R3,736.
INTERIOR DIMENSIONS:
  *Front seat headroom: 3·25 in.
  *Rear seat headroom: 2 in.

PERFORMANCE

ACCELERATION THROUGH GEARS:

M.P.H.	Sec.	M.P.H.	Sec.
0—30	3·3	0—70	14·2
0—40	5·5	0—80	18·1
0—50	7·9	0—90	23·5
0—60	10·1	0—100	32·9

ACCELERATION IN HIGHER RATIOS:

M.P.H.	O/D	Top	2nd	3rd
20—40	—	7·6	4·0	5·5
30—50	—	7·4	4·1	5·5
40—60	10·1	7·6	4·7	5·8
50—70	10·5	8·0	—	6·3
60—80	10·8	9·3	—	7·5
70—90	13·0	9·8	—	9·8
80—100	22·0	12·7	—	—

STANDING-START QUARTER-MILE: 17·5 sec.
GEARED SPEEDS AT PEAK R.P.M. (5,500):
  1st: 33·0 2nd: 60·0 3rd: 87·0 Top: 111·5 O/D: 144
MAXIMUM SPEED IN TOP: 114·0 m.p.h. **IN O/D TOP:** 118·2 m.p.h.
EMERGENCY STOPS (10 at 30-sec. intervals, from 50 m.p.h., in sec.): 2·8, 2·7, 2·4, 2·6, 2·7, 2·5, 2·7, 2·6, 2·5, 2·6. (Average: 2·61 — superlative.)
HANDBRAKE STOP (from 50 m.p.h., in sec.): 7·1
MINIMUM SOUND LEVELS:
  30 m.p.h.: 63   90 m.p.h.: 81   Rough road at 60: 79
  45 m.p.h.: 69   105 m.p.h.: 85   Idling: 49
  60 m.p.h.: 76   120 m.p.h.: 90·5
  75 m.p.h.: 79   Window open at 60: 83
FUEL CONSUMPTION AT STEADY SPEEDS IN O/D TOP:
  30 m.p.h.: 33·0   75 m.p.h.: 24·4
  45 m.p.h.: 32·8   90 m.p.h.: 20·6
  60 m.p.h.: 28·7   105 m.p.h.: 16·5
SPEEDOMETER CORRECTION:
  Indicated:
  20 30 40 50 60 70 80 90 100 110 120 130
  True Speed:
  19 28 36·5 46 55 64 73·5 82·5 91 100 109 118
TEST CONDITIONS: At sea-level; Barometer 30·17; air temperature 70 deg. F.; wind 5–10 m.p.h. down test strip; dry new tarmac surface; sunny and mild; 93-octane fuel.
CAR SUPPLIED BY: Robb Motors, Cape Town.

Engine compartment is crowded, with dry-element air-cleaner just making it on top.

We were able to establish quite clearly that the competition-type disc brakes of the Jag. will go on stopping it quickly and in an effortless straight line all day, if necessary.

Jaguar were pioneers in fitting discs, and their cars were used in the original development work by Dunlop more than 10 years ago. For this reason, it is not surprising that their stopping ability is impeccable, aided by the adhesion of Dunlop Roadspeed tyres, which are standard in South Africa.

Opening the Jaguar bonnet is something to bring joy to the initiated. The engine all but overflows the compartment, and hardly an inch is wasted. From 70-degree twin overhead camshafts to seven-bearing crankshaft it speaks of quality manufacture and sterling ability.

It is a quite remarkable unit and not easily flustered. We found that after a series of 100 m.p.h.-plus performance runs it instantly returned to its normal idling level of 600 r.p.m. when the car was stopped, and when switched off had no tendency to run on.

The Jaguar boot is not big, but it is serviceable enough. There is 13·5 cu. ft. of clear space with a low sill, with the jack and wheelbrace carried in clips on the bulkhead behind the rear seat.

The spare wheel is carried in a compartment under the boot floor, and the tool kit in its special container is also stored in this underfloor position.

The Mk. 2 saloons are built as five-seaters, and they carry five people in great comfort. The front bucket seats will take bulky shapes, and the 56-in. rear seat has lateral space for three adults. Headroom is better than average both at front and back.

Leg reach is adequate for people of average height, but a very long-legged driver would find himself a bit cramped when driving, as the steering wheel is set fairly low.

With the driving seat right back for comfortable cruising, the rear legroom becomes restricted, but not uncomfortably so because the rear seat itself is deep and there is foot space under the front seats.

There are many little features for comfort and convenience: twin foglamps as a standard fitting, reversing lamp, variable instrument lighting, five interior lights, one of which is a map-reading lamp, cigarette lighter, two-speed windscreen wipers, electric windscreen washers, rev-counter, heater and demister, inviting quality upholstery over Dunlopillo cushions, deep-pile carpets, door pockets and ashtrays, and a centre armrest at rear, to name a few.

Leg-level front ventilation is provided by an intake in front of the windscreen, operated under the dash, and its volume is adequate at cruising speeds. Efficient heating and demisting, with a booster fan, is standard.

Altogether, this is a car which makes a strong appeal to people who take their motoring seriously. It has little flashiness, and a rather old-fashioned warmth and comfort about its interior, which is quite rare in modern cars.

Mechanically it is as sound as a bell and immensely satisfying to drive, whether at an about-town dawdle or with 1,000 miles to cover.

Our feeling was: This is motoring at its liveliest and most satisfying level. ●

Gearing is a bit higher in South Africa because the Dunlop Roadspeed tyres have a larger rolling circumference.

★

Boot has 13·5 cu. ft. of space. Spare wheel is underfloor, and jack and wheelbrace are clipped to the bulkhead.

The JAG BELT

One year with 3.4 litres

THINGS are different now. There was a time, you may remember, when the Establishment used to curl its lip ever so slightly upon introduction to a Jaguar owner; a Gentleman wouldn't really be seen in a car that looked so expensive and wasn't. In the hard world of the '60s, where income tax is a universal disease, you can creep up on unsuspecting lord or layabout alike and say "fast car", confident that nine times out of ten his reflexes will prompt the reply "Jaguar". The two thoughts are inseparably linked. If you want to be terribly with-it you can still speak condescendingly of the Jag Belt to identify yourself, by contrast, as an honest, liberal-minded toiler with neither cigar nor expense account, but you run the risk of looking silly. When the wealthy are taxed and most workers can aspire to wealth, £1,500 is within anyone's comprehension as value for money, whether or not he happens to have it in the bank.

by Richard Bensted-Smith

I wouldn't deny that the expense account is a powerful aid to Jagmanship. Lots of things help to make a 3·4-litre, four-seater saloon an attractive proposition at £1,500, but few of them carry as much weight as using someone else's money. In the main, I suppose, motoring editors pick Jaguars for the same reasons as other customers. What else, at anything like the price, has the same combination of performance, safety, comfort and (be honest) prestige? An overworked word if you like, prestige is a tool of commerce—and it has more meanings than simple snobbery. Polished walnut fittings may say nothing to you or me (they don't, to me), but if the drawing-room atmosphere helps their owner to clinch a business deal they will have earned their keep.

Motoring journalists don't usually do that kind of business, but they do consort with a variety of people, from captains of industry to racing drivers. If you want to run with the hare and hunt with the hounds you need spiked shoes *and* brass buttons. Hence 842 ELX, a Mark 2, 3·4-litre dark green saloon with synchromesh and overdrive transmission, reclining seats and normal steering.

The steering wasn't normal to start with, which was entirely my own fault. On paper, the standard steering ratio of 4·7 turns from lock to lock inspires visions of Brockbank-type contretemps with the driver simply running out of arms in an emergency. Carried away with the idea of lightning response, I requested the special-order competition steering box with 1·2 turns less, and suffered for several summer months of heaving myself into London parking slots. There were other, unexpected snags: an inch or so of free play in the middle assumed much greater importance, and an average turn into an average side-street required too little wheel winding to trip the self-cancelling mechanism of the direction indicators. In the end I gave in and paid to return to normal, which is a lot better than it sounds since (a) ELX's turning circle of 36 ft. beats the pants off most cars of its size and (b) the progressive steering ratio has a quicker response over the first 15° of wheel movement right or left.

All cars being a compromise, steering as often as not is an excuse for some other feature of the design. Unquestionably the Mark 2 Jaguar could have lighter and more direct steering—and probably better handling into the bargain—if the light-alloy racing engine took the place of 500 lb. of cast iron between the front wheels, but that sort of speculation always brings you back to the astonishing price. To wear a Jaguar engine out at the same rate as the car, you have to drive it continually harder than most people ever do, which is the penalty of owning a vehicle of such potentially huge performance. There are compensations. Running-in instructions put a limit of 2,500 r.p.m. on the first 1,000 miles, corresponding to an easy 70 m.p.h. in overdrive; if you have tried M1 at a steady 30 m.p.h. you will see the value of instant cruising.

The early running-in period (during which the Jaguar, for one, needs a constant check on the dipstick) is not generally the most economical. By force of circumstances ELX probably made the most unthirsty journey of its life when, with less than 1,500 miles on the recorder, *The Motor* went visiting the German Ford works in Cologne. With three people and their luggage, and many miles of very open motoring, 80 m.p.h. cruising gave us 21 m.p.g.

The liquid consumption record since then has been mixed. After four or five thousand miles the piston rings settled down as they were designed to, so that weekly inspection of the oil level has become a mere formality; the price of 1½ pints of oil per 1,000 miles is less than most people spend on putting too high a grade of fuel into their tanks. The radiator does need looking at, especially after an hour of the sort of walking-pace queue you get on the way to Silverstone —the only conditions which move the thermometer needle more than a millimetre above its normal 75° position. Petrol consumption is less a function of how you drive than of how well you attend to minor details of maintenance. In early days we hit rock bottom for 3½ litres with a spell at just over 13 m.p.g. and several symptoms of a sticking automatic choke. After two visits to the factory a maladjusted carburetter needle was finally pinned down as the villain, since when any check taking into account at least a proportion of country driving has shown a consistent 19 m.p.g.

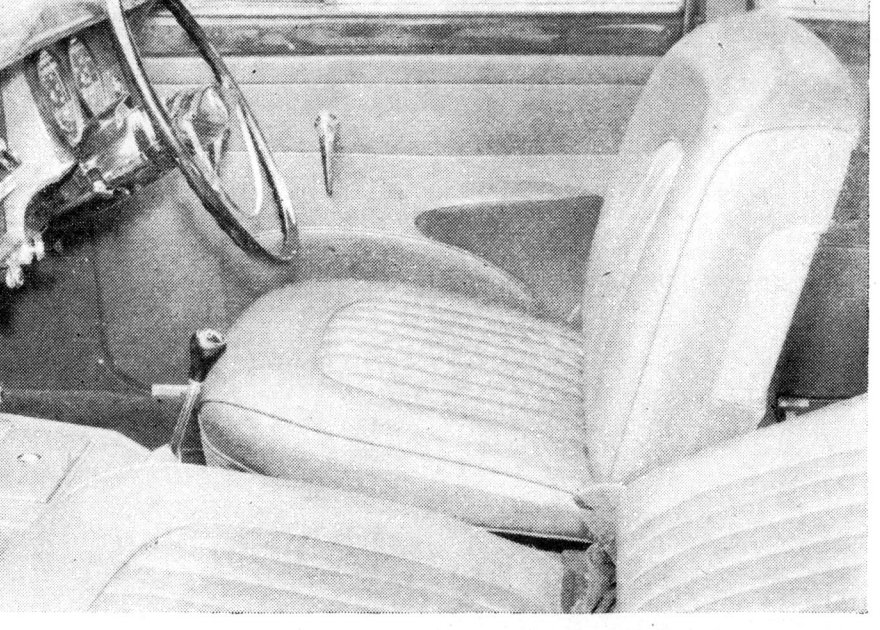

Coughing and Sooting

Occasional fits of coughing with a tendency to stall inconveniently at traffic lights are a reminder to keep the dashpots of the S.U. carburetters well oiled. More trying for my personal routine is the Jaguar's reluctance to spend a life of leisure in town for more than a few days without sooting up sparking plugs. Too many cold starts and short journeys without relief result in misfiring the first time an opportunity does arise to combine, say, three-quarter throttle with 2.500 r.p.m. or more; a demoralizing performance which can take five or ten minutes to restore itself. A promising remedy in Lodge's new Golden sparking plugs failed when the first batch sent out by that unfortunate company were let down by a faulty bought-out component and I have since experimented to no avail with a variety of grades and makes. The answer, I suppose, is not to buy a racehorse for delivering coal, but I shall persevere in the search for perfection without concession. A new set of Golden Lodges is now being tried.

Everybody has his own story to tell of last winter's motoring. An alarming death-rattle which lasted a full minute after one exceptionally cold start seems to have had no ill effect, and (except for periods under the influence of plug or dashpot trouble) starting has always been child's play with a battery obviously quite equal to turning a big, cold engine without fail. A permanent sceptic about electricity, I bow respectfully to a car which surely presents a strong case for an alternator—crawling a few miles through London every day of the winter with heater fan, radio and, on the way home, sidelamps—and has got along very well without one.

The fan is necessary because the heating system of the Mark 2 is archaic. For a car of this quality to produce such a dribble of low-temperature air is both

"One of the greatest virtues of reclining seats, incidentally, is their ability to adjust the other way for a super-upright driving position in falling snow or fog." The steering wheel is shown right in, and right out, and in the picture below, the frayed stitching down the seams of the arm rest is evident.

THE JAG BELT

ludicrous and a mockery of the neat trunking that taps off part of it for the rear passengers. Summer ventilation suffers likewise from the poor airflow at low speed, though on the open road it can be greatly improved by opening the back quarter-lights.

I shan't trouble you further with arctic memories, except to record frozen door locks (generally cured by cigarette lighter), frosted headlamps and a frosted windscreen and rear window, for which by far the cheapest remedy is a shilling's worth of soft plastic kitchen dish-scraper, and an Iceguard device which obligingly lit up every evening in February to confirm my observation that it was freezing. It seemed to work; whether one is justified in concentrating on slippery roads when a warning light says so, I am less sure. On a steep, snowbound garage approach, 3·4 litres of Jaguar with about 18 out of 31 cwt. on its back tyres proved less adept than 850 c.c. of Mini carrying 8 out of 13 cwt. on the front, which is not altogether surprising, but even on normal Road Speed tyres at normal pressures ELX was a most reassuring car to drive about in, on ice and snow. One of the greatest virtues of reclining seats, incidentally, is their ability to adjust the other way for a super-upright driving position in falling snow or fog. Adaptability is made more practical by a seat belt which it is either easy or unnecessary to adjust. Two makes are fitted to ELX, both of the lap-strap-and-diagonal configuration: an Autosafe, which is infinitely adjustable by flipping open the buckle, and a new Britax inertia-reel belt which adjusts itself to a rather greater tension than I find comfortable, though it is undoubtedly safer that way. I hope not to be able to report on how well either type behaves in emergency.

Pirelli Demonstration

In spring, as the winter moved on, I took up an invitation from Pirelli to try the Cintura tyres, without which, as numerous influential persons had been earnestly assuring me from the advertisement pages, I could not afford to be. At the aerodrome test track at Stoke-on-Trent I made a dutiful exhibition of myself, trying to go faster through Pirelli's timed wiggle-woggle in an A40 shod with Cinturas than in the same car on normal tyres. The experiment might have been more conclusive if my reactions and the A40's half-turn of steering wind-up could ever see eye-to-eye. There was a hint of sympathy in the Pirelli men's polite approbation when they totted up my meagre improvement after a score of runs on each type of tyre. Fortunately for my morale, I tried a slightly despairing go at the same test with the Jaguar, now also fitted with Cinturas, and wove its impressively greater bulk through the pylons at almost my best A40 time with only two practice runs. Cinturas, as far as one can tell from a semi-objective test, give the Jaguar considerably more wet-road grip; through the quicker steering response of a braced-tread tyre they also give the driver a sense of precise control which is not sent tumbling by the sudden breakaway of earlier, steel-banded designs on one or two cars in my experience; better roadworthiness has to be paid for,

The leaping Jaguar on the bonnet blends particularly well with the sweeping lines, and the wing mirrors and Cintura tyres provide faster, safer driving.

A varnished wood facia gives a comfortable, almost at home feeling, with rows of switches, a radio, cigar lighter and a cubby hole, just in front of the gear lever.

Keeping to this basic body shape with only small changes over a long period not only helps keep down production costs, but also wind drag.

The JAG BELT

however, and under a variety of conditions from town-pottering to very fast cruising on the concrete section of M1 the ride is now perceptibly harsh and a lot noisier.

Thus can you only try to please most of the people most of the time. Almost perfect cars can be built, in penny numbers, for limited purposes at millionaires' prices. The rest can hope only for constant improvements in the compromise—amongst them, one day perhaps, a lighter Jaguar clutch and a better gearbox. The former certainly exists amongst the component maufacturers' developments, for I have driven briefly a 3·8 Jaguar with an experimental diaphragm clutch which made a measurable, though not immediately, noticeable improvement. Measurable pedal pressure is what matters at the end of a tiring day, and it will be an excellent thing when the lightness of Jaguar steering and braking is matched by its manual transmission. Forgiving the non-synchromesh bottom gear (which one will only be able to do for a year or so longer), there is still not a great deal of excuse these days for even a very powerful car to demand skill or strength, or both, from the driver who wants to change his own gears. Being built to last, like the engine, the Jaguar gearbox frees up with use at about the same rate as the driver learns to change from first to second without getting frustrated in the middle; but not quickly, or enough.

In France, fortunately, gears do not matter very much. Driving prestigiously to Le Mans in June, I rediscovered the great advantage of having a car much faster than you need. Cruising on the great open stretches of N138 between Rouen and Le Mans is scarcely related to mechanical limits when a comfortable overdrive 3,500 r.p.m. represents a shade off 100 m.p.h. and the speed is still within the scope of direct top gear for overtaking. The ride in a laden car travelling fast over a wavy road might be improved by alternative damper setting, and it would certainly be better with independent rear suspension (though quarter-elliptic springs make a quite different device of the live axle), but it is wind noise around poorly fitting window frames which dictates, as much as anything, the choice of touring speed between 80 m.p.h. and 100 m.p.h. To go fast, you must be able to stop. Apart from a slight juddering if they are used hard at high speed, the brakes have given me no qualms in heat, cold or wet—and the extra calipers on the rear of the all-disc Dunlop system make a more effective parking brake than a good many drums.

Performance, comfort, roadholding, braking. The question "how is it done for the money?" remains only partly answered. There is one answer, of course, in designing a machine (like the six-cylinder engine) which will hold its own for many years; another in keeping a basic body shape with only small changes over a long period; a third, in making clever use of bought-out parts which cost little because they are sold also to manufacturers of cars by the hundred thousand, with the proviso that you cannot afford to take chances at any price with a car that may be driven at 120 m.p.h. Good though the finish is by comparison with a popular saloon, it does not take a long time or great carelessness to acquire a pattern of chipped paint down the edges of the doors, the wood veneer on top of the facia has a thumbprint in its varnish and if you peer up past the instruments you can see code numbers scrawled on the bare wood. The point of my elbow has torn the armrest covering away from its stitches.

The leather upholstery, on the other hand, is perfect and a good wash will give the paintwork a deep lustre and take out any marks left on the chromium by owner's idleness. At 12,000 miles the original tyres would probably have been just about needing replacement. The carburetter and sparking plug ailments, a leaking clutch master-cylinder seal and a tailpipe bracket which defies all efforts to stop it clonking under engine torque are the extent of a year's troubles. Direct running costs per mile are now 3d. for petrol, negligible for oil and 3d. for tyres, plus just under a 1d. for servicing and oil changes. In its first year the depreciation from showroom price to trade-in value is probably about £450. You can spend your money an infinite number of ways, according to taste. If this is the way your taste lies, there are very few other means of satisfying it.

Six beats to the leap. The heart of a Jaguar looks and is very impressive; it turns out 210 b.h.p. at 5500 r.p.m. With the clutch and gearbox it weighs 640 lb.

JAGUAR 3·8 LITRE MARK 2 SALOON

ENGINE CAPACITY: 231.25 cu in, 3781 cu cm;
FUEL CONSUMPTION: 18.5 m/imp gal, 15.4 m/US gal, 15.3 l x 100 km;
SEATS: 5; **MAX SPEED:** 120 mph, 193.2 km/h;
PRICE: list £ 1,288, total £ 1,557.

ENGINE: front, 4 stroke; cylinders: 6, vertical, in line; bore and stroke: 3.43 x 4.18 in, 86.9 x 105.9 mm; engine capacity: 231.25 cu in, 3781 cu cm; compression ratio: 8 : 1; max power (SAE): 220 hp at 5500 rpm; max torque (SAE): 240 lb ft, 33.1 kgm at 3000 rpm; max number of engine rpm: 6000; specific power: 58.1 hp/l; cylinder block: cast iron; cylinder head: light alloy, hemispherical combustion chambers; crankshaft bearings: 7; valves: 2 per cylinder, Vee-slanted, thimble tappets; camshaft: 2, overhead; lubrication: gear pump, full flow filter; lubricating system capacity: 5.7 imp qt, 6.9 US qt, 6.5 l; carburation: 2 SU type HD 6 horizontal carburettors; fuel feed: electric pump; cooling system: water; cooling system capacity: 10.0 imp qt, 12.0 US qt, 11.4 l.
TRANSMISSION: driving wheels: rear; clutch: single dry plate; gear box: mechanical; gears: 4 + reverse; synchromesh gears: II, III, IV; gear box ratio (I) 3.376, (II) 1.859, (III) 1.282, (IV) 1, (Rev) 3.376; gear lever: central; final dri limited slip final drive; ratio: 3.54 : 1.
CHASSIS: integral; front suspension: independent, wishbones, coil springs, ar roll bar, telescopic dampers; rear suspension: rigid axle, cantilever se elliptic leaf-springs, trailing arms, transverse linkage bar, telescopic dampe
STEERING: recirculating ball; turns of steering wheel lock to lock: 4.3.
BRAKES: disc, servo.
ELECTRICAL EQUIPMENT: voltage: 12 V; battery: 60 Ah; ignition distribut Lucas; headlights: 2 front and reversing.
DIMENSIONS AND WEIGHT: wheel base: 107.37 in, 2727 mm; front track: 55.00 1397 mm; rear track: 53.37 in, 1356 mm; overall length: 180.75 in, 4591 m overall width: 66.75 in, 1695 mm; overall height: 57.50 in, 1460 mm; grou clearance: 7.00 in, 178 mm; dry weight: 2968 lb, 1346 kg; distribution of weig 56.5 % front axle, 43.5 % rear axle; turning radius (between walls): 16.7 ft, 5.1 tyres: 6.40 - 15; fuel tank capacity: 12.00 imp gal, 14.52 US gal, 55 l.
BODY: saloon; doors: 4; seats: 5; front seat: double.
PERFORMANCE: max speed in 1st gear: 35 mph, 56.3 km/h; max speed 2nd gear: 64 mph, 103 km/h; max speed in 3rd gear: 98 mph, 157.8 km max speed in 4th gear: 120 mph, 193.2 km/h; power-weight ratio: 13.4 lb/ 6.1 kg/hp; useful load: 882 lb, 400 kg; acceleration: standing ¼ mile 16.3 s 0 — 50 mph (0 — 80 km/h) 6.4 sec; speed in direct drive at 1000 rpm: 20.1 m 32.4 km/h.

PRACTICAL INSTRUCTIONS: fuel: petrol, 95-100 oct; engine sump oil: 5.4 qt, 6.5 US qt, 6.2 l, SAE 20 (winter) 30 (summer), change every 2500 mi 4000 km; gearbox oil: 1.2 imp qt, 1.5 US qt, 1.4 l, SAE 30, change every 10 miles, 16100 km; final drive oil: 1.4 imp qt, 1.7 US qt, 1.6 l, SAE 90, cha every 10000 miles, 16100 km; greasing: every 2500 and 5000 miles, 4000 8000 km; tappet clearances: inlet 0.004 in, 0.10 mm, exhaust 0.006 in, 0.15 valve timing: (inlet) opens 15° before tdc and closes 57° after bdc, (exha opens 57° before bdc and closes 15° after tdc; tyre pressure (medium lo front 30 psi, 2.1 atm, rear 26 psi, 1.8 atm.

VARIATIONS AND OPTIONAL ACCESSORIES: tuned competition engine, speed 130.4 mph, 210 km/h; 4-speed mechanical gear box and Laycock Normanville overdrive in IV (I 3.377, II 1.859, III 1.284, IV 1 - ov. 0.777), m plication ratio 0.77, speed in overdrive at 1000 rpm 21.1 mph, 34 km/h, ratio 3.77 : 1; 3-speed automatic gear box (see Mk X); Rudge wire whe Mk II 3.4-litre, engine capacity 210.51 cu in, 3442 cu cm, bore and st 3.27 x 4.17 in, 83 x 106 mm, compression ratios 7 - 8 : 1, max power 210 hp 5500 rpm, max torque 215 lb ft, 29.7 kgm at 3000 rpm, limited slip final d max speed 120 mph, 193.2 km/h; Mk II 2.4-litre, engine capacity 151.86 cu 2483 cu cm, bore and stroke 3.27 x 3.01 in, 83 x 76.5 mm, compression 8 : 1, max power 120 hp at 5750 rpm, max torque 144 lb ft, 19.9 kgm at 2000 4-speed mechanical gear box (axle ratio 4.27 : 1), 4-speed mechanical box and overdrive (axle ratio 4.55 : 1), 3-speed automatic gear box (axle 4.27 : 1), max speed 105 mph, 169 km/h (see Mk X).

BEST LUXURY COMPACT ✱ JAGUAR 3.8 MARK II

Back in 1947 such makes as Armstrong-Siddeley, Frazer-Nash and Lea-Francis apparently had as much going for them as Jaguar, but the others are no longer in the car manufacturing field, while the firm of Jaguar has been going through a spectacular expansion, absorbing such old-established companies as Daimler, Guy and Coventry Climax.

Jaguar went against the post-war trend to austerity-marked economy cars with flamboyant high-powered machinery and had greater success than many who went after the mass market.

The sports sedan began its career in 1956 with a short-stroke 2.4-liter six, and a year later the 3.4-liter XK-140 engine became optional. In 1959 came the first of the 3.8-liter installations, which have dominated sedan racing until Ford became interested. A crab-tracked, bulbous design which combined firm understeering with a high sensitivity to crosswinds, it needed thorough race preparation. Racing experience finally resulted in a wider rear track on the production models in 1960.

The basic appeal of Jaguar cars, and the SS before that, has always been their good looks. The 3.8-liter sports sedan is more controversial than any other Jaguar model in this respect, but it does have an air of speed and stability about it, which has earned for it the nickname of "The Businessman's Express."

Perhaps just as important is the majestic appearance of the twin-cam engine with its polished cam covers and the big pots of the triple SU carburetors. Its 3.8-liter version is also the best—both the original 3.4 and the short-stroke 2.4 have a poorer service record. Attempts to enlarge it to 4.2 liters have been patently unsuccessful, and the S-type (the sports sedan with independent rear suspension) will not have the benefit of improved engine performance over the normal 3.8, as originally intended.

The Jaguar comes with a choice of two transmissions, both of which have some annoying shortcomings. The Borg-Warner automatic won't engage first gear in low range on kickdown at any speed above about 5 mph, and the four-speed (either made by Moss Gears or Jaguar to a basic Moss design) has a synchromesh action of little more than token value.

The strength of the car lies in the performance, and the inconveniences become forgiveable to all who are charmed by the impressive engine and the pretentious frontal appearance, as well as those who buy it because Jaguar remains the cheapest of the good cars. On a value-for-money basis, it's in a very strong position.

ENGINE Type	water-cooled in-line six
Bore x stroke	3.44 x 4.18 in, 87 x 106 mm
Displacement	231 cu in, 3781 cc
Power (SAE)	265 bhp @ 5500 rpm
Torque	260 lbs-ft @ 4000 rpm
TRANSMISSION	4-speed, non-synchro first gear, overdrive
BODY AND CHASSIS	Unit-construction, all-steel body
SUSPENSION	F: Ind., unequal-length, wishbones and coil springs, anti-roll bar. R: Rigid axle, radius rods and cantilever leaf springs.
BRAKES	11-in discs F, 11½-in discs R, 495 sq in swept area.
WHEELBASE AND TRACK	107.4, F 55.0, R 53.4 in
DIMENSIONS	L-181, W-66.75, H-54.5 in
WEIGHT AND DISTRIBUTION	3300 lbs, 56/44%
TIRE SIZE	6.40 x 15
PERFORMANCE 0-60 MPH	9.7 seconds
SPEED IN GEARS	37/67/98/120/OD 125
FUEL ECONOMY	14-20 mpg
BASE PRICE	$5040 POE NY

Released yesterday

THE JAGUAR 3·4 and 3·8S

An all-independently-sprung version of the Jaguar Mk. II range

THE Mk. II Jaguar series, which has been in production since 1959, has now been joined by two additional models to be known as the S-type. The Mk. II range will be continued in 2.4 and 3.4 forms but the 3.8 has been dropped and the S-types will be available only in 3.4 and 3.8-litre forms. No prices have been announced for the new cars as we go to press but they are expected to be more expensive than the equivalent 3.4 and 3.8 Mk. II versions.

At a quick glance the new car is little different in outward appearance from the Mk. II model but more detailed study shows numerous differences. The tail of the car is longer, having a similar shape to the Mk. 10, and the luggage boot capacity has been increased from 12 cu. ft. to 20 cu. ft. as well as having a much better shape for the stowage of cases. The roof line has been lengthened and flattened slightly and the front end styling has been subtly altered with small cowlings over the headlamps, differently shaped bumpers and overriders, and strip-type winkers instead of the separate round lights used on the Mk. II models.

The most important change on the S-type is the introduction of independent rear suspension. This is based on that of the E-type and Mk. 10 with certain changes to enable it to be adapted to the 3.8 bodyshell which has a narrower track. A good deal of additional stiffening has had to be incorporated into the rear end of the unit construction shell to take the Salisbury chassis-mounted differential, and this partly accounts for the fact that the S-type is slightly heavier than the normal 3.8, the dry weight being 3,440 lb. Apart from the normal benefits of i.r.s. this system has allowed the rear seat to be moved back approximately 3 in., giving more leg room. By virtue of the fixed drive line and chassis-mounted differential the rear seat has been lowered, allowing the roof line to be lowered also.

All the seats of the S-type are new, the front seats being separate bucket types with better shape, having adjustment for height and reach as well as reclining backrests. The rear seat is contoured for two passengers with a large central arm-rest, but three people can be carried if required. The facia is similar to the Mk. II but the space above the central tunnel is taken up with a new fresh-air heating and ventilation system similar to that of the Mk. 10 with a control for directing air to the rear compartment. The door locks have been modified to the much improved "zero-torque" type and quarter-lights now have thief-proof catches fitted.

The fuel system has been modified to that of the Mk. 10 with twin 7-gallon tanks in the rear wings with a facia-mounted change-over switch. The fuel gauge registers the contents of the tank in use at the time.

Dunlop self-adjusting disc brakes of the latest Mk. III type are fitted on all four wheels and the tyres fitted as standard are the Dunlop RS5. The usual Jaguar options are available for the S-types. It can be had with 4-speed gearbox, and overdrive if required, or with the Borg-Warner automatic transmission. Power steering is also optional, this being modified so that only 3.5 turns lock-to-lock are required against the 4.7 turns of the manual steering and 4.9 turns of the previous power steering.

We were recently able to spend a couple of days with the latest Jaguar, this particular car having the 4-speed gearbox with overdrive, which operates on top gear only. Power steering was also fitted. As the car was brand new and not fully run-in we were restricted to a rev. limit of 4,000 r.p.m. and were therefore unable to take any performance figures. However, as the car uses the same engine as the 3.8 Mk. II performance should be much the same, although it will be interesting to see whether the theoretically improved traction of the i.r.s. will offset the greater weight. Jaguar claim a top speed of 123 m.p.h., 0-60 m.p.h. in 10.1 sec., and 0-100 m.p.h. in 29.4 sec., which are fairly similar to the figures obtained on the ordinary 3.8.

The i.r.s. definitely imparts a much smoother, softer ride to the Jaguar, although not as soft as that of the Mk. 10. This smooths out the bumps most impressively although they are still heard rather than felt. The car seems to roll slightly more than the rigid axle cars and more tyre squeal is heard during hard cornering, but we did have the tyres on their softest recommended settings. The power steering is commendably light without losing the feel of the road, and with 3½ turns gives much more controllability than previously. Jaguar owners will undoubtedly appreciate this feature as the car can be aimed much more precisely and cornered quicker. Understeer is quite prominent but this turns into roll oversteer as really high cornering speeds are reached, but with the i.r.s. bumpy bends provide no problems with the displacement of the rear end. The engine is the same familiar turbine-smooth unit and the gearbox is also regrettably the same unit, having poor synchromesh and requiring full depression of the clutch pedal. Fuel consumption is in the 16 to 18 m.p.g. region, which empties the 7-gallon tanks rather rapidly.

Despite the fact that we had the feeling that this Jaguar is only an interim model until something really exciting is revealed, the S-type is obviously going to appeal to Jaguar enthusiasts who prefer the more compact dimensions of the 3.8 but crave the all-independence of the Mk. 10. We look forward to longer acquaintance on the road.—M. L. T.

DETAIL front end styling changes with hooded lamps and narrow bumpers characterise the S-type.

INTERIOR changes include the installation of a heating and ventilation system similar to that of the Mk. 10.

PRIDE OF JAGUARS CONTINUED FROM PAGE 85

on hairpins at speed, when I find myself getting a little "ravelled" about the arms.

Everyday motorists who use their Jaguars purely as transport will no doubt find this light steering the outstanding feature of the car. Parking-station attendants are loud in their praise of it, remembering how the Mark I steering nearly pulled their arms out.

I hope the gearbox will be the next item for attention by Sir William Lyons' engineers. It is not a very good box — heavy in movement and slow in synchromesh. That is why I prefer the automatic Jaguar to the manual-gearshift car.

The automatic is first-class, and the hold switch for "intermediate" range makes it delightful in twisty hill country.

For those who consider automatic Jaguars sissy, I must recount Mike Hawthorn's story of his "derby" with a lass in an automatic 3.4.

Mike was driving his famous VDU 881—the modified 3.4 in which he later had his tragic accident. This lass and Mike were doing a TV show for the BBC in their respective Jaguars —and afterwards they had a standing-start quarter-mile dice.

Much to everyone's surprise, there was less than a car's length between them at the finish, Mike just having the edge!

And Now, the 3.8 . . .

The 3.8 is another matter altogether. Whereas the 2.4 is perhaps the "feminine" Jaguar and the 3.4 the confirmed Jaguar man's buy, the 3.8 is the "projectile."

It's so much of a projectile that I feel it calls for a rather special type of owner — one who can really handle the car, has a high sense of responsibility, and who, after an evening at the club, has the courage to go home by taxi.

It is not a beginner's car, and parents should vet their children very carefully before handing over the keys of this 130 m.p.h. beauty. Granted, this applies to many cars that aren't half as safe to drive as a 3.8; but they aren't quite so heady a plaything as the Jaguar.

In case people think the 3.8 is a sports car, I had better say right away it isn't. Certainly, it will outperform most sports cars; but it will transport four elderly people in silence and comfort just as readily.

The 3.8 automatic now holds the touring-car record for my "private" mountain test course and looks like retaining it for some time. It also galloped up the hillclimb in an inspiring manner, hurling itself up the steepest part as quickly as most cars could descend!

The understeering characteristic is still pronounced; but with the help of the throttle and the limited-slip diff. it doesn't call for more than average skill to point the Mark II 3.8 effectively.

The ride is comfortable at all times, the suspension even managing to sort out uneven bitumen corrugations. However, heavier front shockers would prove useful for touring under Australian conditions.

Despite all this praise, I have a couple of personal complaints about the new Mark II's:

Those dreadful sun visors, not only ineffective but hard to operate; the steering wheel, which is hardly in Jaguar tradition and would be more at home in a ten-horse family saloon; and the rev-counter.

This last point was even more apparent in the 2.4 than the 3.8, for I was up in the high figures more in the little car during normal use. The tacho is marked with an abundance of figures, and I would prefer fewer figures and clearer markings.

A minor point, perhaps — but Jaguars are noted for detail, and I hope they'll note this criticism.

Apart from this, I can find little to fault about the Mark II's — and until someone builds something better, my opening remark still stands. • • •

JAGUAR 3.8

CONTINUED FROM PAGE 74

power ratings 5 units each, the effect throughout the rev range is said to be exceedingly worthwhile. With domestic cars using up to two ratios more, we found premium gas satisfactory for all but the most arduous conditions.

The accompanying acceleration curve is graphic demonstration of the Three-Eight's sporting performance. Considerable care is necessary to avoid excessive wheelspin when you start in first with the revs raised high. About 2000 seems best. First gear could usefully be raised, for the 59% step to second requires a leisurely pause because of the weak synchromesh. For the same reason, the angle of the pendant accelerator pedal is annoying as it makes the desirable heel-and-toe downshifts nearly impossible. A solution is to install a Corvette gearbox (we *have* come a long way) as one 3.8 owner has done; another is to take full advantage of the Jaguar engine's remarkable flexibility. The 3.8, which may be comfortably started in second, pulls smoothly and strongly from as low as 14 mph in fourth. And up to some two miles per minute, if you can find the opportunity.

The trunk is large (13½ cubic feet), perhaps at the expense of a too-small gas tank. Engine compartment accessibility is very good for those items which are inspected frequently, except the battery. Inside and out, everything seems to fit properly and work nicely. But then, Jaguar buyers are a demanding lot. The Dunlop disc brakes are smooth and strong and easily a match for this car's power.

The Jaguar's steering takes a little over **two turns** from straight ahead to full lock **in either** direction, at which point the **turning** diameter is a bit under 36 feet. **This is** quicker and sharper than a Corvair or Valiant, for example, yet at high cornering forces the steering seems extraordinarily slow. On the 400-foot Test Circle, with the tires at 34/31 psi, front and rear, (the pressures recommended for continued high speed use), an initial setting of some 55 or 60 degrees grew with increasing rapidity to 95° at 30 mph and 225° or so at the maximum speed achieved of 45 mph (indicated). On the circular graph used in SCI's Road Research Reports, this would put the 45 mark at about 8 o'clock, thoroughly counterclockwise compared to any car we've tested before. This is understeer with a capital U, yet it doesn't keep keen 3.8 owners from racing.

For all of this, the 3.8 is a nimble car. Its engine's roar is somewhat muted but not its bite. If the tire noise could be equally hushed, Jaguar might commandeer Bentley's old motto and call the 3.8 "The Silent Sports Saloon". Or to paraphrase a more modern Bentleyism, people who feel diffident about driving practically any kind of car may find what they're seeking in a Jaguar. Sybarites will find creature comforts, while performance- and status seekers will find just what they're looking for.   —*SCI*

USED CARS ON THE ROAD

No. 216 • 1961 JAGUAR Mk. II 3·8

PRICE: Secondhand £1,050; New—Basic £1,255, with tax £1,779

Petrol consumption	17-21 m.p.g.	Date first registered	17 May 1961
Oil consumption	300 m.p.pint	Milometer reading	11,068

ONE might well answer the familiar expression, that you can't beat a Jaguar for good value, with the reply that you can—with a used Jaguar. High running costs restrict the demand for them among those who cannot afford to buy new, so that prices are artificially low. Counting the reductions in purchase tax, our used Jaguar for this year—a 3·8 overdrive model—has lost £800 of its value in 2½ years, and is being offered at two-thirds of today's price for a new one.

Age and the mileage covered represent a small portion of the potential life of such a car, so that it is not surprising for it still to be in very fine condition. Its magnificent performance, thrusting the car effortlessly past the 100 m.p.h. mark when really unleashed, is only mildly inferior to that of our original Road Test car, and oil consumption has increased little. The automatic cold-starting device is working correctly, and whether hot or cold the engine usually responds to the first touch on the starter button. The combination of uncanny silence when cruising on a light throttle, with a purposeful snarl of power when accelerating hard, remains exactly as with the model when new.

The long and rather heavy clutch pedal movement, and slow but positive gear change, with easily beaten synchromesh, are also characteristic and not faults resulting from wear. A fair amount of gear whine is heard in the indirects, and there is also some axle hum audible when the engine is pulling in top gear. Spring damper efficiency has not deteriorated, and the ride is still firm for fast driving; large deflections of the wheels are absorbed very well and without "float." Some shiver of the whole body structure at high speeds suggests that the wheels are beginning to go out of balance.

Unassisted steering is fitted on this model. Directional stability is exceptionally good, but steering precision around the straight-ahead position is not quite as accurate as the car's performance merits. In all these aspects this Jaguar is identifiable with a new one, and it is only in respect of the brakes that there has been any fall-off from the new car standards. Superb response is still available, giving reassurance for fast driving, but the pedal has a slightly spongy feel and rather longer free travel than normal for power-assisted Dunlop disc brakes. A point which calls for special care is that there is a marked delay before the pads bite when driving fast in wet weather. The handbrake is adequately effective.

A pleasant shade of deep blue, which suits the car and camouflages road filth, is used for the exterior finish; and on close inspection of joints and edges it is clear that there has been an extensive, if not total, respray. In a car only 2½ years old this may mean accident damage; and although there are some features to support this thought, the car obviously has been expertly repaired, and does not seem to be the worse for any mishap which it might have suffered. Thus there is the advantage of practically new and unblemished paintwork. A slight looseness of the driver's door results in rattle over bumps. The chrome mainly is in excellent shape, though some of the zinc fittings such as the door handles have a few pit marks of rust.

Slight sag of both front seats accentuates the high scuttle. The leather upholstery is mildly creased, and the carpets rather worn. There are no marks on the cloth roof linings or polished wood facia and window surrounds, and the overall impression of the interior is of a car that has been well cared for but which makes no secret of the fact that it is not new.

Jaguar spares are not cheap, and the used car buyer does well to examine carefully such matters as the condition of the exhaust. In this case it is not badly rusted, and the general underbody condition is good. All five tyres are Dunlop RS5, half to two-thirds worn; and the toolkit is practically complete, nestling in its container within the spare wheel. The jack is in its clips in the boot, and the handbook is in the driver's door pocket.

So comprehensive and elaborate is a Jaguar's equipment that a few detail faults may be forgiven. The interior light switch in the driver's door is not working, although the other three function correctly, but a greater inconvenience is that the electrically operated windscreen washer is out of action. Two fog lamps and a fresh-air heater are included with the initial equipment. The fog lamps have been rewired through separate switches on the facia, instead of the standard fourth position on the main lighting switch. H.M.V. radio, with variable control for a second loudspeaker on the rear shelf, is fitted, but is not working properly—possibly through a fault in its wind-up telescopic aerial.

For a large and powerful car the fuel consumption is very fair; and fast, effortless travel on a long run really can be enjoyed at (or very near to) 20 m.p.g. This 3·8 is very typical of the value offered by the many Mark II Jaguars now coming on to the used car market at around the £1,000 mark.

As well as having radio, and the very comprehensive array of instuments and equipment standard with the model, this Jaguar features chrome-plated wire wheels. Except for a slightly erratic speedometer needle, the instruments are all working correctly

PERFORMANCE CHECK

(Figures in brackets are those of the original Road Test, 26 February 1960)

0 to 30 m.p.h.	3·8 sec (3·2)	0 to 90 m.p.h.	22·3 sec (18·2)
0 to 40 m.p.h.	6·1 sec (4·9)	0 to 100 m.p.h.	28·0 sec (25·1)
0 to 50 m.p.h.	7·9 sec (6·4)	0 to 110 m.p.h.	36·5 sec (33·2)
0 to 60 m.p.h.	9·9 sec (8·5)	**Standing quarter-mile**	17·6 sec (16·3)
0 to 70 m.p.h.	13·7 sec (11·7)	20 to 40 m.p.h. (top gear)	6·6 sec (6·0)
0 to 80 m.p.h.	17·1 sec (14·6)	30 to 50 m.p.h. (top gear)	6·4 sec (6·1)

Car being sold by Lazenby Garages Ltd., 929-931 Loughborough Road, Rothley, Leicestershire. Telephone: Rothley 2494

USED CARS ON THE ROAD

No. 223 • 1960 JAGUAR 2.4 Mark 2

PRICE: Secondhand £795; New—Basic £1,082, with tax £1,584

Petrol consumption	18-22 m.p.g	Mileometer reading	26,047
Oil consumption	120 m.p.pint	Date first registered	19 October 1960

USED car buyers who long for the grace, space, pace features that go with Jaguar ownership but are fearful of high running costs, tend to look to the 2·4, often thought of as the "small Jag," to give them the best of both worlds. Frequently we are asked for advice on whether they will be able to afford to keep one. To answer these questions, as well as to re-acquaint readers with the model, we have chosen an early 2·4 Mark 2 as this year's Jaguar for the used car series.

With known service history to support its mileometer reading of 26,000 the car is considered to be representative of some of the better examples now on the market, and it was quickly found to be in good tune and thoroughly sound. The short-stroke 2,483 c.c. engine is in especially good fettle, and has the characteristic Jaguar combination of remarkable quietness, changing when throttle and high revs are used for hard acceleration, to a purposeful roar of power.

It is not until high performance is needed that one is really conscious of the difference between 2·4 and the over-3-litre versions. Only then is it appreciated that much more "engine" has been in use for brisk progress, and less kept in reserve. Even so the 2·4 is a lively car with ample acceleration and speed to please most drivers.

Optional overdrive is fitted, giving remarkably quiet cruising around the 70 m.p.h. mark. Rather more effort and noise goes with a motorway 85 m.p.h., and given plenty of time the 2·4 is still capable of a true 100 m.p.h. on the flat in overdrive at 4,550 r.p.m., and gives a mean maximum of 96·3 m.p.h. In direct top, the rev counter needle reaches the 5,500 r.p.m. red segment at 96 m.p.h. The only Jaguar not fitted with S.U. carburettors, the 2·4 has twin Solex. Starting is not quite as immediate as expected, but the engine always fires after a few turns on the starter, and it is quick to warm up.

In both ease of action and efficiency of the synchromesh, the gear box is one of the best we have come across on any Jaguar, and perhaps has benefited from having less torque through it; but it has a tendency to "stick" in bottom gear at low speeds. Clutch take-up is smooth and the pedal load is reasonably light. Slight nose-heaviness is noticed on corners; directional stability and steering response on the straight, even in quite strong cross winds, are good. Ride comfort is excellent, and there is extremely efficient damping.

All-disc servo-assisted brakes replaced the drum brakes of the first 2·4s, and are effective; the car needs pedal pressures slightly heavier than those of the feather-light systems which are becoming the modern trend, but the brakes are excellent for a car that is nearly five years old. The handbrake is good, too.

All of the bodywork is in the shade of light grey which many Jaguar owners have chosen, and barring a few chips and car park touches, the finish is pretty sound and shines well. Chrome condition varies, as some parts are more scratched or worn than others; but the general exterior appearance belies the age. Extensive creasing of the maroon leather seats is the main visible sign of use inside. Fitted mats on the front floor cover practically as-new carpets, and all wooden trim, door leather, and cloth roof linings are in good shape.

Underbody examination revealed complete absence of rust as a result of Undersealing, and the exhaust system appears fairly youthful. All tyres including the spare are Dunlop RS5s with at least two-thirds of the tread depth remaining. The fitted tray of tools nestling in the spare wheel lacks the adjustable spanner, but is otherwise complete. A GB plate on the boot, and yellow, centre-dipping headlamps, tell of some Continental motoring.

Returning to the matter of running costs, one must bear in mind that the 2·4 is no smaller, and only a little lighter, than its more powerful sisters. Not surprisingly, fuel consumption is much the same. There may be a small saving in tyre wear, and about 20 per cent reduction in the cost of insurance, otherwise, running expenses will be practically the same with a 2·4 as with a 3·4. This Jaguar is an excellent car on its own merit.

In addition to overdrive, an extra is the Phillips hand-tune radio; but standard equipment includes fresh-air heater, twin foglamps, dipping mirror, clock and windscreen washer. Every item is in perfect working order, about the only fault being a very stiff driver's window

PERFORMANCE CHECK

(Figures in brackets are those of the Mark 1 Road Test, 21 September 1956)

0 to 30 m.p.h.	5·7 sec (5·0)	0 to 80 m.p.h.	33·3 sec (30.6)
0 to 40 m.p.h.	8·5 sec (—)	0 to 90 m.p.h.	49·9 sec (—)
0 to 50 m.p.h.	12·7 sec (11·5)	Standing quarter-mile	20·8 sec (20·5)
0 to 60 m.p.h.	17·3 sec (15·8)	20 to 40 m.p.h. (top gear)	8·9 sec (8·8)
0 to 70 m.p.h.	23·8 sec (21.7)	30 to 50 m.p.h. (top gear)	9·2 sec (9·6)

Car for sale at: Campbell Symonds and Co. Ltd., Empire Garages, Western Avenue, Ealing, London, W.5. Telephone: ALPerton 1515-7

AFTER THE NEW WEARS OFF
3.8 JAGUAR SEDAN

BY GUSTAF EVAN ENGSTROM, JR., M.D.

A 3.8 JAGUAR SEDAN came into my life on the day before Thanksgiving, 1963. Just one day short of two years later, it had covered a total of 30,000 miles, I think. I say that because the speedometer was broken at that time. And this is an abbreviated version of my experience with a 3.8 Jaguar sedan.

Why did I choose the 3.8 sedan? Frankly, it was an impulse purchase. I had tired rapidly of my previous car, a Triumph TR-4, even though I had owned it for only nine months and driven it less than 12,000 miles. The TR-4 and I weren't *sympatico*. I thought the ride was brutal, civilized amenities totally lacking and the top leaked no matter what. I had just taken it on a 3-day, 1200-mile trip and was ready to *give* it away after that. This eagerness to rid myself of the TR-4, plus the fact that the 3.8 by that time had been superseded by a later model and was available at an advantageous price, were enough to swing the balance.

But I was also attracted to the car itself, make no mistake about that. Black with black leather, chrome wire wheels, beautifully finished walnut trim, little tables, and so forth. It was equipped with only one major extra, a Blaupunkt AM-FM radio. It also had the 4-speed manual transmission (non-synchro first) and overdrive. Fully equipped, it came to $5538.25. I call her *"Chat noir,"* the black cat.

There is a lot to be said for owning a Jaguar 3.8. It looks fierce, goes rapidly, and makes a sound like it means business. The car also has an amazing effect on people. At the opera, theater, parking lots, etc., I am treated with deference reserved for the owners of very few makes of cars. Car-jaded kids turn and goggle, filling station attendants almost beg to be allowed to lift the hood, and dignified old ladies sneak looks at the nameplate and discreetly glance into the interior. No Detroit product and very few imports get such attention. A classic-car enthusiast whose everyday transportation is either a 1937 Aston Martin or a 4.3 Alvis of the same vintage says, "If I *had* to have a modern car, this would be it."

I would sum it up by saying that the car is "conspicuously inconspicuous," the very essence of snob appeal. Perhaps you remember the ad of the Thirties showing a man on the stormswept deck of a "J" boat: "HE drives a Duesenberg." Today, I think there's a good chance he'd drive a Jaguar sedan.

So far as driving is concerned, I've driven few cars I liked better. Steering is light and precise, the brakes are superb and the much-maligned gearbox isn't *that* bad. The ride is firm in the best European tradition and there is a genuine feeling of luxury with the good British hide upholstery and honest walnut trim. It won't beat a U.S.-built Musclecar away from a stoplight but I can think of almost no car better suited for long-distance, high-speed cruising and for all-around driving on anything and everything from turnpikes to country lanes.

I drive the car as I think it is meant to be driven, winding up in the gears and blowing the cobwebs out whenever there is the opportunity. I live and practice in a suburb and all but perhaps 2000 miles have been within 20 miles of home—city, suburbs, stop-and-go, go-anywhere-anytime. It has been maintained by the book but not pampered.

Its greatest shortcoming, from the driver's point of view,

CONTINUED ON PAGE 153

JAGUAR 3.8 SEDAN

**Overall Cost per Mile
for 30,000 miles**

Delivered price	$5538
Gasoline, oil & other "service station" bills	913
Tires (see text)	346
Maintenance & repairs (see text)	1090
Licensing, 2 years @ $10	20
Insurance	353
Regular washing and polishing	130
Total expenditure in 30,000 mi.	$8390
Retail value at end of period	2500
Cost of driving 30,000 mi.	$5890
Overall cost per mile	19.6¢

JAGUAR RANGE REVISED
SMALLER MODELS IMPROVED AND RENAMED—MORE POWER FOR 2·4-LITRE

'68 MODELS

Only the chrome lettering on the boot lid now distinguishes externally between 240 and 340, but the new slim-line bumpers immediately identify the '68 mode

Mark 2 2·4-litre gets SU carburettors and becomes 240. Improved 3·4 becomes 340 and Mark 2 3·8-litre is discontinued. S-types cheaper.

IN their programme for 1968, Jaguar have revised specifications and prices to produce a more even spread of models spanning a range of tax-paid U.K. prices extending from £1,364 (for the 240 in basic form) to the unchanged £2,577 price of the 420G with automatic transmission. No changes are made to any models with 4·2-litre engines—420, 420G and E-type, except for minor revisions on the E-type which are already appearing on the production cars. The chief change here is the removal of the glass fairings over the headlamps to give improved night driving illumination. A road test of the latest E-type is under way, for publication in the 12 October issue. The chief alterations this year concern the 2·4-litre saloon, which is now renamed the 240.

In place of the former B-type cylinder head with its twin Solex carburettors the 2·4 engine is now fitted with the same straight-port-type cylinder head as that of the E-type and a new water jacketed inlet manifold carries twin HS6 1¾in. SU carburettors. Previously, a 2·4-litre Jaguar could be identified quickly from the rear by its single exhaust pipe, but the 240 now has a dual exhaust system as on the rest of the Jaguar range. The induction and head changes together raise the maximum power output from 120 to 133 b.h.p. (gross), now developed at 5,500 r.p.m. instead of 5,750 r.p.m. Torque is increased from 138 to 146 lb. ft, which also comes appreciably higher in the engine speed range, at 3,700 instead of 3,000 r.p.m.

Under the bonnet, the 2·4-litre engine can still be immediately identified from that of a 3·4-litre by its much lower height, and in place of the big pancake-shaped air filter of the larger engine is a very neat rectangular filter across the top of the engine, with twin inlet pipes, and paper element filter, Previously, the 2·4-litre had an oil bath air filter.

Engine dimensions for both the 240 and 3·4-litre (which now becomes the 340) are unchanged, but a detail refinement is a new dual-purpose thermostat, which closes off the radiator bypass when it opens, thus allowing the full flow from the water pump to pass through the radiator. This is an ingenious way of increasing radiator efficiency, without altering the size.

Buyers who specify automatic transmission will be pleased to know that the old design of Borg-Warner DG transmission has been replaced by the better and more modern Type 35. Although automatic transmission used to be listed as an extra for the 2·4-litre, very few were made; but for the 240, automatic transmission will be readily available. Standard transmission will, of course, be the all synchomesh 4-speed gearbox with Laycock overdrive at extra cost.

Marles Varamatic power-assisted steering, previously available only on 4·2-litre models, is now offered as an extra for the 340; power-

Familar outline but with the difference of unusually short cylinders; in fact, of course, it is the 2·4-litre engine with the new cylinder head, manifolding and SU carburettors

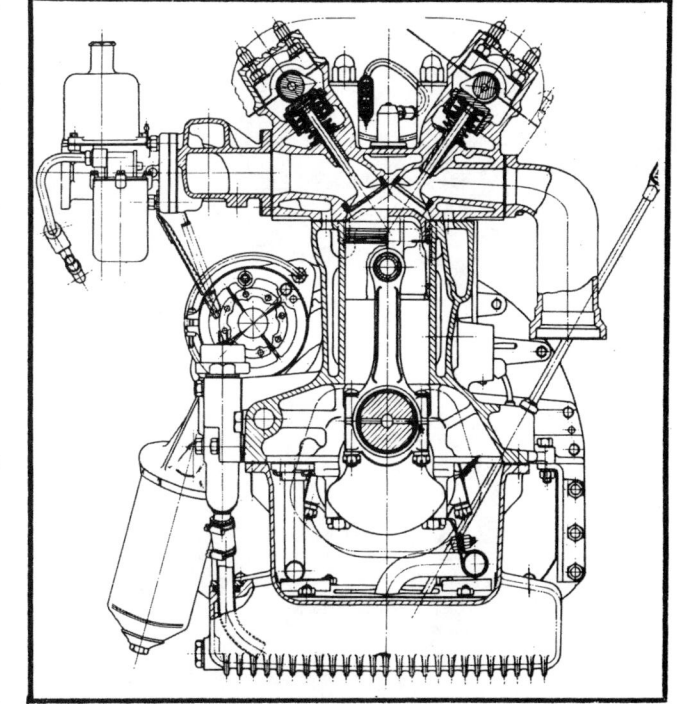

	Basic £	Purchase Tax £ s d	Total (in G.B.) £ s d
240	1,109	255 17 4	1,364 17 4
Overdrive	1,155	266 8 2	1,421 8 2
Automatic	1,186	273 10 3	1,459 10 3
340	1,172	270 6 1	1,422 6 1
Overdrive	1,218	280 16 11	1,498 16 11
Automatic	1,249	287 19 0	1,536 19 0
3·4-Litre S-Type	1,334	307 8 7	1,641 8 7
Overdrive	1,380	317 19 5	1,697 19 5
Automatic	1,413	325 10 8	1,738 10 8
3·8-Litre S-Type	1,415	325 19 10	1,740 19 10
Overdrive	1,461	336 10 8	1,787 10 8
Automatic	1,492	343 12 9	1,835 12 9

assisted steering is not available for the 240. Servicing intervals on both the 240 and 340 are extended from 2,500 miles to 3,000. Both these revised models are readily recognised by the slim-line bumpers now fitted.

Prices of the 340 are unchanged except for the automatic transmission model, which is nearly £40 cheaper. The 240 now costs £23 more than the previous 2·4-litre, but again the saving on automatic transmission makes the price for the 240 automatic £16 less than was previously listed.

S-types

Although the 3·8-litre engine is no longer available for what used to be the Mark 2 range, Jaguar still offer the choice of 3·4- or 3·8-litre engine for the S-type. The price gap between the S-types and the 420 models has been extended by making the same cost savings on the S-type as were introduced earlier on the smaller models. So S-types now have Ambla upholstery, tufted carpets, and chromed grilles fitted in place of the fog lamps which used to be standard. The revised price structure now shows a saving of more than £100 for the 3·4 or 3·8 S-types with manual transmission, and for those who specify automatic, here again the Borg-Warner 35 box is used instead of the old model DG; and in this form the car is more than £140 cheaper than before.

The new cylinder head for the 240 is "borrowed" from the E-type; and the graph below shows how the engine improvements have increased efficiency. Below, left: S-type in its 1968 form, with chrome grilles in place of fog lamps

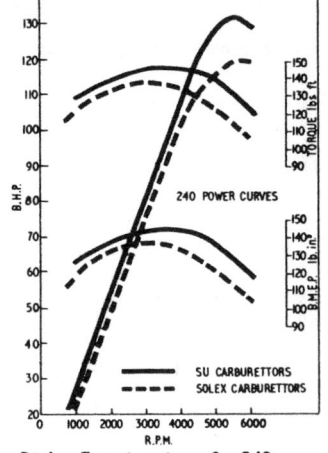

Transverse side-entry air filter box with paper element does not obscure the beautiful appearance of the 240 engine. Right: Exterior view of a 340, showing restyled wheel nave plates, common to 240 as well

ENGINE SPECIFICATION

	240	340
Cylinders	6, in line	
Cooling system	Water; pump, fan and bypass-closing thermostat	
Bore	83mm (3·27in.)	83mm (3·27in.)
Stroke	76·5mm (3·01in.)	106mm (4·17in.)
Displacement	2,483 cc (151·5 cu. in.)	3,422 cc (210 cu. in.)
Valve gear	Twin o.h.c.	
Compression ratio	8·0 to 1	
	Min. octane rating 97 RM	
Fuel pump	SU electric	
Oil filter	Tecalemit full-flow, renewable element	
Max. power	133 b.h.p. (gross) at 5,500 r.p.m.	210 b.h.p. (gross) at 5,500 r.p.m.
Max. torque	146 lb. ft. at 3,700 r.p.m.	215 lb. ft. at 3,000 r.p.m.

Autocar Road test
NUMBER 2165

JAGUAR 240 2,483 c.c.

AT-A-GLANCE: Cheapest model in Jaguar range with short-stroke, twin-cam engine. Top speed well over 100 mph (with overdrive) and brisk acceleration. Excellent all-synchromesh gearbox. Powerful all-disc brakes; no fade. Good stability and plenty of grip from optional radial tyres. Quiet car except at high revs. Remarkable value.

MANUFACTURER:
Jaguar Cars Ltd., Brown's Lane, Coventry.

PRICES:
Basic	£1,109	0s	0d
Purchase Tax	£255	17s	4d
Inertia reel seat belts (pair)	£15	9s	9d
Total (in G.B.)	£1,380	7s	1d

EXTRAS (inc. P.T.):
Overdrive	£56	10s	10d
Dunlop SP41 tyres	£12	12s	4d
Heated rear window	£18	8s	9d
Reclining seat kit (fitting extra)	£7	4s	0d
Radio with remote control aerial	£44	10s	10d

PERFORMANCE SUMMARY:
Mean maximum speed	106 mph
Standing start ¼-mile	18.7 sec
0-60 mph	12.5 sec
30-70 mph (through gears)	12.3 sec
Typical fuel consumption	20 mpg
Miles per tankful	240

IT is now over 12 years since the small-engined Jaguar 2.4 was announced, and it is almost as long since we carried out a full road test on one. Although our experience of other Jaguar models, both for test and as staff cars, has not been lacking over the years, this "junior" version of the Mk. 2 has been tested only in the "Used Car" series. Last autumn it was revised to become the 240, and as such it should enjoy a long new lease of life. Outwardly it may not look all that different from the original, but in the way it performs it is bang up to date with only a few detail points that fall short of 1968 standards.

Both the character of the cars and the way they are promoted put the Jaguar accent on performance. Although the 240 uses the same famous XK twin-cam engine as all the rest, it does have a reduced stroke and a lot of weight to carry around. Gross power output has risen from 112 bhp in 1955 to 133 bhp for the latest version, but with two-up and ready for the road the total weight is now over 32 cwt. What were once very brisk acceleration and a high top speed, can now be matched by several cars with engines of under 2 litres. Nevertheless, from behind the wheel there is never a moment's doubt that this is a full-blooded Jaguar with all the virtues of a thoroughbred.

Comparing figures with our test of the 2.4 Mk. 1 published on 21 September 1956, one can see a big margin of improvement. Top speed is up from 102.5 to 106 mph and almost 8 sec have been pared off the 0 to 80 mph acceleration time. Flexibility has not suffered at the expense of top-end power though, and the

There is an armrest on each door as well as in the centre of the back seat. The front seats are shown here in their most forward position

JAGUAR 240 . . .

240 can still accelerate smoothly from 10 to 30 mph in overdrive top in 12.7 sec (12.4 sec for the 1956 2.4). From 60 to 80 mph, again in overdrive, now takes only 14.1 instead of 21.5 sec, and it is here that the 240 engine is so much better than the original. The earlier 2.4 would run out of breath above 80 mph and it took a long run to reach maximum speed; by comparison the 240 spurts up to—and cruises at—100 mph with admirable ease and lack of fuss.

Of course, extra performance with extra weight can only mean extra fuel consumption and the 240 barely covers 20 miles on a gallon of premium. Using the top end of the engine rev range soon reduces this to around 18 mpg, although with a light foot we did manage 24 mpg for a 100-mile trip during the test. Cruising at a steady 70 mph in overdrive returns 21.6 mpg, so our overall figure probably reflects our driving style more than usual.

One of the latest changes has been from twin Solex to twin SU carburettors. During the test period we experienced everything from snow storms to an unusually mild spell, but never had a moment of starting trouble. There is a manual rich-mixture control worked by a knob sliding in a vertical quadrant on the right of the facia; it has a bright red warning lamp above it which lights as soon as it is moved from the off position. In cold weather the engine takes a long time to warm up and there is tendency to stall if the choke is pushed off before the temperature gauge has reached 70 deg.C. Normally the needle remains steady at 75 deg. Similarly, the heater takes a very long time to produce hot air after a cold start.

Probably to keep the temperature down in hot climates, the cooling fan is very powerful. It would be better both for warming up and to reduce the noise level if some alternative system of driving it were used instead of a fixed pulley. Above about 4,000 rpm a lot of fan whirr intrudes inside the car, which is annoying to a driver who likes to use the gearbox and make the most of the performance.

Otherwise the 240 is a very quiet car. The engine emits only a very faint whine from its camshaft chains and virtually no mechanical clatter. Induction and exhaust are well silenced without even a faint throb. Each of the indirect gears has its own characteristic whine too, but these are all subdued and introduce a pleasant note to the mechanical ear. Wind noise at speed is not offensive and mostly stems from the front quarterlights. Tyre squeal is now a thing of the past (it was a different story in 1956) nor was there any roar from the optional Dunlop SP 41 radials fitted to the test car. On some sharp irregularities there is the characteristic thump of radials, but Jaguar have done a good job in getting the rubber suspension bushes to absorb most of this. Without testing the adhesion, some of our drivers were not sure they were on radials until they looked, which is a big compliment.

It is only recently that Jaguar have approved ▷

The 240 engine now has twin SU carburettors and the cam covers are ribbed in polished aluminium and black crackle. The battery fillers can be reached through the hinge of the bonnet

Autocar road test Number 2165 Make: Jaguar Type: 240 (2,483 c.c.)

TEST CONDITIONS: Weather: Cloudy. Wind: 0-7 mph. Temperature: 8 deg. C. (47 deg. F.). Barometer: 29.65 in. Hg. Humidity: 70 per cent. Surfaces: Damp, concrete and asphalt; dry for brake tests.

WEIGHT: Kerb weight 29.4 cwt (3,296lb—1,494kg) (with oil, water and half-full fuel tank). Distribution, per cent: F, 54.8; R, 44.2. Laden as tested: 32.8 cwt (3,674lb—1,667kg).

Figures taken at 5,800 miles by our own staff at the Motor Industry Research Association proving ground at Nuneaton.

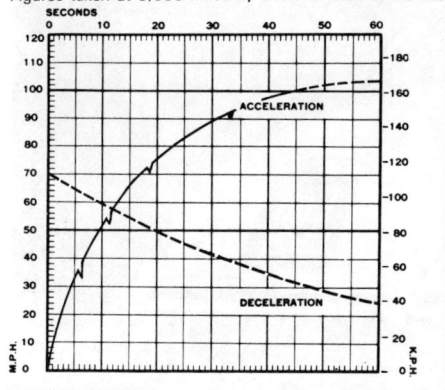

MAXIMUM SPEEDS

Gear	Mph	kph	rpm
O.D. Top (mean)	106	170	4,860
(best)	107	172	4,910
Top	93	150	5,500
3rd	73	118	5,500
2nd	53	85	5,500
1st	35	35	5,500

Standing ¼-mile 18.7 sec 74 mph
Standing Kilometre 34.7 sec 93 mph

TIME IN SECONDS	4.1	6.3	9.3	12.5	16.4	22.8	31.0	44.8
TRUE SPEED MPH	0	30	40	50	60	70	80	90	100
INDICATED SPEED		31	41	52	63	74	85	95	106

Mileage recorder 1.5 per cent over-reading
Test distance 1,274 miles.

Speed range, gear ratios and time in seconds

mph	O.D. Top (3.54)	Top (4.55)	3rd (5.78)	2nd (7.92)	1st (12.19)
10-30	12.7	8.7	6.7	4.5	3.3
20-40	11.2	8.3	6.1	4.3	—
30-50	12.1	8.3	6.3	—	—
40-60	12.9	8.4	6.6	—	—
50-70	14.0	8.7	7.2	—	—
60-80	14.1	11.5	—	—	—
70-90	22.4	14.1	—	—	—
80-100	35.7	—	—	—	—

FUEL CONSUMPTION

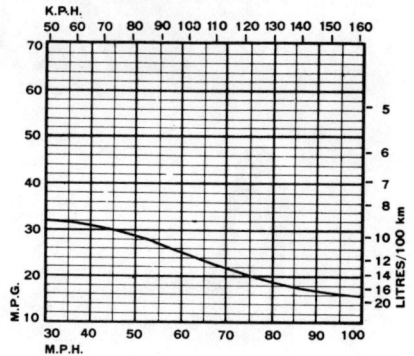

(At constant speeds in OD top—mpg)
- 30 mph 32.2
- 40 .... 31.7
- 50 .... 28.8
- 60 .... 25.0
- 70 .... 21.6
- 80 .... 19.1
- 90 .... 16.7
- 100 .... 14.2

Typical mpg 20 (14.1 litres/100km)
Calculated (DIN) mpg .. 19.6 (14.4 litres/100km)
Overall mpg 18.4 (15.4 litres/100km)
Grade of fuel .... Premium, 4-star (min. 97RM)

OIL CONSUMPTION
Miles per pint (SAE 10W/30)......... 320

BRAKES (from 30 mph in neutral)

Load lb	g	Distance ft
25	0.25	120
50	0.47	64
75	0.68	44
90	1.02	29.5
Handbrake	0.35	86

Max. Gradient: 1 in 3

Clutch Pedal: 30lb and 5.5in.

TURNING CIRCLES
Between kerbs   L, 37ft 10in.; R, 36ft 7in.
Between walls   L, 39ft 9in.; R, 38ft 7in.
Steering wheel turns, lock to lock: 4.75

HOW THE CAR COMPARES:

Maximum Speed (mean) mph
- Jaguar 240
- BMW 2000
- Fiat 2300
- Rover 2000 TC
- Triumph 2000

0-60 mph (sec)
- Jaguar 240
- BMW 2000
- Fiat 2300
- Rover 2000 TC
- Triumph 2000

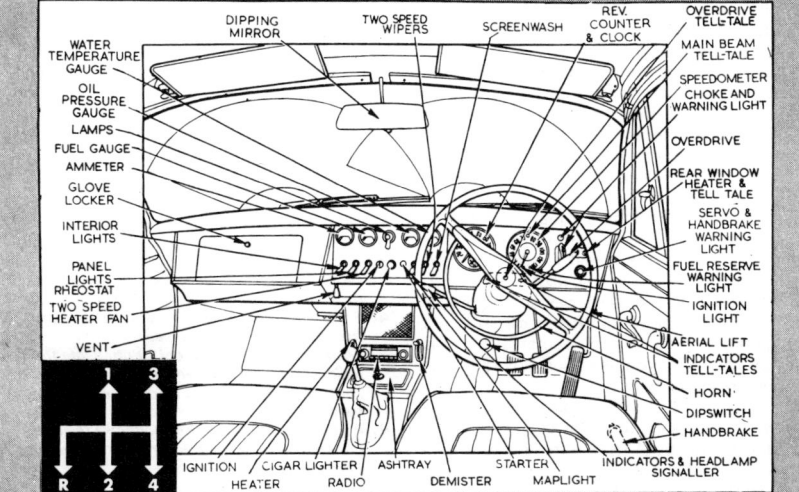

Standing Start ¼-mile (sec)
- Jaguar 240
- BMW 2000
- Fiat 2300
- Rover 2000 TC
- Triumph 2000

MPG Overall
- Jaguar 240
- BMW 2000
- Fiat 2300
- Rover 2000 TC
- Triumph 2000

PRICES

Jaguar 240	£1,365
BMW 2000	£1,778
Fiat 2300	£1,354
Rover 2000TC	£1,415
Triumph 2000	£1,198

On the latest models the bumpers are much slimmer and the 240 has the same dual exhaust system as the 340. Rear quarterlights open to improve the ventilation. The petrol filler cannot be locked

JAGUAR 240 . . .

the fitment of radial tyres to the Mk. 2 models, and they are a very cheap extra. Unlike on some of the more powerful Jaguars there is no limited-slip differential with the 240, yet it is only by vicious use of the clutch with high revs that the back wheels can be made to spin on wet roads and it is impossible to kick the tail round under power.

The steering is heavy *and* low geared, which makes parking in tight spaces very wearisome. There are 4¾ turns between locks on a 37ft. dia. turning circle; power assistance is not available on the 240. Despite its gearing and a small amount of lost motion, the response of the recirculating ball system is quick and very positive at speed. The marked nose heaviness (nearly 17 cwt on the front alone) makes it hard to turn the wheels, but it gives excellent stability and in a very gusty cross-wind on M1 the 240 was much less deflected than most other cars. More than this, we found we could relax at the wheel in these conditions, in distinct contrast to our experience in a lighter car which we had driven over the same route earlier that day.

With this weight distribution one expects the 240 to understeer and it does, more and more the faster one corners. On the MIRA track in the wet it was the front end which eventually lost adhesion first, and it was only in a very tight turn at high revs in second that we managed to get any other behaviour to show itself. Here, just as the tail tried to swing out, the inside rear wheel lifted enough to spin and so cut the speed back to the steady understeering state. All these exploits were at speeds far above the comfort limit for normal use on the road, but even under these extremes such as might occur in a rare emergency, the 240 did nothing that was not progressive and entirely predictable with no loss of control.

Brakes and braking systems have come a long way since the early 2.4 Jaguars appeared, with their servo-assisted drums front and rear. Apart from fade (which we did not measure in 1956) the responsiveness and pedal effort have been improved out of all recognition. The 240 has large discs all round backed up by a powerful vacuum servo. It took 95 lb on the pedal to give a better-than 1g stop with the wheels just on the point of locking; in 1956, 100 lb gave only 0.7g and that was the best measurement recorded. After 10 stops from 70 mph in quick succession, the brakes were behaving exactly as at the beginning, although they did fluctuate a little in between.

The handbrake likewise has been considerably improved since early examples of those working on rear discs, and it now holds the car very securely facing either way on a 1-in-3 hill. Used on its own from 30 mph it can easily lock the rear wheels. Unfortunately, the lever is difficult to get at on the right of the driving seat because of the armrest on the door, and several times when getting out we accidentally kicked the release button with our heel; a release guard is urgently needed and perhaps the armrest could be moved back, or contoured to be less obstructive. A bright red warning on the facia lights when the handbrake is on, or when there is a drop in the hydraulic level.

Because of its long movement (5½in.) the clutch feels heavier than it is. In fact it takes only 30 lb to release and it is a man-sized unit well able to take the full torque of the engine without excessive slip. We restarted on the 1-in-3 once by keeping the revs up and deliberately easing the clutch, and again by letting the clutch in quickly and allowing the engine to slog.

Probably the one feature most criticized on early Jaguars was the gearbox. All that has changed now and the latest all-synchromesh unit is hard to fault on any score. The lever, with its large knob, is ideally placed and moves sweetly between all ratios. If the clutch is not quite floored, then the movements become stiffer as the synchromesh works harder, but we never crashed a gear.

The test car had the optional Laycock overdrive which works on top gear only. It cuts the revs at 80 mph by 1,000 rpm and forms an essential part of the 240's high-speed cruising ability. Without overdrive the rev counter would be in the red at only 96 mph, even allowing for the slightly higher final drive ratio when overdrive is not specified. A lever on the steering column is used to engage and disengage overdrive, and with the accelerator wide open the changes are smooth enough. On a trailing throttle the take-up—particularly down from overdrive to direct—is jerky unless the clutch is used. A red tell-tale lights up in an indicator on top of the column when overdrive is engaged; this would be better in another colour (red should be reserved for danger signs) as it is the same size as, and very near, the ignition warning lamp. During the test an oil seal failed in the overdrive and we lost all the oil without any signs of trouble until severe vibration indicated damage.

Inside there have been a lot of small changes since that first Mk 1 2.4. The rev counter and speedometer are now directly in front of the driver, with the supplementary gauges and all the switches grouped in the centre—standard Jaguar layout. The seats now have Ambla covering and adjustable backrests are another cheap extra. But all the polished wood veneer and the cloth headlining are still there to delight the traditionalist. Ignition key and starter button continue to be separate. By modern standards the areas of glass look rather small and the windscreen seems shallow, especially when peering through the wiper arcs on a dirty

1. *The clarity of the instrument markings is an example to others. There is no crash padding*
2. *Only the 240 badge of the boot lid gives the game away. The crest on the radiator still reads 2.4.*
3. *Auxiliary lamps are no longer fitted*
4. *Back seats are well shaped and the armrests hold the passengers snugly in place*
5. *There is a well-stocked tool kit fitted into the hollow of the spare wheel under the mat in the boot. The low sill makes loading heavy luggage easy*
6. *A guard is needed for the handbrake release button. The other short lever is for the reclining backrest (a £7 extra)*

1 2
3 4
5 6

winter's day.

It is not difficult for any size of driver to get comfortable because the steering column adjusts for length and the seat cushion is high enough to give a good view. The nearside front wing is just out of sight, but the leaping Jaguar mascot forms a useful aiming point. In the back there is enough room for legs, but nowhere to stretch. The ride is soft and accommodating without plunge and wallow, and without those short, sharp jogs from which many cars suffer. Seat springs seem well tuned to the suspension.

In 1956 we summed up the 2.4 by saying it was one of those cars whose capabilities are appreciated as the mileage mounts. The latest 240 is still such a car, rather ponderous around town perhaps, but superb on the open road and ideal for journeys. As far as value goes, it offers much more today to a standard higher than ever. The basic list price now is only £133 more than it was 12 years ago, which is a remarkable achievement considering how costs in every other field have escalated so much more.

SPECIFICATION: JAGUAR 240 (FRONT ENGINE, REAR-WHEEL DRIVE)

ENGINE
Cylinders	6, in line
Cooling system	Water; pump, fan and thermostat
Bore	83mm (3.27in.)
Stroke	76.5mm (3.01in.)
Displacement	2,483 c.c. (151.5 cu. in.)
Valve gear	Twin overhead camshafts
Compression ratio	8-to-1: Min. octane rating: 97RM
Carburettors	2 SU HS6
Fuel pump	SU electric
Oil filter	Tecalemit full flow, renewable element
Max. power	133 bhp (gross) at 5,500 rpm
Max. torque	146 lb. ft. (gross) at 3,700 rpm

TRANSMISSION
Clutch	Borg and Beck diaphragm spring single dry plate 9in. dia.
Gearbox	4-speed, all synchromesh; overdrive on Top
Gear ratios	Top 1.0 OD Top 0.78
	Third 1.27
	Second 1.74
	First 2.68
	Reverse 2.68
Final drive	Hypoid bevel, 4.55-to-1 (4.27-to-1 without overdrive)

CHASSIS and BODY
Construction	Integral, with steel body

SUSPENSION
Front	Independent, wishbones, coil springs, telescopic dampers
Rear	Live axle, radius arms, Panhard rod, cantilever half-elliptic leaf springs, telescopic dampers

STEERING
Type	Burman recirculating ball
Wheel dia.	16.8in.

BRAKES
Make and type	Dunlop discs front and rear
Servo	Lockheed vacuum
Dimensions	F. 11in. dia.; R. 11.37in. dia.
Swept area	F. 242.0 sq.in.; R. 252 sq.in. Total 494.0 sq.in. (301.0 sq.in./ton laden)

WHEELS
Type	Pressed steel disc, 5-stud fixing 5in. wide rim.
Tyres—make	Dunlop
—type	SP41 radial-ply tubed
—size	185-15in.

EQUIPMENT
Battery	12-volt 51-Ah
Generator	Lucas C40L 25-amp d.c.
Headlamps	Lucas sealed filament 150/90-watt (total)
Reversing lamp	Standard
Electric fuses	8
Screen wipers	Two-speed, self-parking
Screen washer	Standard, electric
Interior heater	Standard, water valve type
Heated backlight	Extra
Safety belts	Extra, anchorages built-in
Interior trim	Ambla seats, cloth headlining
Floor covering	Carpet
Starting handle	No provision
Jack	Screw pillar
Jacking points	2 each side, under sills
Windscreen	Zone toughened
Underbody protection	Rubberised compound on all surfaces exposed to road

MAINTENANCE
Fuel tank	12 Imp. gallons (no reserve) (54.6 litres)
Cooling system	20 pints (including heater)
Engine sump	13 pints (7.4 litres) SAE 10W/30. Change oil every 3,000 miles. Change filter element every 3,000 miles
Gearbox and overdrive	4 pints SAE 90EP. Change oil every 12,000 miles
Final drive	2.75 pints SAE 90EP. Change oil every 12,000 miles
Grease	10 points every 6,000 miles
Tyre pressures	F. 30; R. 27 p.s.i. (normal driving)
	F. 33; R. 30 p.s.i. (fast driving)
	F. 33; R. 30 p.s.i. (full load)

PERFORMANCE DATA
Top gear mph per 1,000 rpm	16.95
Overdrive top mph per 1,000 rpm	21.8
Mean piston speed at max power	2,760 ft./min.
Bhp per ton laden (gross)	81.0

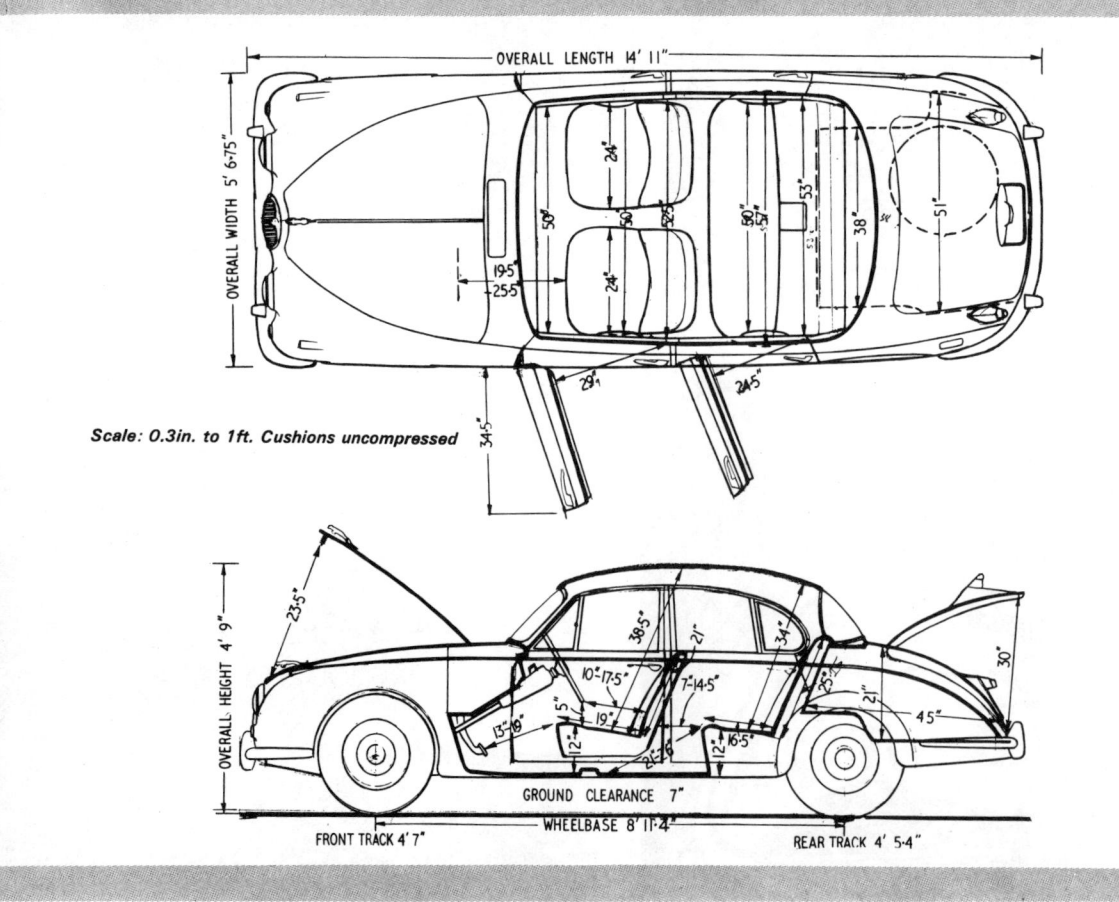

Scale: 0.3in. to 1ft. Cushions uncompressed

MOTOR ROAD TEST No. 4/68 ● Jaguar 240

Value for money

Smooth, quiet engine and excellent performance. Very good roadholding and adequate ride. Poor fuel consumption and heavy steering at parking speeds.

So many crises have beset the allegedly foundering British economy during the past couple of years that a real effort is needed to recall the name of the first of them all: The Squeeze. It was The Squeeze that forced Jaguar to take a second look at what was (relatively speaking) the Cinderella of their range, the 2.4 model, since company directors were finding it much easier to produce the £1,300 or so it cost than the £1,700 to £1,900 for a larger model. This smallest-engined car in the Jaguar line-up had sold steadily over the years but with its exploits largely unsung—the present test, for example, is our first since the original which appeared in 1956, soon after the car was introduced. Jaguar's slight shift in promotional emphasis culminated in the announcement at last year's London Show of a new version of the 2.4, called the 240, costing £23 more and incorporating both improvements and economies. Fitted with a cylinder head similar to that of the E-type, and larger, 1¾-in. carburetters, the revised model now develops 133 (gross) b.h.p. at 5,500 r.p.m. instead of 120 b.h.p. at 5,750 r.p.m., while maximum torque is up from 138 lb.ft. at 3,000 r.p.m. to 146 lb.ft. at 3,700. Other changes include twin pipes for the exhaust system, a new type of wax thermostat and, externally, the "slimline" bumpers introduced on the S-type.

Price: £1,109 plus £255 17s 4d tax equals £1,364 17s 4d. Heated backlight £18 8s 9d with tax, radial ply (Dunlop SP 41) tyres £12 12s extra with tax, seatbelts £9 16s 8d with tax; total as tested £1,405 14s 9d.

It is a pity that the praises of the 2.4/240 have been a trifle neglected in the past because the car has much to offer. To begin with, its extremely smooth engine is derived from the larger one by using a short throw crankshaft so that it can operate at high r.p.m. at lower stresses. Next there is the price, £1,365 for the standard car without overdrive (as tested), and what you get for it. Almost the only cost-cutting expedient adopted in the latest face-lift—and one which few owners are likely to notice—is the use of Ambla leathercloth for the upholstery in place of real leather: in all other respects the 240 remains luxurious and very well equipped.

If this is Jaguar's economy package it manages to be both jumbo-sized and lavish. It has the usual elegant and well-placed set of instruments, rows of impressive-looking switches, bags of walnut (which some of us don't like very much) and a toolkit which has more tools in it than are to be found in some complete households. More important, the 240 is surprisingly wieldy on twisting roads, has excellent handling and adhesion, and will carry four people in comfort and quietness at high speeds over long distances. Always provided that the filling stations are not too far apart—the overall fuel consumption of 17.1 m.p.g. is one of the few drawbacks. Another is heavy steering at parking speeds. These faults are to some extent offset by the patrictic feeling (experienced by more than one of our testers) engendered by looking down that long bonnet from the driver's seat at the Jaguar mascot, and by the sensation of being encased in a great deal of solidly-constructed motorcar.

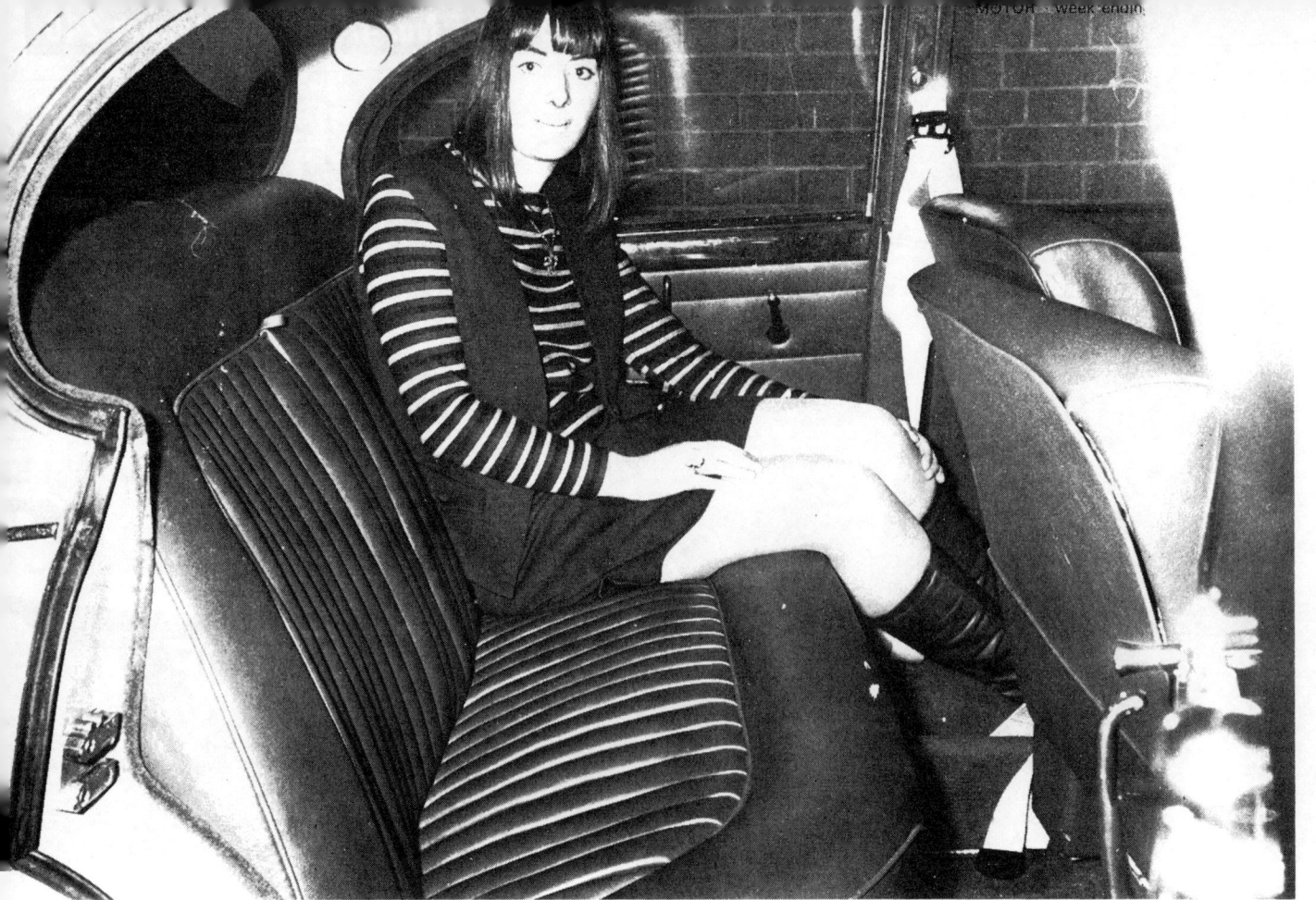

The rear seat is comfortable with adequate legroom.

Performance

Performance tests carried out by *Motor's* staff at the Motor Industry Research Association proving ground, Lindley.
Test Data: World copyright reserved; no unauthorized reproduction in whole or in part.

Conditions:
Weather: Cool and dry.
(Temperature: 46°F. Barometer: 29.1 in. Hg.)
Surface: Dry concrete and tarmacadam.
Fuel: Premium 98 octane (R.M.), 4 star rating.

Maximum speeds
	m.p.h.
Direct top gear	98.0
3rd gear	74.3
2nd gear } at 5,500 r.p.m.	50.0
1st gear	32.5
"Maximile" speed: (Timed quarter mile after 1 mile accelerating from rest)	
Mean } see text	104.8
Best	105.8

Acceleration times
m.p.h.	sec.
0-30	3.6
0-40	5.8
0-50	8.0
0-60	11.7
0-70	15.4
0-80	20.5
0-90	28.5
0-100	40.1
Standing quarter mile	18.4

m.p.h.	Top sec.	3rd sec.
10-30	—	6.4
20-40	8.8	6.5
30-50	9.1	6.7
40-60	9.8	7.2
50-70	10.7	7.4
60-80	11.2	7.8
70-90	14.0	—

Brakes
Pedal pressure, deceleration and equivalent stopping distance from 30 m.p.h.
lb.	g	ft.
25	0.30	100
50	0.63	47½
75	0.92	33
100	0.98	31
Handbrake	0.38	81

Fade test
20 stops at ½g deceleration at 1 min. intervals from a speed midway between 30 m.p.h. and maximum speed (=64 m.p.h.)

	lb.
Pedal force at beginning	35
Pedal force at 10th stop	35
Pedal force at 20th stop	35

MAXIMUM SPEED / ACCELERATION

- Jaguar 240 — £1,363
- Ford Zodiac Mk IV — £1,282
- BMW 1600 coupe — £1,298
- Lancia Fulvia sedan GT 1.2 — £1,298
- Volvo 144S — £1,415
- Rover 2000 — £1,357

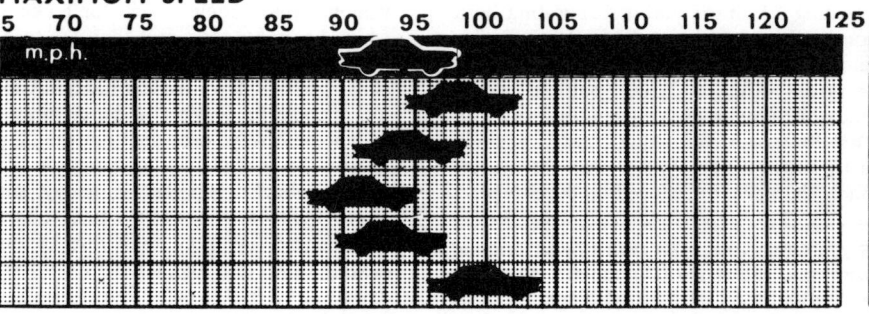

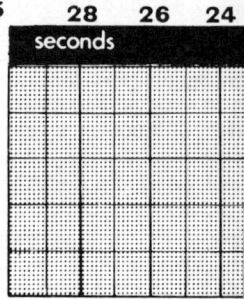

Jaguar 240 continued

The front seats have a good range of fore-and-aft adjustment and the steering wheel is adjustable for reach. Note the fire extinguisher (an optional extra) mounted at the front of the driver's seat.

Performance and economy

One of the many pleasantly old-fashioned features of the Jaguar is its choke, which is described in the handbook as a "Mixture Control" and consists of a lever moving in a quadrant on the facia marked "Start" at the top and "Run" at the bottom. With the lever pressed fully upwards—an extra resistance is encountered halfway along when moving from the fast idle condition to actual operation of the choke—and a touch on the separate starter button, the engine always sprang instantly to life in the coldest of weather. The choke could then be returned almost at once to the fast idle position and the engine was ready to pull cleanly without hesitation although sudden opening of the throttle caused spitting back through the carburetters.

The power unit was outstandingly smooth and quiet, better in this respect than most other Jaguars we have tested over the past few years. A subdued breathy hum is its normal song, changing to a satisfying muffled growl if the throttle is opened at, say, 50 m.p.h. or so in top. Above about 3,500 r.p.m. a whine from the camshaft chains becomes apparent when the car is being accelerated hard through the gears. Flexibility is another outstanding feature of the engine, which will pull smoothly from little more than 10 m.p.h. in top.

Performance is good: 60 m.p.h. is reached from a standstill in 11.7 seconds, and the 30-50 m.p.h. time in top gear is 9.1s. On our test car with no overdrive and the standard 4.27:1 final drive ratio (4.55:1 with overdrive) the engine reached its red-line limit of 5,500 r.p.m. at 98 m.p.h. in top gear—our Maximile figures were obtained by considerable over-revving, albeit for no more than a few seconds. The resultant combination of acceleration with effortless cruising is ideal for speed-limited Britain or the United

Continued on the next page

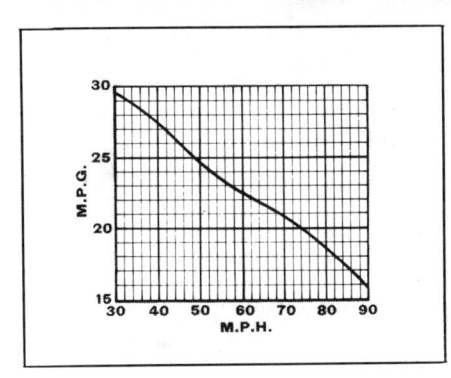

Overall . . . . 17.1 m.p.g.
(=16.5 litres/100 km.)
Total test mileage . . . . 1,820 miles
Tank capacity (maker's figure) . . . . . 12 gal.

Fuel consumption
Touring (consumption midway between 30 m.p.h. and maximum less 5% allowance for acceleration) . . . . 20.7 m.p.g.

Steering
Turning circle between kerbs: ft.
Left 34
Right . . . . $35\frac{1}{2}$
Turns of steering wheel from lock to lock . . 4.9
Steering wheel deflection for 50ft. diameter circle . . . . 1.5 turns

Weight
Kerb weight (unladen with fuel for approximately 50 miles) . . . . $28\frac{1}{4}$ cwt.
Front/rear distribution 57/43
Weight laden as tested . . . . 32.0 cwt.

Speedometer
Indicated	10	20	30	40	50	60	70	80	90
True	10	20	29	$38\frac{1}{2}$	48	$57\frac{1}{2}$	67	76	85

Distance recorder . . . . $1\frac{1}{4}$% fast

Clutch
Free pedal movement . . . . $\frac{3}{4}$in.
Additional movement to disengage clutch completely . . . . 2in.
Maximum pedal load . . . . 35lb.

Parkability
Gap needed to clear a 6ft. wide obstruction parked in front:

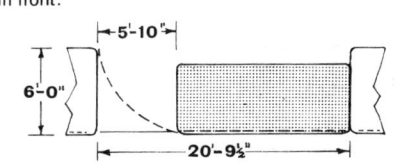

5'-10"
6'-0"
20'-9½"

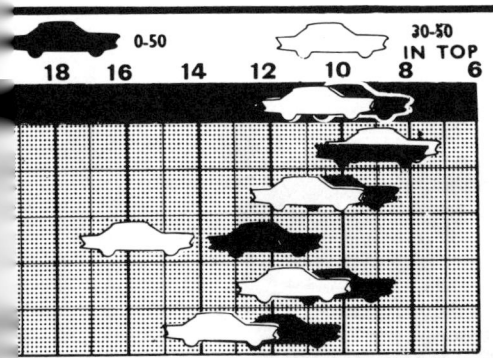

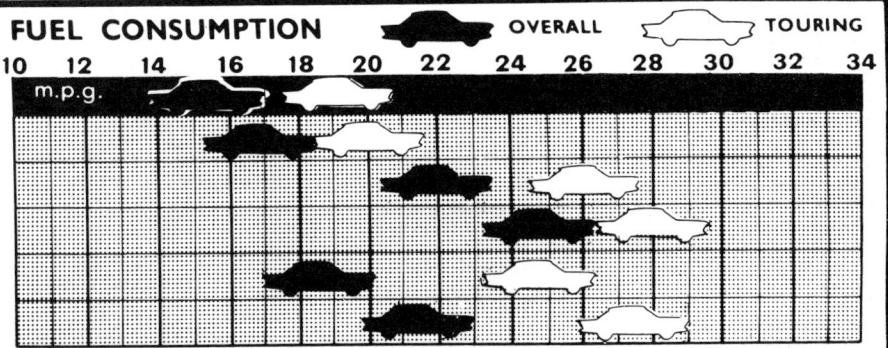

FUEL CONSUMPTION

0-50 | 30-50 IN TOP | OVERALL | TOURING

149

Jaguar 240 continued

States. Prospective owners intending to use their cars for long distance commuting on the Continent, however, would be well advised to invest in an overdrive (£56 10s. 10d. extra), especially as there is an injunction in the handbook not to exceed 5,000 r.p.m. for long periods without lifting off every now and then.

With a kerb weight of 28 cwt. the 240 is about half a ton heavier than many more recent designs offering the same passenger space and comfort; it also has an indifferent drag factor, so that the 2,483 c.c. engine must do the work that is now quite satisfactorily done by engines of 2-litres capacity or less. The inevitable penalty of such middle-aged corpulence is a high, 17.1

New "slimline" rear bumpers leave a certain amount of exposed skirt at the rear.

All four armrests are provided with ashtrays. Warm air for the rear passengers emerges from ducts on the transmission tunnel (partly covered by seatbelts).

m.p.g. overall fuel consumption (20.7 m.p.g., touring) whereas virtually all the 240's more modern competitors achieve an overall consumption above 20 m.p.g.; hard driving can reduce the fuel consumption to around 15 to 16 m.p.g.

Transmission

The pleasant gearchange has a longish travel but is light, with good synchromesh—except on second where it tended to be weak when the gearbox was hot. A long-travel clutch with all the bite at the top (not shown up by our test figures, which measure the

The boot has a flat floor and accepted 9.4 cu ft of our test boxes. The lavish Jaguar tool kit fits within the spare wheel.

Safety check list

Steering assembly
Steering box position	Under engine, protected
Steering column collapsible	Yes
Steering wheel boss padded	No
Steering wheel dished	No

Instrument panel
Projecting switches	Yes
Sharp cowls	No
Padding	None

Windscreen and visibility
Screen type	Zone toughened
Pillars padded	No
Standard driving mirrors	Interior only
Interior mirror framed	Yes
Interior mirror collapsible	No
Sun visors	Padded

Seats and harness
Attachment to floor	By slides
Do they tip forward?	No
Head rest attachment points	No
Back of front seats	Unpadded
Safety harness	Lap and diagonal
Harness anchors at back	No

Doors
Projecting handles	Yes
Anti-burst latches	No
Child-proof locks	No

14" x 11" x 5"
17½" x 13" x 6"
21" x 15" x 7"
24" x 18" x 8"

Specification

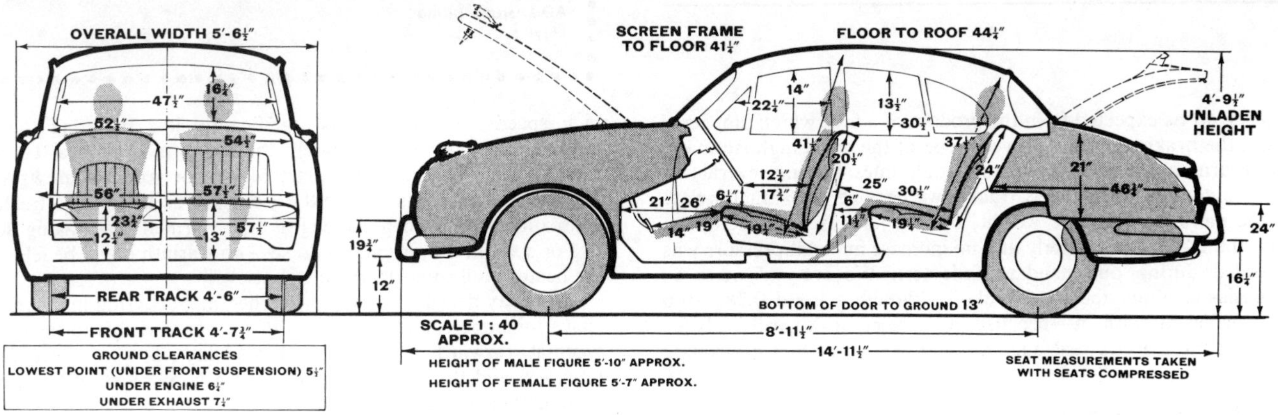

OVERALL WIDTH 5'-6½"
SCREEN FRAME TO FLOOR 41¼"
FLOOR TO ROOF 44¼"
4'-9½" UNLADEN HEIGHT
REAR TRACK 4'-6"
FRONT TRACK 4'-7¾"
BOTTOM OF DOOR TO GROUND 13"
SCALE 1 : 40 APPROX.
8'-11⅜"
14'-11½"
GROUND CLEARANCES
LOWEST POINT (UNDER FRONT SUSPENSION) 5½"
UNDER ENGINE 6½"
UNDER EXHAUST 7½"
HEIGHT OF MALE FIGURE 5'-10" APPROX.
HEIGHT OF FEMALE FIGURE 5'-7" APPROX.
SEAT MEASUREMENTS TAKEN WITH SEATS COMPRESSED

Engine

Cylinders	6
Bore and stroke	83 mm. x 76.5 mm.
Cubic capacity	2,483 c.c.
Valves	d.o.h.c.
Compression ratio	8:1
Carburetters	2 SU HS6
Fuel pump	SU electrical
Oil filter	Full flow
Max. power (gross)	133 b.h.p. at 5,500 r.p.m.
Max. torque (gross)	147 lb. ft. at 3,750 r.p.m.

Transmission

Clutch	Borg and Beck 9 in. s.d.p.
Top gear (s/m)	1.0:1
3rd gear (s/m)	1.33:1
2nd gear (s/m)	1.97:1
1st gear (s/m)	3.22:1
Reverse	3.50:1
Final drive	4.27:1 hypoid bevel
M.p.h. at 1,000 r.p.m. in:—	
Top gear	17.8
3rd gear	13.5
2nd gear	9.1
1st gear	5.9

Chassis

Construction	Unitary

Brakes

Type	Discs
Dimensions	Front 11 in. dia.; rear 11.4 in. dia.
Friction areas:	
Front	15.8 sq. in. of lining operating on 240.1 sq. in. swept area of disc
Rear:	15.8 sq. in. of lining operating on 252 sq. in. swept area of disc

Suspension and steering

Front	Independent by wishbones and coil springs with an anti-roll bar
Rear	Live axle on cantilevered semi-elliptic leaf springs with radius arms and a Panhard rod
Shock absorbers:	
Front / Rear	Girling telescopic
Steering gear	Burman recirculating ball
Tyres	185-15 Dunlop SP41
Rim size	5J

Coachwork and equipment

Starting handle	No
Jack	Screw pillar
Jacking points	Four
Battery	12-volt positive earth, 51 amp hrs. capacity
Number of electrical fuses	8
Indicators	Self-cancelling flashers
Screen wipers	Electric two-speed self-parking
Screen washers	Lucas electrical
Sun visors	Two
Locks:	
With ignition key	Doors
With other key	Glove compartment and boot
Interior heater	Fresh air
Extras	Automatic transmission, overdrive, reclining seats
Upholstery	Ambla
Floor covering	Carpet
Alternative body styles	None

Maintenance

Sump	13 pints SAE 10W-30
Gearbox	2½ pints SAE 90EP Shell Spirax
Rear axle	2¾ pints SAE90 hypoid
Steering gear	SAE 140 gear oil
Cooling system	20 pints (2 drain taps)
Chassis lubrication	Every 6,000 miles to 8 points
Minimum service interval	3,000 miles
Ignition timing	12° b.t.d.c.
Contact breaker gap	0.014-0.16 in.
Sparking plug gap	0.025 in.
Sparking plug type	Champion N5
Tappet clearances (cold)	Inlet 0.004 in.; Exhaust 0.006 in.
Valve timing:	
Inlet opens	15° b.t.d.c.
Inlet closes	57° a.b.d.c.
Exhaust opens	57° b.b.d.c.
Exhaust closes	15° a.t.d.c.
Front wheel toe-in	Parallel to $\frac{1}{16}$ in.
Camber angle	½°-1° positive
Castor angle	$-\frac{1}{4}°$ to $+\frac{1}{4}°$
King pin inclination	3½°
Tyre pressures:	
Front	30 lb./sq.in.
Rear	27 lb./sq. in.

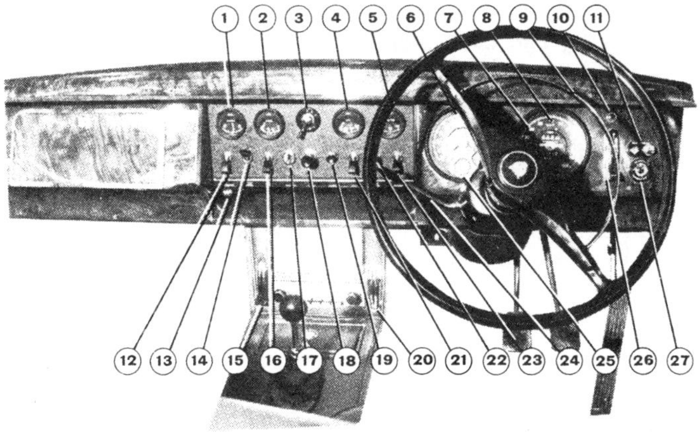

1
2
3
4
5
6
7
8
9
10
11
12
13
14
15
16
17
18
19
20
21
22
23
24
25
26
27

1, ammeter. 2, fuel gauge. 3, lights switch. 4, oil pressure gauge. 5, temperature gauge. 6, rev-counter. 7, speedometer. 8, main beam warning light. 9, horn ring. 10, choke warning light. 11, heated backlight switch and light. 12, interior light switch. 13, heater air inlet control. 14, panel light switch. 15, heater temperature control. 16, heater fan switch. 17, ignition lock. 18, cigar lighter. 19, starter button. 20, heater distribution control. 21, maplight switch. 22, wiper switch. 23, washer switch. 24, indicator/flasher stalk. 25, clock. 26, choke. 27, brake fluid level/handbrake warning light.

movement from its uppermost position) meant that short, dabbing movements of the foot gave the smoothest changes, although the position of the take-up point seemed to vary from time to time. Well chosen ratios allow, for example, an easy restart on a 1-in-3 hill and a third-gear maximum of over 74 m.p.h.

Handling and brakes

Perhaps the worst part of the Jaguar was its steering, and even this had a Jekyll-and-Hyde quality—with apparent faults in some situations which are quite absent in others. Despite a steering-box ratio which gives nearly five turns from lock to lock, the effort required when parking is excessive, and the kindest thing to say of the steering under these conditions is that it develops the muscles. But despite, again, those five turns from lock to lock, things are quite different when the car is on the move: all our testers were agreeably surprised at the nimble, responsive handling and the way in which the car could be hustled fast along winding country roads, especially as this agility was allied to outstanding adhesion and roadholding. When "earholing" on MIRA's test track it was virtually impossible to make the car breakaway in the wet, let alone the dry, the normal behaviour being mild understeer with moderate roll and not much squeal from the SP41 tyres. It took a sharp second-gear turn, a brutal right foot and a greasy patch on public roads to make the tail swing out, and even then it did so in a gentle and controllable way. Nor was wheelspin easy to provoke, while the live rear axle, on cantilevered semi-elliptic springs with upper radius arms and a short transverse rod, did not hop or tramp.

Jaguar 240 continued

Insurance
AOA Group rating 4
Lloyd's 5

As is to be expected from the product of a firm which pioneered discs, the braking system proved to be of the same high standard, giving firm, progressive retardation at all times. During several fast laps—again on MIRA's test track—we became very much aware of the car's considerable weight, but the brakes never failed to slow the 240 down properly and no increase in pedal pressure was recorded during our standard fade test. We are not quite so enthusiastic about the hand brake—although a good 0.38g stop was achieved with it—because it is very awkwardly placed between the driver's seat and the door, underneath a protruding armrest.

Comfort and controls

Compared to the S-type and 420 Jaguars with all-independent suspension, the ride of the 240 is not outstanding, although it is quite acceptable by ordinary standards, being firm and well-damped with a certain amount of pitch and bounce on rough surfaces. Reactions to the front seats varied from toleration, to outright dislike. The main complaints were of a too upright backrest and insufficient lumbar support—the optional reclining seats would be worth having.

With plenty of fore-and-aft adjustment in the seat and a steering wheel adjustable for reach (but not rake) few of our staff had any difficulty in adopting a comfortable driving position, all the major controls such as the wheel, gearlever and pedals being well located in relation to each other. Although the horn ring and flasher/indicator stalk are within fingertip reach, and the old-fashioned rotary lights switch on the facia is not too hard to find, the other minor controls such as the wiper and washer switches tend to get lost in the uniform row of other switches on the dashboard and become difficult to find in the dark. And for the British market it would be better if all the switches were turned round so that "down" meant "on".

Insulation from road noise was generally excellent, although there was a certain amount of radial-ply thump on Cats-Eye studs and shallow ridges. Wind noise was also very moderate, even at high speeds, and there was significantly less buffeting than in many other cars. Unfortunately there was a good deal of wind roar as soon as the front quarter-lights or front windows were opened, and this was frequently necessary as there is no fresh air ventilation. Even when the engine had warmed up—taking some five or six miles of driving to do so—no warmth could be felt from the heater until some 15 minutes later.

The body design shows its age when manoeuvring in a confined space for none of the extremities of the car is visible to the driver. Normal forward and rearward visibility is quite acceptable, however, and the headlamps provided a good blaze of light, both on main beam and when dipped.

Fittings and furniture

The old Jaguar cliche—"How do they do it for the price?"—is particularly relevant to the equipment of the 240. Take the instruments, for example: set in a rather garish walnut facia—which sometimes gave the road ahead an orange glow through its reflection on the windscreen—are a large (5 in.) speedometer and a matching rev-counter which incorporates a clock, with separate gauges for charge, fuel contents, oil pressure and water temperature, and warning lights for the choke, low fuel level and brake fluid level/handbrake. Then there are the lights: interior, reversing, and in the boot and glove compartment. And all four doors are fitted with armrests (including the one which gets in the way of the handbrake) each of which contains an ashtray, with a fifth such receptacle on the transmission hump in front of the gearlever; naturally there is a cigar lighter on the facia to help create the ash in the first place. For oddments there is a lockable glove compartment, a small central under-facia parcel shelf and another one behind the rear seats, while for big luggage the large boot accepted 9.4 cu.ft. of our test boxes and has a flat floor for easy loading.

So far we have considered items which Jaguar no doubt regard as the bare minimum: the final masterly touch comes with the toolkit which contains a jack ratchet, grease gun, brake cleaning set, nave plate remover, pliers two feeler gauges, screw driver (with Philips head), four open-ended spanners, pressure gauge, adjustable spanner, spare sparking plug, plug spanner and T-bar and cam timing gauge—a spare fanbelt is also provided.

Servicing and accessibility

Despite a rather crowded bonnet, the main service points of the engine are easy to get at except the distributor, which is buried down the side of the engine at the front. The spare wheel—with the toolkit in a shaped plastic box inside it—is stowed under the boot floor, and the pillar jack is easy to use. Maintenance is needed every 3,000 miles and is exacting in its requirements, there being eight points to grease (every 6,000 miles) while the gearbox and back axle oils have to be changed every 12,000 miles.

M

1, radiator filler cap. 2, crankcase breather. 3, brake fluid reservoir. 4, fuel filter. 5, oil filler cap. 6, dipstick.

MAKE: Jaguar. **MODEL:** 240. **MAKERS:** Jaguar Cars Ltd., Coventry, England.

Maintenance summary

Every 3,000 miles: Change engine oil, replace filter, top up carburetter dampers and adjust slow running, lubricate and check distributor, clean and check plugs, check brake fluid, gearbox and back axle levels.
Every 6,000 miles: Tune carburetters, clean fuel filter, check and adjust timing chain, check fanbelt tension, lubricate all grease nipples and rear wheel bearings, check brake disc pads.
Every 12,000 miles: Replace gearbox and back axle oil, lubricate front wheel bearings, renew sparking plugs and air cleaner element.

3.8 JAGUAR SEDAN

CONTINUED FROM PAGE 138

is bad-weather work. With limited-slip and snow tires, it slips and slides all over the road when the roads are wet.

What does it cost to drive? As someone said, "It all depends on how you look at it." And I say, only half joking, "It's such a nice car that I wish I could afford it." It is certainly not inexpensive, but I doubt that it's any more costly to operate than many other cars in its price class. A breakdown of expenses is included in this article, but by themselves they mean little. I look at it in terms of having had one major failure, a couple of moderate-to-high expenses, and a whole raft of failures and faults of the sort that drive men berserk.

Briefly, here is a list of the salient points of my experience with the 3.8 sedan:

Dealer Service. My local Jaguar dealer was so unsatisfactory that as soon as the car was off the warranty I switched to a local independent garage, with the result that I had lower bills, better work, on-time completion and a "loaner" to drive in the interim—all in very pleasant contrast to the treatment received at the hands of my dealer. As a footnote, I might add that the local dealership has changed hands and that I have had no experience with the new one.

Major failure. Clutch. Broken pressure plate at 29,000 miles. Replaced at a cost of about $200.

Tires. The original Dunlop RS-5s were gone after 14,000 miles, a big disappointment to me. I replaced the original 6.40-15 Dunlop with Pirelli Cinturato 185-15s (equivalent to 6.70 section) and the result was amazing—ride and cornering were both improved. There has been no visible tire wear since then though one tire had to be replaced after 15,000 miles because a wheel was out of line. All these, plus a spare, plus a pair of American snow tires, totaled $346.

Gas consumption has run 14-17 mpg on Sunoco 260. It has used about a quart of engine oil per 1500 miles. Very good for a Jaguar.

Depreciation. Previous reports by owners in R&T have emphasized the importance of depreciation in overall cost per mile. My 3.8, which sold for $5500 two years ago, now has a retail value of roughly $2500. It is hard to find direct comparisons with other cars, as the Jaguar fits into a price gap between the top of the Buick, Olds, Pontiac line and the bottom of the Cadillac, Lincoln, Imperial prices. Considering everything, the Jaguar depreciates at roughly the same rate as its price competitors, better than some, worse than others.

Other major items of expense have included brake jobs (all four relined once for $68 at 14,000 miles, the fronts again for $34 at 26,000) and two new batteries (one under guarantee, one at my expense, $34 at 25,000 miles).

Routine maintenance. Maintenance "by the book" is expensive on the 3.8 but I believe that it is necessary on such a car. My 2500-mile checkups ran $25–30, the 5000-mile service twice that. At 10,000 miles, the tab came to about $70.

Miscellaneous. A whole horde of minor annoyances have added up to the major frustration Jaguar owners suffer. Here is my list: The tachometer broke five (5) times in the first 30,000 miles, all for a half-dollar-sized piece that takes three hours to replace. The speedometer broke three (3) times in the same period. The clock was replaced twice under the 12,000-mi warranty, after which I gave up. Really, a Jaguar's clock *ought* to work. The power steering pump has been a constant annoyance, never satisfactorily repaired, and now leaks more fluid than the engine burns. I've resigned myself to simply adding fluid as needed. The windshield washer was replaced once at 5000 miles and has been inoperative again since 12,500 miles. The cooling system has developed one leak after another. I finally gave up trying to have this corrected too and simply add water (or anti-freeze) as needed. The entire cooling system is somewhat less than marginal at best and many times, even with a full radiator, I've limped home through summertime rush-hour traffic with the needle on the peg and accompanied by appropriate smells and gurgles. For winter driving, the heater-defroster system is only slightly better than a prehistoric VW I once owned and one is advised to bundle up well. For summer driving, I haven't added air-conditioning, much as I'd like to, simply because I'm sure the system wouldn't stand it. Finally, the 3.8 is a car of many squeaks and groans—undefinable, untraceable, unremediable. And the walnut trim, so attractive when new, needs refinishing after less than two years.

My next car? No, it isn't going to be a Jaguar. Two reasons. First, my experience with the dealer was sufficient to discourage me—at least temporarily—even though the dealership has changed hands, as I mentioned. Second, I'm not a slave to marques and frankly I want to sample something different.

And what sort of compromise have I arrived at? First, I have an Olds Toronado on order. My purist friends are aghast. But I should add that I'm also dickering for a 1937 4¼-liter Bentley Vanden Plas phaeton. After the Jaguar, I'm going to separate my work car from my play car!

ROAD TEST
by John Bolster
Jaguar 340

High performance at modest price

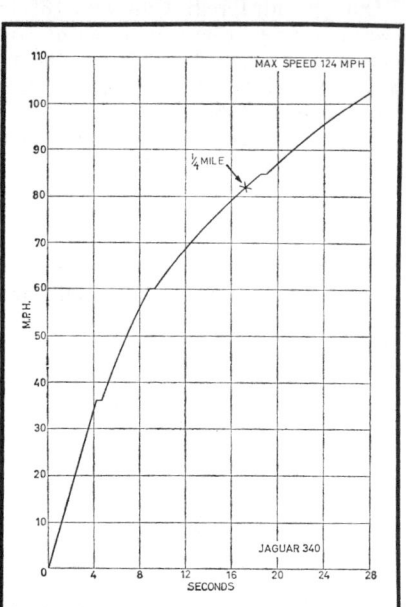

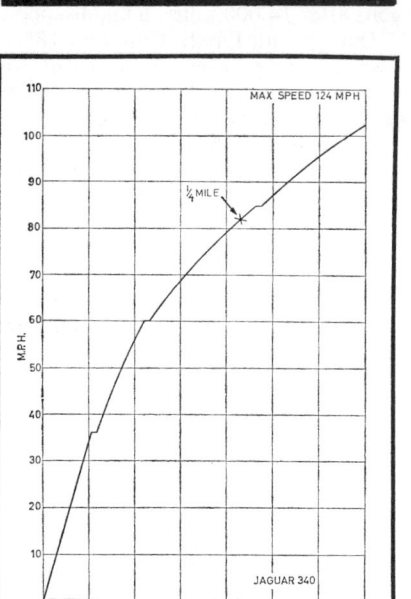

THE Jaguar 340 derives from the famous Mk 2 model, a type which many Jaguar enthusiasts prefer because it combines very high performance with compact overall dimensions. Now that the S and 420 models satisfy the luxury market, the simpler 340 brings 120 mph motoring within the reach of the man of medium income.

Having many traditional Jaguar features, the 340 is beautifully finished and, as always, the interior displays a large area of polished walnut. Perhaps the absence of padding round the instrument panel may dismay the safety-conscious, but genuine tree-wood gives an air of richness even if a close inspection reveals that the upholstery no longer comes off the backs of cows. The absence of built-in fog and driving lamps is the only other indication that this is a less expensive Jaguar.

The independent front suspension delights the engineer because provision is made for setting up all the angles and dimensions correctly. The steering swivels and track rods carry several grease nipples but the suspension pivots, front and rear, have rubber bushes. The back axle has slightly splayed longitudinal cantilever springs underneath, with short radius rods above, and there is a link between the central abutment of the right cantilever and the axle, just short of the differential housing. This improves the lateral stability of the cantilever springs and might rather loosely be termed a Panhard rod.

The engine is the latest version of the immortal twin-cam 3.4-litre, giving 210 bhp. This is a very sensible size to choose, for it is notably more economical than the 3.8 and 4.2-litre versions while still offering most of their performance. The gearbox is the greatest improvement over earlier Jaguars. It has excellent synchromesh on all four gears and, though the ratios are now a little wider, this evidently suits the characteristics of the engine for the acceleration figures are better than ever.

It is easy to enter and leave the car and there is a larger area of glass than on earlier Jaguars, though the scuttle is still fairly high. The test car had the optional adjustable seat backs, but even so the leg room was insufficient for a tall driver and one would have appreciated better lateral location. The steering column is immediately adjustable when on the move—a very useful feature.

The automatic starting device on the SU carburetters works instantly from cold. The engine is as smooth and powerful as one would expect, with that wonderful muted hum at full revs which reminds one of many Le Mans victories. The clutch approaches perfection, not only in its smooth action and its absolute grip but because the angle of the pedal and its progressive response are just right. The gearlever is a delight to handle and though the indirect gears are audible, this is scarcely noticed inside the car. The rear axle achieves total silence, a virtue which is becoming increasingly rare, curiously enough. Having been a vociferous critic of earlier Jaguar transmissions, I am all the more delighted to give this one full marks, and it is typical of the car that the excellent Laycock-de Normanville overdrive works remarkably smoothly. In some cars, this component gives a jerky change, so it must be careful and intelligent installation which allows it to work so well on the Jaguar.

When I first took over the car, I disliked the steering. It seemed too low geared, yet it was also heavy at parking speeds. The manufacturers advised me to inflate the tyres —the optional Dunlops SP41 radials were fitted—to 35 lb pressure for high-speed work, and this transformed the steering. Not only did it become far lighter, but the improved responsiveness made it seem higher geared. Personally, I also preferred the ride with the harder tyres and for really fast driving I would like more damping, especially on the front suspension. When a rigid rear axle is used, it seems best to avoid ultra-soft suspension settings. The radial ply tyres do not cause the usual road noise problems, except for a slight rumble at low speeds on certain surfaces.

The 340 is an understeering car and it runs straight at high speeds, even on bumpy or cambered roads, though violent gusts of wind cause some deflection at 110 mph or more. We know, from watching them racing, that Jaguar saloons do lean on corners, but there is little sensation of rolling inside the car. The rear axle does not patter, even under violent acceleration, but wheelspin is easily provoked and the fastest getaways are made at moderate engine revs.

As the acceleration figures in the data panel show, the 340 is an extremely potent car. I generally changed up as the rev counter needle touched the red mark, but I admit that I trespassed momentarily on this area to attain a genuine 60 mph in second gear in recording the best 0-60 figure. Normally, I changed into third at 57 mph. It is interesting that this 4-door saloon takes some 4 secs less to go from a standstill to 100 mph than did the hairy old XK120 two-seater which I tested for AUTOSPORT many years ago. This is progress.

Naturally, a car of this calibre can best be exercised on the Continent and so, at the first opportunity, I took my breakfast on the Townsend Ferry and then let the Jaguar have her head. One can cruise at 100 or 110 mph with little mechanical sound, and the brakes can be used repeatedly from such speeds without any apparent warming up. Incidentally, the hand brake works better than those of other recent Jaguars but, as I have written before, the release button needs a guard, for one can kick it when entering or leaving the car as it is situated on the right of the seat.

If the overdrive is ideal for fast Continental touring, it is also useful for fuel economy in England. In spite of its latent 120 mph performance, the Jaguar normally consumes petrol at a rate of 20 mpg or better, even with quite brisk handling. When I was taking the performance figures and really using the gearbox, the consumption only rose to 17 mpg. The 340 is therefore not only cheap to buy but cheap to run as well.

The heating and demisting system is not outstandingly powerful but it heats the rear compartment and is a great comfort on a long journey. The optional demisting of the rear window is well worth the extra cost. For fast driving at night on the Continent, more potent headlamps would be appreciated. The test car was fitted with a radio and a very neat aerial that could be retracted without leaving one's seat, though neither the range nor the volume of the set were impressive at the time of my test.

The Jaguar 340 is a very fast car which is a delight to drive, but it is as practical for everyday use as any ordinary vehicle. To the driver and his passengers, it feels, sounds, and looks like a very costly car, and everything about it is rich except its price. Once again, I am astonished at the Jaguar miracle of value for money.

SPECIFICATION AND PERFORMANCE DATA

Car tested: Jaguar 340 4-door saloon, price £1442 6s 1d. Extras: overdrive £56 10s 10d; electrically heated rear window £18 8s 9d; Britax automatic seat belts £15 9s 9d; SP41 tyres £2 6s 1d; reclining seat kit (fitting extra) £7 4s; radio with remote control aerial £44 10s 10d; all including PT.

Engine: Six-cylinder, 83 mm × 106 mm (3442 cc). Twin-overhead camshafts driven by 2-stage roller chain, operating inclined valves in light-alloy head. Compression ratio 8:1. 210 bhp at 5500 rpm. Twin SU carburetters. Lucas coil and distributor.

Transmission: Single dry plate clutch. Four-speed all-synchromesh gearbox with Laycock-de Normanville overdrive and remote control central gearlever. Open propeller shaft and hypoid rear axle. Overall ratios 2.933, 3.77, 5.0, 7.44 and 11.46:1.

Chassis: Combined steel body and chassis. Independent front suspension by wishbones and helical springs with anti-roll torsion bar. Re-circulating ball steering. Rear axle on cantilever springs with trailing arms and Panhard rod. Telescopic dampers all round. Disc brakes with vacuum servo. Bolt-on disc wheels fitted Dunlop 6.00/6.40-15 ins SP41 tyres.

Equipment: 12-volt lighting and starting. Speedometer. Rev counter. Clock. Ammeter. Oil pressure, water temperature and fuel gauges. Heating and demisting with heated rear window. 2-speed windscreen wipers and washers. Flashing direction indicators. Radio. Cigar lighter.

Dimensions: Wheelbase, 8 ft 11¼ ins; track (front), 4 ft 7 ins, (rear) 4 ft 5¼ ins; overall length, 15 ft 0¾ in; width, 5 ft 6½ ins; weight, 1 ton 8 cwt.

Performance: Maximum speed, 124 mph. Speeds in gears: direct top, 115 mph; third, 85 mph; second, 60 mph; first, 36 mph. Standing quarter-mile, 17.2 s. Acceleration: 0-30 mph, 3.5 s; 0-50 mph, 6.9 s; 0-60 mph, 8.8 s; 0-80 mph, 16.6 s; 0-100 mph, 26.4 s.

Fuel consumption: 17 to 22 mpg.

50,000 Miles In a 3.8 Mark II or if you don't like cars, don't buy a Jag

by Milt Rosen

At social gatherings talk among men invariably gets around to cars. When I say I own a Jag someone—you can count on it—will ask: "How can you afford the servicing?"

Depending upon my mood, liquid intake and the attitudes of the listeners my answer sometimes is a facetious "make a lot of money." But because that usually and perhaps unfairly ends the discussion I am more likely to offer my two part solution to the dilemma of how to be a Jaguar owner and remain solvent.

Part one requires that to own a Jaguar and to enjoy it to its fullest, the owner should possess a fondness for automobiles. He cannot look upon his car merely as transportation. A sudden squeak or change in performance calls for investigation *now*, not when the next scheduled oil change or grease job is due. Overly prompt, by ordinary standards, maintenance can save replacement of parts. The Jag is an expensive car with expensive parts and these parts have to come to you from a long ways away.

Part two, and of equal importance to the Jaguar owner, is the cultivation of a local mechanic who will take care of your car and who does not look in the "book" to see what Jaguar Ltd. says that care is worth. Choose this gentleman cautiously, making sure that when you drive your expensive car into his place of business, dollar signs do not suddenly appear in his eyes. Once you've found him, you have to get to know his limitations. By all means let him perform Jaguar's suggested schedule of checks at regular intervals. He can save you more than $50 on a 20,000-mile checkup. He'll change your fan belt for a lot less than the $4.50 charge listed by agencies. But, it can be a traumatic experience to drive into the station to check on some work and see your parts scattered around and your mechanic standing there scratching his head. For the more major jobs pay the cost and go to where the experience is. With luck and adherence to part one above, you may never have to.

Our Jaguar 3.8 came to California in anything but a routine fashion. It was bought in Oslo, Norway just before our family's return from a two-year sojourn in Sweden. It made its way to New Haven aboard a ship carrying 300 other cars, all Saabs. It's a fast car, faster than a lot of people credit it with being including a Corvette driver on the trip out to California who played games while I was trying to pass. He quit at 100 mph with as puzzled a look as one could project from the window of one of those. It's also not prone to overheating as I've so often seen reported. While still relatively new and stiff it absorbed 100-degree weather that stretched from Oklahoma to the California coastal range. A quick

The car today, 3000 water and 50,000 land miles away from Oslo, looks little the worse for wear. White paint job was prematurely necessitated by a contractor's errant sand-blasting gun. Rosen says: "If you don't like cars, don't buy a Jag."

Author Rosen poses slightly out of wife Audrey's focus with their new (in 1963) 3.8 in front of the Kon-Tiki museum in Oslo. Body-mounted driving lights indicate European purchase.

50,000 MILES

check on the cost of an air-conditioner for this car, incidentally, convinced me that sweating was not really uncouth. Gas consumption averaged over 20 miles to the gallon.

Ours is a 1963 Mark II and is a comfortable 4-people car. It was originally what Jaguar calls "Opalescent Silver Grey" in color but after a sand-blasting on a Los Angeles street, it is now Buick White. Red leather (real) upholstery and wool carpets, walnut trim, a full set of instruments (not idiot lights) including tachometer and clock, fuel and brake warning lights, are standard equipment. It has a 4-speed manual transmission with a Laycock electric overdrive that the 230-CID, 220-horsepower dohc six handles with ease. Hemispherical combustion chambers and dual SU carburetors help, of course. A limited slip differential is also standard, there is no synchronization in first or reverse and no power anything.

After trips during which I have driven rented cars with automatic transmissions and power everything, I look forward to driving my own unautomated machine again. I like the straight, quick stops that the 4-wheel disc brakes provide from any speed. I like the third gear, good from 10 to 70 miles an hour, for the sudden maneuvers required by city traffice. I like the steering and suspension which actually allows the car to go in the desired direction whether braking or accelerating. I like the reserve of power available for normal driving situations and there is plenty there for the abnormal too. In Nevada, where this sort of thing is legal, I found that 125 mph equals a lazy 4400 rpm in overdrive.

Much thought has gone into the little things that American manufacturers ignore. The four jacking points welded to the chassis simplify tire changing as well as making the job less dangerous than it is with a bumper jack. All the controls and switches are centrally located, well-lighted and easy to reach. The panel they are on is hinged, offering access to all the wiring, fuzes and flasher unit. All of the behind the dashboard mish-mash is covered by removable panels. Large pockets are located in each of the doors and there are neat little drop tables behind each of the front seats.

On the other hand, I'd be less than honest if I ignored annoying and troublesome features. They do exist. For example, the rear-view mirror is poorly supported and vibrates badly enough at times to make looking behind you an excursion into vertigo. The pressure in the hydraulically assisted clutch sometimes suddenly and inexplicably vanishes. Generally you're waiting in first gear to make a left turn at an intersection when this happens. The car stalls and you're left stranded unless you get the car out of gear, start it and then the pressure is back again perhaps for months. (The brakes are also hydraulically assisted but in the five years we've owned the car, they've never given us any difficulty.

The tachometer impeller, made from a hunk of plastic worth all of 50¢, lasts less than 2500 miles. It's located, of course, in the most inaccessible region of the engine compartment and to replace it requires a highly developed sense of touch because you can't see what you are doing. The trunk is small by American standards; put two large suitcases in it and there is room left only around the edges. I called the 3.8 a four people car but to fit their luggage in the trunk would require the services of a packing specialist. We solved this problem by buying a rack for the top of the car.

The beautiful walnut trim needs lots of care especially in desert and subtropic areas with intense sun. Taking care of it is simplified, though, by the ease with which each panel may be removed and replaced. In typical long-stroke English fashion, the car consumes a lot of oil. A quart every 250-400 miles is usual but the large capacity (8-9 quarts) allows some margin for errors. When my wife drives in, the station attendant now asks: "Fill the oil and check the gas?"

But it is remarkable how easy it is to forget these annoyances when you're out on the highway, cruising along smoothly and quietly. The faster you go, the lower the car rides. The steering remains precise and quick and eight hours behind it is a breeze. My Jag has 50,000 miles on it and before it's traded it will have lots more. Something I like better may come along, assuming the makers of luxury imports will face up to the mandatory American safety and emission requirements, and when it does I suspect it will be a later model of the car I now have. You see, I like that car! •

SEAT BELTS SAVE LIVES!

JAGUAR 240

● LONG TERM ASSESSMENT

Eight months and 10,000 miles behind the polished wood facia, and twin cam engine of Jaguar's smallest—the 240. Last of the long line first launched in 1955, the 240 is still a highly competitive car, particularly on performance and value, and has served us excellently. With the enormous demand for the XJ6, one wonders how long can Jaguar afford to go on producing it.

LOOKING back over 14 years of driving and testing the world's cars I have many wonderful memories of fabulous machinery and magnificent journeys; and among them is all the excitement and pleasure when, way back in September 1956, after nearly a year since its announcement, the Jaguar 2.4-litre arrived for Road Test. I remember precisely the places I went to in it when my turn came to try it, and how in those pre-motorway days we used the jet bomber station at Wittering for maximum speed runs and repeatedly hurtled off the end of the main runway, between the landing lights and on to the grass at 102 m.p.h., trying to get a true top speed; and I recall clearly the tremendous impression the car created at the time.

In 1956 the 2.4 seemed incredibly fast and quiet, and so very low and sporting; improvements to the current 240 specification have made it faster still, though of course many cars have caught up and the difference from others is nowhere near as marked as then. By contrast, too, it seems now rather high-built, particularly around the scuttle; but after having the use of one for eight months and 10,000 miles I have again found much of the pleasure and satisfaction in driving it that I enjoyed with the first Mark I model. It is remarkable also that although the value of money must have halved in the meantime, the 240—which now reaches 90 m.p.h. in the same 31 sec. time which the Mk. I took to 80 m.p.h., has the wider track, later and better body style and disc brakes all round instead of drums (all changes that came in with the Mk. II)—sells at a total price that has risen only £36, from £1,465 to £1,501.

Our particular car is slightly a hybrid, in having the old Mk. II style bumpers, but it is a 240 in all important respects of specification, and was taken over as a "near-new" ex-demonstrator with 5,703 miles indicated, and only just qualified for its "F" year letter, first registration being in August 1967. First impressions on taking over were that it seemed heavier, slower and rather noisier than expected. I think the feeling of heaviness was because of the way power steering had spoilt us on other models, particularly after extensive experience of the M.D.'s Daimler Sovereign earlier in the same year; and the other two aspects of performance and noise are related. The 240's kerb weight of nearly 30 cwt. is a lot for 2.4 litres, so that fairly high engine revs are needed for lively performance. From about 2,000 r.p.m. there was excessive fan noise, and one of the first things we did was to remove the huge pressed steel fan and fit a Kenlowe.

This improvement was reported in an article published in our 29 August 1968 issue, explaining how maximum speed had gone up from 106 to 111 m.p.h., but the most marked gain was the reduction in noise level, to such extent that I would certainly recommend this rather costly

Our Jaguar has the old Mk. II, bumpers, but in all important respects is up to the 240 mechanical specification. The appearance of the engine compartment is very beautiful and keeps remarkably clean in wet weather. One could not wish for a clearer or more attractively laid out facia, but safety demands have meant the end of the polished wood capping and the snarling bonnet mascot will disappear from the Jaguar range when 240 production finally ends. In the centre picture below, the Radiomobile tape recorder unit can be seen to the left of the console

modification to anyone. The Kenlowe sensors had to be changed twice, the first one tending to bring in the electric fan too soon, and the second one much too late. With the latest thermo-switch the cut-in point is just right, and occurs just as the thermometer has risen from its normal 70 deg. C. to about 80. The arm carrying the electric fan also slipped after high speed running on rather poor Belgian roads, and our garage used the link rod of a Cortina's wipers to make a side bracket for it, completely curing the problem. With the extra cooling provided by the heater in winter, and lower ambient temperatures, it takes a prolonged traffic snarl-up for the electric fan to cut in, and all the rest of the time the thrash is being saved.

Because there are no child safety locks on the 240's rear doors I made another small modification by simply pushing out the securing pins and removing the interior handles. We keep one of the handles loose in the door pocket for locking or unlocking the rear doors, the slight inconvenience being well worth the peace of mind when our little boy is in the back, and the shanks do not protrude far enough to be dangerous. An incidental gain is that the car is less easily broken into even if one of the windows has been left open, but when it was used as a chauffeur-driven car for our directors I gather the arrangement was frowned upon; "What do we do if there's a fire?"

Good instrumentation proved its value in the first month, when it was noticed that the charging had become intermittent. The ignition warning light did not come on, there being still a small output from the dynamo, but the ammeter often showed a substantial discharge. So while the car was at Gloucester after collection of a Cotswold motor caravan, Kingscote and Stephens checked it over and fitted a new voltage regulator, after which it charged perfectly.

The car ran well until, when some performance testing was being done at 8,000 miles, clutch spin was noticed, and oil was found to be leaking from the overdrive—the same fault that we had reported in the original Road Test of the 240 in January last year.

The oil level was still well up and the car perfectly driveable so an appointment was made for Henlys to have a look at it a little later on. It was reported that the clutch slip was purely due to a fault in the hydraulic linkage, preventing full engagement, and the overdrive leak was rectified as well. The repaired clutch was very smooth and pedal action light, as is normal on the 240, and uncomplainingly absorbed the full power of the engine when engaged abruptly and repeatedly during performance testing. Our figures are a little below the Road Test standard, but our "fanless" 240 is rather more economical and appreciably faster.

On this and later visits for routine maintenance I was most impressed by the service given by Henlys' north London depot. Usually I dislike big service emporiums, but Henlys do exactly what is asked, ring up when the car is ready (and it *is* ready when they say), the receptionists are not white-coated and impersonal; there is a pleasant club atmosphere about the reception room.

Costing £61 7s. 3d. extra, overdrive is an essential part of the car's effortless high speed cruising. It engages and disengages smoothly and can even be used quite often in town at 25-30 m.p.h., though there has always been a pronounced whine from the overdrive at low speeds; it disappears at about 50 m.p.h. At 90 m.p.h. in overdrive, engine revs are only 4,200—not that the engine is incapable of much higher revs without distress. On one journey I was sitting in the back of the car when a driver (who should best remain nameless) temporarily forgot the overdrive while running pretty near maximum speed. None of the three of us in the car noticed that the unfortunate engine was doing 6,500 r.p.m. at 110 m.p.h., the rev. counter needle off the scale. Happily it cannot have been for long, and the engine evidently suffered no ill effects. It is a magnificent power unit in every way—smooth, responsive, untemperamental.

On this and another occasion when Innes Ireland drove the 240 out to the Nurburgring, both drivers used much higher speeds than I, as the "owner elect," have found sensible. To me, the 240 is a 90 m.p.h. car, since it settles so quietly and effortlessly at this speed, and will still return more than 20 m.p.g. after a motorway journey in which the speedometer seems to have been on 90 most of the time. But Innes' constant flat-out driving pretty nearly emptied the 12-gallon fuel tank about 160 miles from Ostend. Normally I can get 20-21 m.p.g. on a journey while driving at and up to 90 m.p.h.; it is only in London traffic that there is a disappointing drop to 16 or 17 m.p.g. Like all the instruments, including the speedometer, the fuel gauge is very accurate. The yellow warning tell-tale begins to flash when three gallons remain, and there are still nearly two gallons in the tank when the light is glowing steadily.

Oil consumption has been mild for a Jaguar, averaging 650 miles per pint; Less on a long, fast trip—more in local running with many cold starts.

When he drove the car, Innes Ireland complained about the sponginess of the steering on corners, and of course you have to pull on a fair amount of lock in really hard cornering to counter the understeer. I learnt a lot about cornering on that trip to Germany, and although I could never take the car round corners at 70 or 80 m.p.h. as he does, I began to see how he uses the power at the right time to push it through without apparent effort. The Jaguar is wonderfully stable at speed, and it never matters much whether there is a cross wind or not.

An unusual fault occurred when turning the car round in the paddock at Nurburgring—I actually broke the end off the steering wheel front trim. It glued back in place all right with Araldite.

At 12,000 miles I reported a slight water leak, and clock not working, when the car went to Henlys for service. It came back with the clock repaired and although the water leak had not been serious, Henlys evidently thought so and had replaced the radiator. As with all previous repairs, there was no charge for this although the car was then well over a year old, and the mileometer indicating 12,136. Penalties of a big engine are £1 14s. 6d. for the 12 pints of oil which the sump takes, and £2 3s. 9d. for winter's 8¾ pints of Smiths anti-freeze. In subsequent mileage there were only trivial matters: new wiper blade rubbers became necessary, and I had to replace the sidelamp tell-tale after knocking it off when cleaning the car. The spare was obtained easily, but what a job fitting

JAGUAR 240 LONG TERM...

it, sliding a little spring on from underneath! While driving along in Holland the cold start warning lamp suddenly fell in my lap, the lens having fractured where it passes through the wooden panel. This, too, was difficult to replace, requiring removal of the thin gauge metal panel beneath the facia. It is much easier to get at the central part of the facia, where simply undoing two thumb screws allows the centre part of the panel to hinge forward—an excellent Jaguar feature.

In my previous car, an Auto Union Audi Super 90, I greatly enjoyed having a Philips tape recorder unit in conjunction with Radiomobile radio, but it is very bulky and out of the question to fit in the Jaguar with its centre console. The solution was a change to the Radiomobile tape recorder which, although using the same components as the Philips, performed even better and is much more compact. In my view this is far better than tape players which play only pre-recorded cassettes, as one tires of hearing the same music after a while. With this recorder I change my choice of canned listening from time to time, and there really is no detectable loss of quality between recorded or direct radio.

Auxiliary fog and spotlamps were also fitted, the spot being mounted on the left and wired through the dipswitch. In this rather fog-free winter the fog lamp, mounted on the right, has never had to be used. I was unable to obtain back mounting Bosch lamps, so had to jury-rig a pair of brackets for them. In fact I have found the standard Lucas headlamps excellent, and it would be no hardship to go over to twin foglamps, as required by the new law coming in on 1 January next year, and forego the spot.

The one really dated aspect of the Jaguar is its heating, which can be left full on from the start of winter until the spring and it hardly ever gets too hot. However, even in bitter cold during February, I never had to resort to wearing a coat in the car, and provided the front quarter vents are kept closed and the rear ones open a fraction, there is adequate ventilation. The front quarter vents have in fact never been opened except to clean their frames. In wet weather the rear window mists up badly even with the blower kept on to provide through-flow of air, but occasional use of an anti-mist cloth keeps it clear. An advantage of the ventilation layout is the ease with which all incoming air is closed off in one quick movement of the vent lever when a really dense diesel-pong is sighted ahead.

Starting has always been prompt, and usually just one short touch on the starter button is enough. After a night of severe frost, with the car standing out at Amsterdam, the engine still fired on the third try. The rich mixture control has to be used quite ruthlessly for a mile or so when it is really cold, otherwise the engine just spits back through the carburettors when the throttle is opened. Piston slap is audible for a few seconds when stone cold. The gear change, normally so smooth and light, is also very sticky for the first few gear

PERFORMANCE CHECK

Maximum speeds

Gear	mph R/T	mph Staff	kph R/T	kph Staff	rpm R/T	rpm Staff
O.D. Top (mean)	106	111	170	179	4,860	5,240
(best)	107	112	172	180	4,910	5,280
Top	93	90	150	145	5,500	5,500
3rd	73	70	118	113	5,500	5,600
2nd	53	50	85	80	5,500	5,900
1st	35	30	56	48	5,500	5,500

Standing ¼-mile, R/T: 18.7 sec 74 mph
Staff: 18.8 sec 70 mph
Standing kilometre, R/T: 34.7 sec 93 mph
Staff: 34.8 sec 89 mph

Acceleration, R/T	4.1	6.3	9.3	12.5	16.4	22.8	31.0	44.8
Staff:	4.6	6.9	9.5	13.6	18.4	25.5	35.2	49.2

Time in seconds —0

True speed mph	30	40	50	60	70	80	90	100
Indicated speed MPH, R/T:	31	41	52	63	74	85	95	106
Indicated speed MPH Staff:	32	42	53	64	75	86	96	106

Speed range, Gear Ratios and Time in Seconds

mph	O.D. Top R/T	O.D. Top Staff 3.54	Top R/T	Top Staff 4.55	3rd R/T	3rd Staff 5.78	2nd R/T	2nd Staff 7.92	1st R/T	1st Staff 12.19
10-30	12.7	12.9	8.7	9.4	6.7	6.7	4.5	4.9	3.3	3.7
20-40	11.2	12.7	8.3	8.9	6.1	6.5	4.3	5.0	—	—
30-50	12.1	12.8	8.3	9.3	6.3	5.8	—	5.3	—	—
40-60	12.9	14.1	8.4	10.2	6.6	7.3	—	—	—	—
50-70	14.0	16.7	8.7	11.2	7.2	8.8	—	—	—	—
60-80	14.1	19.8	11.5	13.1	—	—	—	—	—	—
70-90	22.4	27.9	14.1	16.5	—	—	—	—	—	—
80-100	35.7	37.2	—	—	—	—	—	—	—	—

Fuel Consumption
Overall mpg. R/T: 18.4 mpg (15.4 litres/100km)
Staff: 19.1 mpg (14.8 litres/100km)
NOTE: "R/T" denotes performance figures for Jaguar 240 tested in AUTOCAR of 4 January 1968

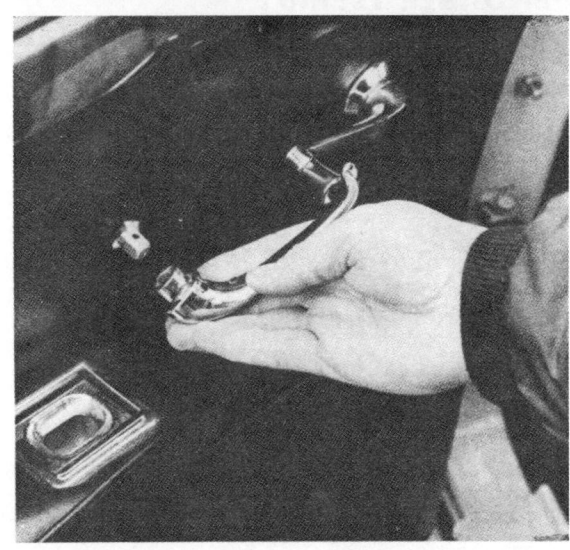

The wing mirrors fitted have convex glasses and are a real aid in traffic or when reversing; auxiliary lamps are non-standard, by Bosch. The front apron collects a black mass of insects in summer and tends to be damaged by flying stones. An effective modification was removal of the interior handles on the rear doors for child safety, and one handle is kept loose for locking

COST AND LIFE OF EXPENDABLE ITEMS

Item	Life in Miles	Cost per 10,000 Miles
		£ s d
One gallon of 4-star fuel, average cost today 6s 4d	19.1	165 18 0
One pint of top-up oil, average cost today 3s 6d	650	2 6 0
Front disc brake pads (set of 4)	24,000	1 3 0
Rear brake pads (set of 4)	30,000	18 6
Dunlop SP tyres (front pair)	30,000	6 8 0
Dunlop SP tyres (rear pair)	35,000	5 10 0
Service (main interval and actual costs incurred)	3,000	51 5 4
Total		**233 8 10**
Approx. standing charges per year		
Depreciation		300 0 0
Insurance*		16 7 6
Tax		25 0 0
Total		**574 16 4**

Approx. cost per mile = 1s 2d
*Insurance is a Cornhill quotation based on business and private use, 65 per cent n.c.b., residence in Herts., and £15 excess.

---

changes after a cold night. Once warmed through, however, the engine retains its heat for a long time.

Wear of the disc brake pads has been extraordinarily mild, the pads being good for well over 20,000 at the front and 30,000 miles at the rear, judging by the present state of wear. They are very consistent brakes, really effective from high speed, and unaffected by water. The handbrake is also very effective, but one of the first things you learn about the 240 is to mind your heel when you get out—it is all too easy to knock the unguarded release button. Tyre wear has also been good on the Dunlop SP radials fitted. Even without using the spare, the tread wear points to mileage of 30,000 for front tyres, and even as high as 40,000 at the back. The slight penalty is a lot of tyre thump at low speeds, and of course they accentuate the steering heaviness. Grip in the rain, and even on snow, is very satisfactory.

One needs to like a Jaguar to get the best out of it—it responds so much better to considerate driving, with engine speeds matched to gear changes. Furthermore, although the mechanical side takes little maintenance I am afraid it is a labour of love to keep the car looking smart. In summer, the smooth shape of the rounded frontal styling, plus, of course, the inevitably high speeds, mean that the front is forever plastered with dead insects. In winter, the problem becomes one of keeping rust at bay from the chromium after driving on salt covered roads. However, the underneath is well protected, and has held corrosion away. My choice of colour would have been the beautiful royal blue which suits the model so well, but our honey beige—unkindly called "accident safety yellow"—certainly does not show the dirt. Small blemishes, mainly from flying stones on the front apron, were retouched with some incredibly quick-drying cellulose supplied with the car; otherwise rust spreads rapidly.

One sees the Jaguar at worst in town, when it always seems rather bulky and lacks the nippiness that is so helpful in traffic; but on a long journey the effortlessness, comfort and quietness, coupled with deceptively high speeds, really pay off, and arrival at journey's end always finds one still enjoying travel. The lack of wind noise must also contribute greatly to the apparent ease and lack of fuss.

Finally comes the matter of personal taste—do you like polished wood facia and window surrounds in an age of plastics and austerity? Some don't. In a car with good safety belts I am afraid I do, and refuse to be persuaded that padded pvc is the thing; and the symmetrical, easily remembered layout of switches and the beautiful clarity of the instruments add to the quality and luxury. My spell with a 240 Jaguar has provided a lot of motoring enjoyment, the only regret being that I was not able to make even more long journeys in it.

Brief Specification Resumé
6-Cylinder twin overhead camshaft engine, 2,483 c.c. Bore and stroke 83 x 76.5 mm. (3.27 x 3.01in.). 2 SU carbs.; C.R. 8.0 to 1; 133 b.h.p. (gross) at 5,500 r.p.m.; max. torque 146 lb. ft. at 3,700 r.p.m. Four-speed all-synchromesh gearbox, optional overdrive on top. Independent front suspension, coil springs, semi-trailing wishbones; rear, live axle on cantilever leaf springs, radius arms and Panhard rod. Dunlop disc brakes and servo, 11 in. front, 11.4 in. dia. rear.

Jaguar 240

Early this year we completed 24,000 miles with our staff Jaguar 240, and have now parted with the car. It was felt that, although the model is now obsolete, readers would be interested in a further report on the 240, particularly in view of the relatively low running costs recorded. The car is an early 240 with Mk. 2 bumpers.

COST AND LIFE OF EXPENDABLE ITEMS

Item	Life in Miles	Cost per 10,000 Miles
		£ s. d.
One gallon of 4-star fuel, average cost today 6s 6d	20	162 10 0
One pint of top-up oil, average cost today 3s 6d	900	1 18 6
Front disc brake pads (set of 4)	35,000	16 0
Rear brake linings (set of 4)	35,000	16 0
Dunlop SP41 tyres (front pair)	35,000	5 10 0
Dunlop SP41 tyres (rear pair)	40,000	4 16 0
Service (main interval and actual costs incurred)	3,000	73 11 10
Total		249 18 4
Approx standing charges per year		
Depreciation		200 0 0
Insurance*		16 7 6
Tax		25 0 0
Total		491 5 10

Total cost per mile — 1s 00d
*Insurance is a Cornhill quotation based on business and private use, residence in Herts, £15 excess and 65 per cent n.c.b.

162

Long term assessment

The second 12,000 miles with a Jaguar 240

By Stuart Bladon

WHEN I reported on use of a Jaguar 240 for 12,000 miles last April, I fully expected to have to hand the car over shortly afterwards, but in fact I had the use of it throughout 1969 except for a few short breaks when I handed it over to colleagues. In the first week in January this year I reluctantly said goodbye at the end of 19 months and 24,000 miles of consistent, dependable and satisfying service.

During this time, of course, the model went out of production, but this appears to have stimulated demand and values. Last December I left the 240 with a used car dealer while we had one of their cars on test, and a trade buyer offered £1,000 cash for it on sight; it was difficult to convince him that the car was not for sale. Retail prices for 1968 models are still often above £1,200, which is the price to which I had written down the value in April. Although first registered in August 1967, this car would still fetch £1,000 without difficulty, so depreciation has been only a further £200 in the second year.

In other respects, running costs have been even more remarkable due to the long life of all components, and when I parted with the car at 24,132 miles it was still on its original tyres, brake pads, exhaust system and battery. The life of the Dunlop SP41 tyres has been almost incredible for a big car. Three of them were less than half worn, one just over half worn, and the one which had spent much of the time as the spare read only 1mm less than new. This indicates that the estimated tyre life figures given in the previous report were pessimistic even at 35,000 per set. I always kept the pressures at 34psi front, 32 rear, which is slightly higher than the pressures recommended for fast driving.

After carrying out the 24,000-mile service Henlys reported that the brake pads were still only "about half worn". No doubt there will be some letters from incredulous readers about such long tyre and brake life, but the car is still running around on its original equipment brakes and tyres, proving the figures.

At about 16,000 miles the oil consumption began to rise alarmingly, and it was found that there was leakage from the sump return pipe. After this was rectified the consumption settled down to about 800-900 miles per pint and would have been a little less if there had not been a slight leakage from the rear main bearing oil seal, revealed by an occasional drop of oil on the ground if the car was left parked overnight on a slope. In a general check-up at the same time a new speedo cable was fitted to cure a slight tapping noise, and the front suspension sandwich rubbers were renewed.

All wheels were rebalanced at 18,000 miles to cure a slight high-speed vibration. Then, near 20,000 miles, it became progressively more difficult to get into gear, particularly into reverse, and the only major repair necessary had to be undertaken—renewal of the clutch diaphragm. A new clutch had been fitted once already, at 14,400 miles, and it could be that we damaged the plate by having to take performance figures before the new linings had a chance to bed in. At any rate, there was no charge for replacement, as this was still within the guarantee period of the second unit.

Jaguar do not mean you to do your own maintenance on this car, as I discovered when I endeavoured to adjust the SU carburettors. The air intake trunking has to be removed (I wish I had known that only the outer bolts had to be undone and that the inner ones were purely locating studs), and then a bare arm must be insinuated through a labyrinth of carburettor and throttle linkages to reach the mixture nut, by which time one is left with no grip or strength with which to turn it.

I succeeded in the end, but vowed to leave it in future to our own capable service unit or to Henlys, who always proved extremely efficient. Henlys' charges were on the high side (£23 for a full 24,000-mile service, plus lubricants and materials), but I appreciated the way in which jobs requested were always done thoroughly, completed when predicted, and the owner kept in touch with progress.

The distributor, buried away beneath the inlet manifold, must be even more difficult to work on, but no doubt all these jobs are easier with the appropriate specialist tools.

Apart from these points, and the above average amount of work involved in keeping the car looking smart, including keeping the walnut facia and window sills beautiful with furniture polish, the car needed very little attention. It could be left for a week or ten days during work on Road Test cars, and after a sharp rattle from the electric pump on switching on the ignition, it would always respond to the first try with the starter. Only after severe frost were two or three touches on the starter button necessary before the engine would fire.

In very cold weather the 240 was not seen at its best. The engine took a long while to heat through, and everything felt very stiff and reluctant during the first couple of miles. The weak heater, also, was the most out-of-date feature of the car, and although it would win through in the end it took some miles to take the chill off the interior.

On long trips, in contrast, the 240 was a joy to drive, with its effortless high speed cruising, comfortable ride and seats and low noise level. A lot was owed to substitution of the engine-driven fan by a Kenlowe electric, a really worthwhile modification. On a journey such as the run from London to Scotland, the power and noise are saved all the way, for 400 miles, and the electric fan does not cut in until the car gets caught up in the Glasgow traffic. The combination of high gearing in overdrive, no fan, and low wind noise due to the good aerodynamic shape, made this one of the quietest cars I know at 90–100mph.

I also had a lot of pleasure out of being able to choose my own music for a long journey, playing it back from the Radiomobile cassette tape recorder/player unit, through the radio. This equipment I have transferred to the Sunbeam Rapier H120 which I have inherited from the Editor to replace the Jaguar.

Among the many things which I liked about the Jaguar 240 were the attractively styled interior with its neat, symmetrical layout of switches, and the superb clarity of the instruments; good visibility, ensured by efficient two-speed wipers which park neatly at the base of the screen and are controlled by a switch conveniently alongside a switch for vigorous electric windscreen washers; and the pleasing air of quality and refinement about the whole car. Perhaps above all, though, it was the wonderfully smooth and responsive twin-overhead camshaft engine which made the car such a delight to drive.

All this and much more, of course, is available in the more powerful XJ6 range, and one fears that there will never again be a compact Jaguar to replace the 240. In its absence, it is not surprising that there is healthy demand for well-preserved used examples. ☐

When our term with the car ended at [24,]000 miles, its paintwork and chromium were [st]ill in excellent condition and underbody rust [wa]s slight—only a patch under the sill on each [side] behind the front wheels, and beneath the fuel [tan]k. Performance tests were thwarted by bad [we]ather (above, left) but after a tune at Henly's, [p]erformance seemed to be fairly well up to [sta]ndard and oil consumption was at the lowest [leve]l in the history of the car so far. Above: Near Barnard Castle, on a trip to Scotland

USED CAR TEST

No. 323
1968 JAGUAR 340

PRICES
Car for sale at Dorking at £1,125
Typical trade value for same age and model in average condition £900
Total cost of car when new including tax £1,654
Depreciation over 3¼ years £754
Annual depreciation as proportion of cost new .... 14 per cent

DATA
Date first registered ...... 5 April, 1968
Number of owners 1
Tax expires .... 30 September 1971
MoT expires .... 11 June 1972
Fuel consumption .... 16-20 mpg
Oil consumption .... 150 mpp
Mileometer reading 39,671

PERFORMANCE CHECK
(Figures in brackets are those of the 3.8-litre automatic Road Test, published 5 April, 1963; the 3.4-litre automatic was never road tested in *Autocar*.)

0 to 30 mph		4.9 sec (3.6)
0 to 40 mph		6.9 sec (5.3)
0 to 50 mph		8.9 sec (7.2)
0 to 60 mph		13.3 sec (9.8)
0 to 70 mph		16.3 sec (12.9)
0 to 80 mph		22.1 sec (16.9)
0 to 90 mph		30.0 sec (21.3)
0 to 100 mph		41.1 sec (28.2)
Standing ¼ mile		18.9 sec (17.2)

In top gear:

20 to 40 mph		7.0 sec (—)
30 to 50 mph		6.9 sec (7.4)
40 to 60 mph		8.8 sec (7.1)
50 to 70 mph		10.3 sec (7.6)
60 to 80 mph		12.3 sec (8.4)
70 to 90 mph		14.1 sec (9.6)
80 to 100 mph		18.4 sec (11.3)
Standing Km		34.6 sec (—)

TYRES
Size: 6.40 H—15in. Dunlop RS5 on all wheels
Approx. cost per replacement cover . £10.78
Depth of original tread 8.2mm; remaining tread depth. 8mm on front tyres, 8.2mm on rear tyres; 3mm on spare.

TOOLS
One or two items only missing from fitted tray; jack apparently unused. Handbook with car.

CAR FOR SALE AT:
F. W. Mays Ltd, South Street, Dorking, Surrey.
Tel: Dorking 2244.

The engine compartment is clean, but much of the paint has flaked off the radiator header tank. There are no obvious signs of oil leakage, but a lot of oil is finding its way out somewhere beneath the engine, mainly at the front. Right: The interior is very clean and tidy, apart from slight cracking of the shiny coat on the polished wood. Fixed Britax lap and diagonal safety belts are fitted

WHEN Jaguar first introduced the 3.4-litre version of the Mk. 2 series, following the original 2.4-litre, its performance seemed quite staggering by the standards of the day. Some 15 years later, a used example taken straight from the dealer's showrooms still rates as a fast car. It has the reasonably compact body which, with its timeless styling, was Jaguar's best seller for

At the front, the 340 was almost identical with the 240, except for the radiator grille name badge. The dark blue finish suits the car very well. One of the wheel hub plates has received an ugly graze against the kerb

such a long period, and so one of these has a lot to offer for someone looking for a fast, comfortable and distinguished five-seater saloon.

The long-stroke 3,442 c.c. engine always had a good reputation for longevity, and at nearly 40,000 miles it remains in very good shape. Inevitably it does not come up to the standard of the original 3.8-litre Mark 2 automatic—we never tested the 3.4 except in Mark 1 form. However, it is still quite creditable to accelerate from rest to 100 mph in just over 40 sec, and the engine, in ordinary swift but not flat-out driving, is impressively quiet. Under full-throttle there is a lusty power roar, and a trace of thrash if it is taken up to 5,000 rpm.

On one morning of the test, after a very warm night, the engine would not start, and was only coaxed into life by using the under-bonnet starter switch and holding the auto choke out of action. Fairly heavy oil consumption was somewhat characteristic with this engine, and as considerable signs of oil leakage underneath are seen, the consumption rate of a pint every 150 miles is not excessive.

Borg Warner Model 35 automatic transmission is fitted, and apart from developing considerable stiffness in the selector, it remains up to the standard for the model. It is almost essential to use the selector for acceleration from low speeds, otherwise the transmission tends to stay in top gear. The accelerator has to be pressed really hard into the carpet to operate the kick-down. Gear changes go through smoothly and a reasonable engine speed of 4,500 rpm allows maxima of 40 in low, and 67 mph in intermediate.

Power-assisted steering is fitted, taking the rather considerable effort out of steering this nose-heavy car; but the system is not good. Control is slightly vague, and there is often too much reaction, following initial woolliness. It is particularly difficult to control the car neatly when pulling out to overtake. The suspension is still very comfortable although it was noticed that the dampers have weakened, allowing some pitching on undulations. The low level of road noise is very good, and the whole car is pleasantly solid-feeling and rattle-free.

Braking efficiency seemed to improve during the test, lack of initial response diminishing after further mileage. The brakes now work well enough, but we think that a check on pad and disc condition would be advisable. The handbrake offers a lot of leverage and holds securely on a steep gradient, even though it can be backed up, of course, by the positive transmission lock.

Fairly new Dunlop RS5 tyres are fitted. From experience we know that on this model radials such as the SP41 give much better handling as well as lasting longer. It is a little too easy to make the tyres squeal on corners; but to have so much tread depth on the tyres of a used car is worth a good bit.

This Jaguar impresses as an honest and generally very fit car, with signs of having been well cared for. The power steering and old-fashioned ventilation are below the sort of standards expected today, but in other respects it remains a very pleasant car to drive, with that quiet, effortless high-speed stride which characterized the model.

Condition Summary

Bodywork

In the log book there is a note of a colour amendment from silver blue to dark blue. However, after careful examination we are satisfied that the paintwork is the original finish and that the change noted was the result of an error at the time of first registration. There are some minor chips and marks on the paintwork

condition. The same is true of the chromium, which is almost unmarked, and even the chromed exhaust tail pipes are still shiny.

Inside the Jaguar, the polished wood facia and window surrounds have begun to crack slightly and are in need of a good rub with furniture polish. There are no marks on the maroon pvc door trim, and the seats, upholstered in maroon Ambla are comfortable and show no signs of wear or cracking. The floor carpets have been well cleaned, and have faded and worn slightly. Tobacco smoke has stained the cloth roof lining.

General underbody condition is good, with rust confined to the underneath of the fuel tank and a light covering on the rear quarters and under the sills. Quite a lot of mud has collected under wings and around the fuel tank at the back. New front exhaust pipes have been fitted, and although the twin silencer boxes and tail pipes look considerably older and are fairly well rusted, they are sound.

Equipment

All the comprehensive instrumentation is in perfect working order, except for the clock, which is out of action. The instrumentation is beautifully clear, and the speedometer gives a steady reading slightly less than the true value—showing only 97 at 100 mph. New windscreen wiper blades are needed.

Accessories

Britax lap and diagonal safety belts are fitted, of static type. The only other addition to the 340 is an early Radiomobile push-button radio, giving excellent quality of tone. Reflective number plates are fitted.

About the 340

The 1955 Earls Court Show produced several exciting and entirely new designs such as the Citroen DS, whose general body appearance continues with remarkably little alteration today; and another destined also to become a classic was the Jaguar 2.4-litre with its smooth, rounded frontal shape. It had not been in production long when the long-stroke 3.4-litre engine was offered in the same body, initially for the American market. From then on, there was always a 3.4-litre Jaguar, continuing in the much improved Mark 2 shape which first appeared in 1959, until it was finally displaced by the XJ6 in 1968.

In September 1967 the 3.4 name gave way to the 340, and "slimline" bumpers effected the first significant styling change for nearly 10 years. A new by-pass-closing thermostat was introduced to increase radiator efficiency, and a big improvement for buyers of the automatic model—as the one tested here—was replacement of the Borg-Warner DG transmission by the more modern Type 35 with D2, D1, L selector. Marles Varamatic power-assisted steering became available, but as explained in the text, it suits this model less well than it does the larger all-independent 420 and Daimler Sovereign on which it was first introduced.

Power output of the 3,442 c.c. engine was claimed to be 210 bhp (gross)—the same value as at original introduction. Re-reading the description of the 340, published in *Autocar* of 28 September 1967, it is interesting to recall that the price remained unaltered except for the automatic transmission model, which became £40 cheaper. "Were those the days?" The year before, fog lamps had been deleted, and Ambla replaced leather upholstery, also with a reduction in price. Production ended in September 1968 at chassis number IJ52265. The 240 survived the 340 for a little while, production continuing until early the following year.

PROFILE
Jaguar Mk I & II

Classic car, classic driver and classic drift — Mike Hawthorn slings his 3.4 Mk I around Silverstone, followed by Tommy Sopwith. Mk I and IIs were darlings of both road and track

THE OLD ONE-TWO

Jaguar's Mark I and II saloons have become sought after classics. Mark Gillies and Richard Sutton are your guides to these 'compact' machines, the first monocoque cars made by Jaguar

There is *something* special about a Jaguar. All Browns Lane products have that touch of class which sets them apart from other makes. The Mark I/II family has that something.

For a start, when you sit behind the wheel of one you know that this is a fifties car. The smell of leather, the polished burr walnut dash, the organ type throttle pedal tell you so. Even the array of toggle switches and the separate ignition key and starter button are typical of that era.

But then, when you start the car up, you are not so sure it is from the fifties. That glorious straight-six engine is so quiet and docile, and feels so modern. On late model cars, with all-synchro 'box and lighter clutch, the change is slick and sure, again so thoroughly modern. It takes the Moss 'box, with its long travel and heavy but sure action, to remind you of how old the design really is.

On the move, the engine is superb. It has power and torque aplenty, but will just potter along in traffic at zero revs, as content as can be. Open up, and there is a turbine-like rush, the road in front of the proud leaping cat mascot gets more blurred, and soon you are travelling at highly illegal speeds — quietly, comfortably and without drama. Come to a corner, stamp on the reassuring all-round disc brakes, and you slow down without drama. Boot the car round, the tail slides, but in the progressive manner of a thoroughbred. Positive, if slightly low geared, steering aids your progress.

The car feels more like a good sixties design as you drive it harder. What were other motor manufacturers doing in the fifties, when Jaguar produced a car that feels at least 10 years ahead of its time?

It is no small wonder that these cars have acquired a classic status, in the last five years or so. It is of such avowed excellence that the word classic is appropriate, a rare accolade these days.

The initial aim of this new saloon was to make a small car to back up the Mk VII. Not only would it introduce the marque to new customers, but, more importantly, it would increase the production of components shared with Jaguar's other models.

Design parameters were set out by Bill Heynes, styling was by Sir William Lyons, and the body designers were left to engineer the spaces in between. In keeping with the compact bodyshell, it was planned to fit the car with a small version of the XK unit, a move which proved to be a mistake as most Jaguar customers put performance before economy, and the 2.4-litre Mk Is and IIs not only proved pretty gutless, but weren't very fuel efficient either . . .

Modified engines

The 2.4-litre engine shared the same bore as the 3.4 unit, but had a shorter (76.5mm) stroke, giving 2483cc. The new block was 3ins lower than the standard unit and weighed 50lbs less. The familiar twin ohc layout was retained as was the chain camshaft drive, but twin Solex downdraught carbs somewhat strangled the unit. Power output was 112bhp, produced at 5750rpm.

The unitary bodyshell was the first for the firm, and was very strong — a bit over-engineered if anything. Like many early monocoques, it had outriggers and cross members which can fool MoT examiners into thinking that it is a separate chassis car! Front suspension was carried on a sub-frame, which being rubber mounted insulated the bodyshell from road shocks. The steering box was also mounted on this sub-frame. The front suspension itself was by coil springs and unequal length wishbones, there was an anti-roll bar, and the tubular shock absorber ran down through the coil spring. The subframe worked remarkably well, even if it was a very complex engineering solution.

The rear suspension was also pretty complicated. The axle was live, but unusually it was supported on cantilever leaf springs, with the trailing ends attached via rubber bushes to fixed extensions on the axle casing. Trailing arms and Panhard rods provided further axle location. Brakes were Lockheed drum vacuum assisted, and a four-speed manual 'box was standard fitment with overdrive an option.

Styling was ultra-modern when the car came on sale in 1955. Wing line and grille showed a marked affinity to the XK140, and, like all fifties Lyons cars, it was smooth and swoopy and curvaceous, although in this case a heavy looking back end spoiled its purity. Spats, a small rear window and hefty pillars gave a slightly tail heavy appearance, and while the interior was state-of-the-art, small windows made it feel a bit claustrophobic.

English press acclaim was inevitable, as it lived up to the Jaguar reputation, which was by now pretty formidable. Road-holding and ride were superb, the engine willing and flexible, smooth and silent, the gearbox good. And there was nothing that compared with it for double the £1269 price tag.

166

While road testers in this country eulogised madly, but justifiably, American journalists liked it, but felt it lacked power. To the English press, a saloon car capable of over 100mph was a revelation — the Mk I was able to reach 101.5mph in overdrive, and 0-60mph in 14secs was pretty good, too. While *Road & Track* felt 'that sports car owners will appreciate the combination of road manners and high speed stability incorporated in the 2.4', some magazines wanted a larger engine for the car.

Never slow off the mark, Jaguar responded to this criticism by installing the 3442cc XK unit in the Mk I. Power of 210bhp was going to be far more acceptable to the American market.

The actual announcement of the 3.4 came in February 1957, but before Jaguar had envisaged. The firm had sent 200 cars across to the USA to build up stocks over there, but the news was leaked.

The 3.4 incorporated a number of subtle chages from its smaller brethren. For a start there was a larger grille with narrower slats, the rear spats were of the cutaway type, there weren't any fog lamps, and the model was identified by the new '3.4' badging and twin exhaust tail pipes. Under the skin, there was a 10ins instead of 9ins diameter clutch, a V-type mounting under the gearbox, a Salisbury 4HA rear axle, the Panhard rod mounting (which had proved vulnerable) was altered, and an automatic option was offered for the first time.

The up-engined car caused American testers to change their collective tunes. *Road & Track* said that the net result of the larger engine was 'a sedan which can at least hold its own with all but one or two of our most potent power packed domestic machines.'

Back in England, *The Motor* wound a 3.4 auto up to 119.8mph which was very quick for 1957, and 0-60mph came in 11.2secs: that type of performance was better than all but the most overtly sporting machinery. Mind you, the brakes didn't match up to the improved performance, and with all that extra power *The Motor* felt that the 3.4 behaved in rather an unruly manner. Press reports were in fact tantamount to saying that the car was dangerously underbraked, and hastened the option of four wheel discs. The narrow track axle has always been held responsible for the less than perfect high speed manners.

Changed tunes

Those optional discs were superb, *R & T* claiming them to be the 'finest brakes we have ever found as a production item on a sedan.' Like *The Autocar*, their road testers felt that nothing compared with it, and the car's handsome lines were further improved that season by the availability of wire wheels — they also helped cooling if you bought a car with drum brakes.

October 1959 saw the arrival of the Mark II, a car that has been called 'the last true sports saloon to come from Jaguar.' It was a superbly executed refashioning of the original Mk I concept, given more power, better handling, looks and disc braking as standard equipment.

Externally, there was a new radiator grille with central rib, spotlights were now mounted where the air intake grilles had been, sidelights were moved to the top of the wings, there were slimmer windscreen pillars, and the big increase in glass area was achieved by means of side window frames extending beyond the door openings and by a new wraparound rear screen. Subtle front suspension revisions were made, and, more importantly, the new rear axle was wider, with a 4ft 5.357ins track. Inside, the speedo and rev counter had been moved directly in front of the driver, with the central dash space now taken up by minor instruments and a whole row of toggle switches. The seats were new, while the steering wheel was fitted with a 'half ring' horn button.

As far as the performance driver was concerned, though, the best news of all was the introduction of the 3.8-litre version of the compact saloon — the bored out engine of 3781cc gave 220bhp at 5500rpm, a useful boost of 10bhp over the 3.4.

Bill Boddy writing in *Motor Sport* reckoned that the 3.8 Mk II was one of the 'best saloon cars in the world... that such a car can be sold for just over £1800 is a commercial miracle understood only by Sir William Lyons.'

1954 styling mock up had forward opening bonnet/wings

Mk I has small rear window, and bulky tail appearance

JAGUAR MkII 3.4
SPECIFICATION
Engine	In-line 'six'
Capacity	3442cc
Bore × stroke	83mm × 106mm
Valves	Twin ohc
Compression	8:1 (7:1 or 9:1 optional)
Power	220bhp @ 5500rpm
Torque	240lbs ft @ 3000rpm
Transmission	Four-speed manual with optional overdrive/three-speed auto
Final drive	3.54:1 (standard or auto)/3.77:1
Brakes	Servo discs all round
Suspension front	Ind by wishbones and coil springs, anti-roll bar
Suspension rear	Live axle, semi-elliptic leaf springs, Panhard rod, radius arms
Steering	Burman recirculating ball — optional power assistance
Body	All steel monocoque
Tyres	640-15

DIMENSIONS
Length	15ft 0.75in
Width	5ft 6.75ins
Height	4ft 9.75ins
Wheelbase	8ft 11.35ins
Weight	29½cwt

PERFORMANCE
Max speed	119.9mph
0-60mph	11.9secs
Standing ¼-mile	19.1secs
Fuel consumption	16mpg

It was certainly some performer. *The Autocar* managed a 125mph top speed and achieved a 0-60mph time of 8.5secs, which is quick even by today's standards, and was certainly better than 80% of sports cars then on the market. No small wonder that they became favourite getaway cars for criminals.

With the car's wider rear track, the 3.8 was far more stable than the somewhat skittish Mk I. It was still regarded as an understeerer, and testers found that the steering was heavy at parking speeds, and the five turns lock to lock gearing made it difficult to apply corrective opposite lock.

Later in 1960, power-assisted steering became available as an option on UK market cars, but, far more importantly, that year saw the acquisition of Daimler by Jaguar. It was bought mainly because Browns Lane was proving too small to cope with the record production levels now being maintained.

It also saw a Daimler version of the Mark II, the V8 250. This used a pushrod ohv Daimler V8 of 2547cc as fitted to the SP250 sports car, shoe horned into a Mk II shell, and mated to a Borg Warner Type 35 auto 'box.

This car fitted well into the existing range in terms of both performance and price — it cost £1568, and with a 0-60mph time of 13.8secs and a top speed of 112.3mph, outperformed the manual 2.4 Mk I.

The Mk II story wasn't all rave reviews. The heating, as with most fifties and sixties English cars, was fine — unfortunately, the ventilation and cool air circulation made for grilled owners in countries where sunshine is not something to celebrate.

Poor heating

On the positive side, though, the auto gave away little to the manual cars and was felt to complement the car's character well, a view which is reinforced when you drive one of these cars today. Moreover, the 3.4 car was most practical in this country, as it was cheaper to run and insure than the 3.8, and gave away very little in terms of performance. On the other side of the pond, however, it was the 3.8 which sold, and very few 3.4s and 2.4s made the journey across the Atlantic.

The Mk I/II had a fabulous competition history to back up its excellent reputation as a road car. From 1956 to 1962, the compact Jaguar saloons won all major British saloon car events, and the antics of Hawthorn, Salvadori, Parkes, Duncan Hamilton, Jack Sears and Tommy Sopwith, tyre squealing and drifting their cars, provided superb entertainment.

The first major defeat in Britain was at Brands Hatch in '62 where Salvadori was beaten by Kelsey's Chevrolet. On the continent, things looked brighter for the firm. From 1959 to 1964, these cars won the touring category in the Tour de France, and in 1962 Lindner and Nocker posted a warning to all the other firms due to contest the following year's inaugural European Touring Championship by winning the Nürburgring Six Hour race. Nocker then won the ETC the following year, but little was made of it.

Rallying was never the car's real *forte*, but the Morley Brothers won the Tulip event in 1959 in a 3.4, and several useful performances were posted in long distance road rallies.

The Mk II in its various guises was phased out in 1967, and its replacement 'compacts', the 240 and 340, were destined for short production lives, as the XJ saloon would be announced in 1968.

Both cars were slimmed down and rather more austere versions of the Mk II. In a way, with the XJ just a year away, and the three car Mk II range existing alongside 3.4 and 3.8 S-types and the 420G, some form of rationalisation was necessary anyway.

The cars had slimmer bumpers, which meant a new valance at the rear, dummy foglight grilles, new hubcaps, and the car came with the Ambla (or plastic) seats that had been fitted to late model Mk IIs. Perhaps more significantly, the 240 had a slightly revised engine, which meant a 13bhp increase in power output to 133bhp. A new cylinder head with straight ports, twin 1¾ins SU carbs and a dual exhaust system were the reasons for the increase. New ribbed cam covers marked this engine out.

At £1364, the 240 cost but £20 more than the original Mk I 2.4, an incredible piece of costing when you think about it. At the same time as this revamp, the Daimler was left alone . . .

240s and 340s still drew acclaim from the press. While they found the steering to be heavy and low geared, and the shallow screen and high scuttle dated, they found the car's road-holding and handling to be unexpectedly good, despite developments in independent suspension systems. Performance was much improved, with the 240 transformed from being a 'gutless wonder' into a 106mph motor car.

The car's prestige still counted for a great deal — even if a Triumph 2000 or Rover 2000TC could perform as well, the Jaguar name had a great deal in its favour. The 340 was highly rated by John Bolster, whose closing words had a familiar ring to them: 'Once again I was astonished at the Jaguar miracle of value for money.' Even today, the Mk Is and Mk IIs still provide fabulous value for money, and have to be regarded as the last true Jaguar sports saloons. It was as well that they died in 1968-69, because by then they were becoming dated and expensive to produce — a graceful demise is always better than a long and lingering death, more usually the fate of British cars.

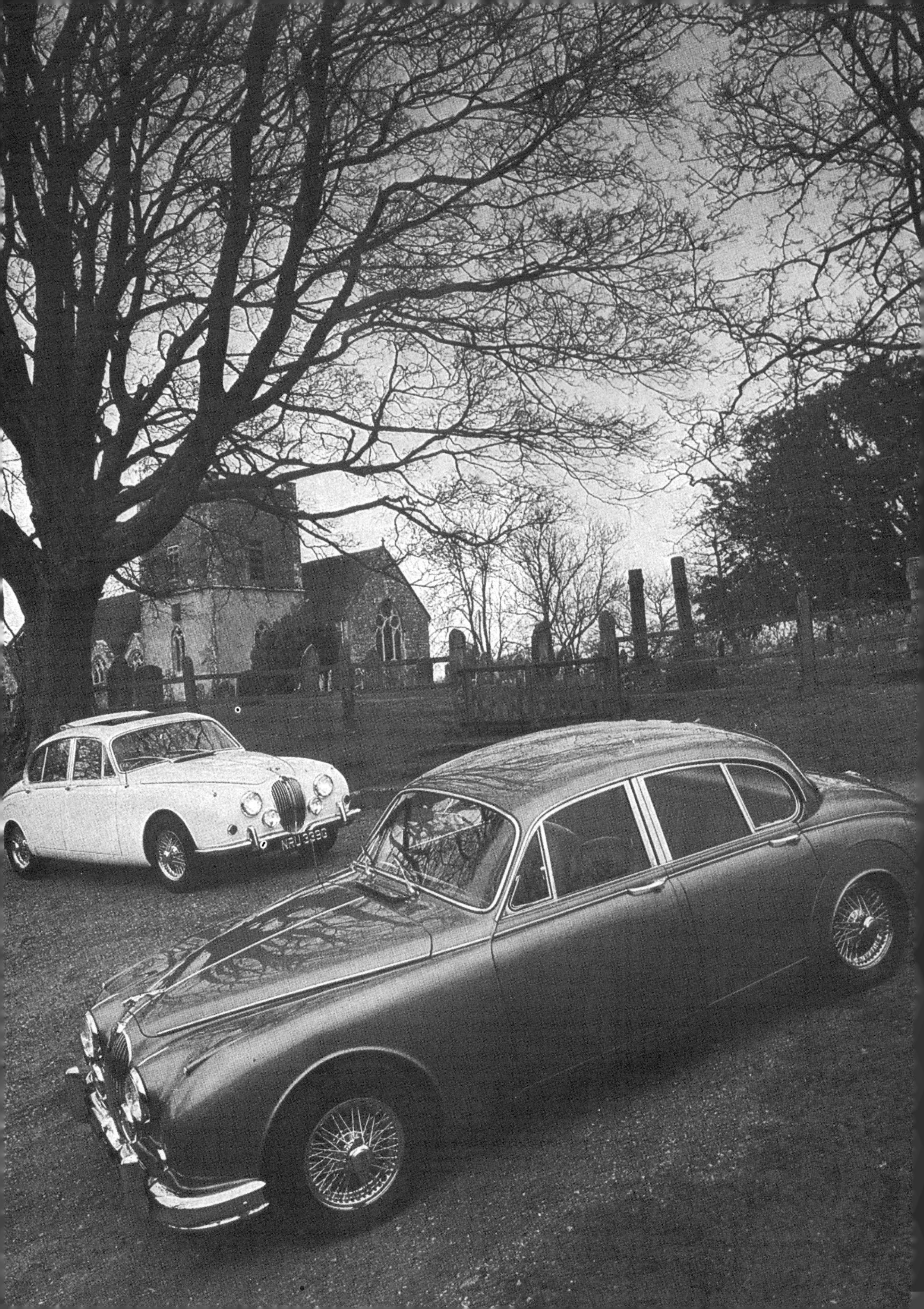

Production history

The compact Jaguar saloons had a long life span so there were a number of running changes on all the different models.

Early on in the Mk I's life it was given stiffer dampers and an adjustable end to the Panhard rod, from June '56 there was a Metalastik crank damper from engine BB 2500, November '56 saw a steel sump while the car also gained a 4.27:1 final drive in non-overdrive form that year.

September '57 saw standardisation of the wider grille in the 2.4, which meant new front wings and a modified intake for the air cleaner behind the grille from chassis numbers 907974 and 942465 and there was the option of disc brakes and wire wheels from the end of the year. For February 1958 there was a larger brake servo (up from 5.5ins to 6.875ins), disc braked cars received a cast iron master cylinder and a new progressive tapered rubber bump stop was fitted to all Mk Is.

Improvements for '58 numbered fitment of RS4 tyres instead of the previous RS3s, the option of a Powr-Lok differential, improved dampers, and in July, the distinctive 'cut-glass' overdrive switch was replaced by a metal switch.

The Mk I was phased out in September 1959, but there were still a number of changes to come. January '59 saw quick change brake pads for those cars with discs, 72 spoked wire wheels replaced the 60 spoke versions (but they were still an option), a ½in wide fan belt was standardised, the vacuum reservoir was positioned between manifold and servo, and there were larger diameter upper and lower ball joints. April saw lead indium bearings and June an electric rev counter instead of a cable driven version.

As Browns Lane managed that lot in three years, there were going to be plenty of changes for the Mk II! February '60 saw stiffer dampers, March 60lbs oil pressure gauge, April a modified telescopic rear view mirror, May a round paper oil filter on 3.4s and 3.8s, in June the indicator switch moved to the left, overdrive switch to the right and November saw a modified oil filter, swivel sun visor and a spring lid on the gearbox cowling ashtray. And there was more to come.

The following year's mods kicked off in February with a stronger anti-roll bar and forged instead of pressed steel upper wishbones, June saw a modified oil pump and a larger sump, August deflectors for the front hub bearings and a self adjusting handbrake mechanism, while December saw larger diameter propshaft UJs and a modified crankshaft rear oil seal. Earlier, short body SU fuel pumps had been adopted.

New York Show car in '60 had gold plated brightwork

Less glamorous ladies, but Mk IIs were good chase cars

Mk I/II motive force was the classic twin ohc XK unit

Changes in '62 included an 'O-ring' seal on the back of the exhaust cam cover (June), drilled camshafts (May), sealed beam headlights (June), and an optional high output dynamo (February, standardised from April '63).

For the next season there was a revised water pump (September), gas cell dampers, and a 3ins diameter propshaft with sealed-for-life components. By now, fewer and fewer changes were made to the car, although yet to come were a modified radiator block and fan cowl (January '64), improved wheel swivel grease seals (March '64), an all synchro 'box (September '65, from chassis numbers 119200/127822 (2.4 rhd and lhd), 169341/180188 (3.4 rhd and lhd), and 234125/224150 (3.8 rhd and lhd)), Ambla upholstery became standard, with leather an option, and foglights were replaced by dummy grilles (September 1966), while July '67 saw Variomatic power steering and a revised boot lid lock.

As the 240 and 340 models lasted only a short time while the XJ saloons were introduced, there were no changes in specification.

Production figures

MkI 2.4	Sep 1955 – Sep 1959	16250 RHD
		3742 LHD
MkI 3.4	Mar 1957 – Sep 1959	8945 RHD
		8460 LHD
MkII 2.4	Oct 1959 – Sep 1967	21768 RHD
		3405 LHD
MkII 3.4	Oct 1959 – Sep 1967	22095 RHD
		6571 LHD
MkII 3.8	Oct 1959 – Sep 1967	15383 RHD
		14758 LHD
240	Sep 1967 – Apr 1969	3716 RHD
		730 LHD
340	Sep 1967 – Sep 1968	2265 RHD
		535 LHD
Total production		**128623**

Buyer's spot check

Mk I and II Jaguars are among the most rust-prone mass-produced cars ever made! This is not entirely due to inherent rust trap design (although that is a considerable weakness) and certainly not because of poor metal quality. It is largely the people who bought used MkI/IIs who are responsible.

The unitary construction of these cars is naturally rather primitive. The Mk I was Jaguar's first attempt at this form of construction and, creditable as it was, the structure did little to inhibit condensation and damp traps. Moreover, the insides of sills and box sections were barely painted.

Initially, it is likely that these cars were well looked after. Inflation was low in those days and Jaguars were relatively rare and desirable.

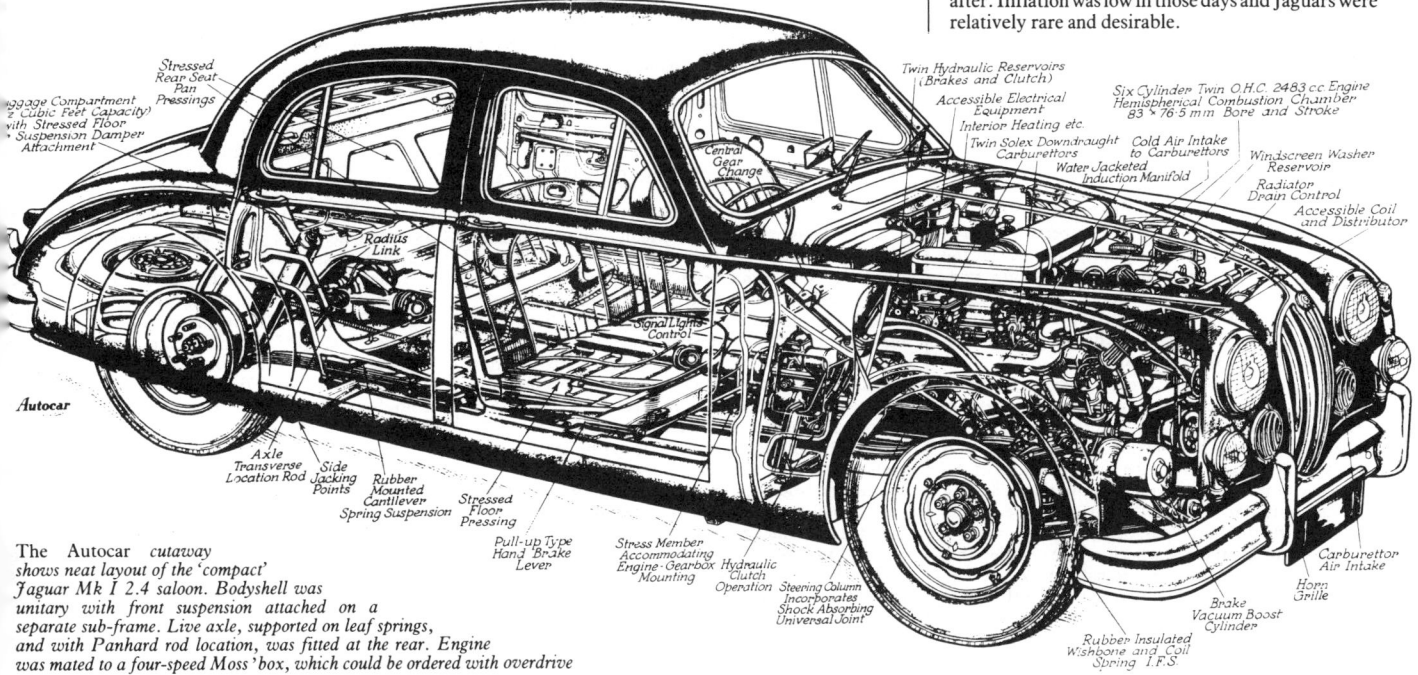

The Autocar cutaway shows neat layout of the 'compact' Jaguar Mk I 2.4 saloon. Bodyshell was unitary with front suspension attached on a separate sub-frame. Live axle, supported on leaf springs, and with Panhard rod location, was fitted at the rear. Engine was mated to a four-speed Moss 'box, which could be ordered with overdrive

169

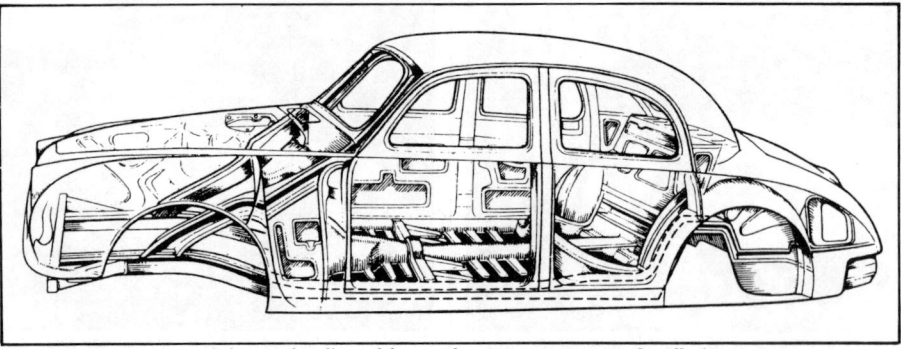

Compact monocoque was made from good quality steel, but complex structure ensures a number of bad rust traps

Browns Lane fire in 1957 saw huge production losses

Unfortunately, severe depreciation occurred during the late sixties and early seventies. This brought these cars into the reach of a far larger and less well off market at the time the average car was beginning to require substantial bodywork repair. The massive box sections of the MkI/II were difficult and expensive to make good. Such repairs were seldom carried out satisfactorily, and by the mid-seventies the classic market was flooded with rotted-out MkIIs.

In spite of the considerable dangers of buying secondhand, the cars' huge following prompted the growth of numerous specialists and several parts remanufacturers. This has made restoration far cheaper and easier in recent years, with the result that there are now more good condition, properly rustproofed, examples on the market than ever before. But the risks are still high.

The condition of the car's bodywork is of primary importance. Look first at all the rear spring hangers, Panhard rod assembly and under the rear seat. If there are any signs of corrosion in these three areas, then you might do well to forget it. To repair these areas is expensive and often a fair indication that the whole car is rotten.

Next departments for inspection are the inner and outer sills. These are fundamental structural areas and their condition is important. The inners go particularly badly around the front and rear jacking points, which themselves are very vulnerable. Don't be afraid to prod around. What at first seems solid very often exfoliates away with a little encouragement. All parts of the floor pan should be carefully examined for rust and even cracks. The boot floor around the spare wheel well is especially vulnerable.

Behind the bottom of the front wings are two fan-line structures known as crows feet which are among the first panels to disappear. These are there purely to stop the front wings from flapping around, and replacement is easy. The front cross-member which sits under the radiator is a more serious matter: it frequently rots and, in serious cases, the rust can effect the Y-shaped girders which run from near the bumper mounts to the front outriggers.

A lot of water and dirt is thrown up into the front wheel arches, so these areas are always under attack. The brake servo vacuum tank is also mounted under the front offside wing and it can rust through.

The skirt behind the front bumper often rusts through, as does the area around the indicators and the auxiliary air vents. On MkIIs the sidelight pods which are leaded into the surface of the front wings are rust-prone and often lift from their apertures. Even if these pods look good, ensure that they have been replaced, and not just repaired.

The sides of the front wings collect dirt and are subject to rot, especially at the 10 o'clock position where the wing surfaces can suffer appallingly. The splash panels and box section closing panels also quickly deteriorate. Moving around the sides of the car, check the door skins and bottoms, the area along the rain gutter and the door pillars. Dropped doors indicate structural problems here.

Spats covering the rear wheels are very susceptible to rust and are usually non-original. Check carefully behind them, as the arches are more difficult to replace. Again, the inner splash panels are prone to rot through, as is the rear under valance.

Finally, check the boot edges, nose of the bonnet and radiator grille, and areas around the windscreen and rear window. It may well be necessary to remove these items to repair and avoid leaks.

In summary, virtually every panel on a MkI or II is vulnerable bar the roof (which is easily dented!). Restoration costs can be enormous. The very good news is that there are repair sections or replacement panels available for all these areas, and numerous companies carry out competitively priced work.

Always a good reason for buying a post-war Jaguar is the car's exceptional mechanical longevity and the easy and cheap availability of associated parts. Any properly maintained XK engine will last for 100,000 miles without requiring any costly attention.

A good XK engine should start instantly and give between 35 and 45psi at 3000rpm when hot. The engine should not be noisy. A slack top timing chain is quite common, but is easy and inexpensive to rectify. A noisy bottom chain is more serious. Tappets need routine adjustment. Pulsing can be put down to poor carburettor balance, misfiring often down to old spark plugs (they don't last very long) and poor oil pressure is often caused by a blocked pressure relief valve or worn oil pump.

Don't be put off by smoke from the exhaust. All XK engines use a lot of oil, especially the 3.8 versions and a thin blue smoke on acceleration is prefectly normal. The condition of the exhaust is fundamental. The slightest leaks can cause substantial power losses, and splitting is common.

All gearboxes, both automatic and manual, fitted to MkIs and IIs are reliable but expensive to repair. Early manual cars are fitted with the Moss 'box — a rather agricultural but very strong unit which will probably have worn second gear synchro and will make quite a row when reversing and on acceleration in first gear. Unless it becomes impossible to select second gear, or the 'box is excessively noisy, do not worry. The later fully synchromesh box is rather a slow changer, so don't let that bother you, but all the synchros should work well. Like the Moss 'box, parts are rare and expensive. The earlier Borg Warner DG transmission may slip. If so beware — parts for these are also scarce.

On such a heavy performance car clutches wear quickly. Unfortunately, plate replacement is a massive job and is therefore often put off until too late — flywheels are easily, and often, damaged.

Suspension poses no problems, but all rubber mountings should be in good order. The car's stance, or how it 'sits', is important. A low nose where the front wheels have damaged the paintwork on the arch rims is a fair indication of tired springs. A lop-sided rear end can be caused by broken springs.

Steering and brakes are generally reliable, but watch out for leaks in the power steering mechanism. Brakes should be efficient, but check for seized caliper pistons and scored discs. The handbrakes on MkIs and IIs are notoriously poor, as they operate separate pads via a complex Heath Robinson affair.

The condition of chromework is very important. Over £1000 can quickly be spent rechroming, and it is possible to spend twice that. Most chrome trim items are still available.

Restoring any Jaguar interior can also be frighteningly costly. The 240/340 cars at least score here as their Ambla upholstery is long-lasting and still young. Complete retrims of MkIs and IIs will cost £1500, and that's without replacing the mohair headlining or lacquering the woodwork. A new fitted headlining alone will cost £150, and allow over £200 to have the 24 pieces (on the MkII) of woodwork refinished.

In summary, the only attractive features of buying a MkI or II in need of work is the model's mechanical longevity and parts cost, and the availability of all repair sections, from jacking points to sidelight nacelles. The drawbacks are the huge potential outlays.

There is a golden rule when buying any monocoque Jaguar. If buying an average condition car with a view to restoring, ensure that at least some of the car's areas are in excellent order — there is no point in buying a car which is basically sound, but will eventually need welding and a respray, which has only mediocre chromework and an interior which will eventually need attention and a drivetrain which needs detail work. Such a car is common for around £1500, and buying is a waste of time. If you're going to have to replace everything, then buy a wreck in the first place.

Clubs, specialists & books

If you are considering buying, running or restoring a compact Jaguar or V8 250, you are very strongly advised to join any of the three active car clubs catering for these models.

Longest established of the three is the Jaguar Drivers Club. Formed in 1956, it now has a membership of over 10,000 and continues to grow. The club is organised into registers, one of the largest of which is for MkI and II models. The club has a very comprehensive calendar of events including a MkI/II Inter Area Racing Challenge, and publishes a monthly magazine, *Jaguar Driver*.

The newly-formed Jaguar Enthusiasts Club was started last December by a JDC breakaway group, most of whom were members of the MkI/II Register committee. Although only months old the Enthusiasts now number over 700, produce an excellent monthly publication, *Jaguar Enthusiast* and offer countless benefits to all Jaguar owners and especially MkI and II fans. The club is working hard to set up a comprehensive spares scheme.

The Daimler and Lanchester Owners Club was formed in the early sixties and has over 2500 members most of whom are V8 250 owners. The club provides excellent technical advice and insurance services via its magazine, *Driving Member*, coupled with an unrivalled spares store.

Annual membership of the JDC costs £15 (plus a £ joining fee): applications to Jaguar Drivers Club Ltd Jaguar House, 18 Stuart Street, Luton, Beds LU 2SL (tel: 0582 419332). JEC membership is £12 from Graham Searle, 37 Charthouse Road, Ashvale, Surrey (tel: 0252 316696). DLOC membership costs £1 from John Ridley, Boxtree Cottage, Brightwalton Green, Berks RG16 0BH (tel: 048 82 563).

Classic Jaguar restorers are very fortunate to have such a large number of specialists available either to assist with parts supply or carry out full blown restorations. The following lists those of good repute and represents a fair geographical spread:

For general restoration: Gantspeed Engineering Chapel Lane, Mareham-le-Fen, Boston, Lincs Classic Spares, 503 Southbury Road, Enfield, Middx Jaguar Specialists, 36 Druid Street, London SE1 Elm Restorations Ltd, Elm Mills, Station Road Skelmanthorpe, nr Huddersfield, West Yorks

Advanced Garage Engineering Ltd, 95A Mitcham Lane, Streatham, London SW16; M.R. Buckeridge, Airedale Works, Otley Road, Shipley, West Yorks; A.J.M., 19 Billington Road, Elmesthorpe, Leics; Symonds Motor Engineers — Jaguar Service, Unit 2, Brittania Industrial Park, Dashwood Avenue, High Wycombe, Bucks; Stanleys, Westwood Farm, Bretton Gate, Peterborough; London Jaguar Centre, 10 Penhall Road, Charlton, London SE7; Car Craft, The Workshop, Railway Sidings, Meopham, Kent; Southern Classics, Chertsey Mead Garage, Bridge Road, Chertsey, Surrey; Vicarage Classic Car Company, The Old Vicarage Workshop, Knowbury, nr Ludlow, Shrops; Phillips Garage, 206 Bradford Street, Deritend, Birmingham; Marina Garage Ltd, Marina Close, Sea Road, Boscombe, Bournemouth, Dorset; Hooked on Classics, 28 Hyde Road, Kinson, Bournemouth, Dorset; Greggs Autos, 3 Marshall Road, Hayling Island, Hants; Hordle Garage, Crossroads, Hordle, Lymington, Hants; A. W. Hannak, Central Garage, Snaith, nr Goole, Humbs; West and Kean, The Square, Wolvey, nr Hinckley, Leics; Swallow Engineering, 6 Gibcracks, Basildon, Essex; D. K. Engineering, Unit D, 200 Rickmansworth Road, Watford, Herts.

For purely mechanical work: Forward Engineering, 780 Kingsbury Rd, Erdington, Birmingham; Classic Power Units, 18 Trevor Close, Tile Hill, Coventry; Autospeed Engineering, 31 Rutherford Close, Progress Road, Leigh-on-Sea, Essex.

For spare parts: G.H. Nolan, 1 St Georges Way, London SE15; WV Engineering in Kent Ltd, 30 Ivatt Way, Westwood Industrial Estate, Peterborough; David Manners, 17 Hagley Road West, Birmingham; Chris Coleman, 17 Devonshire Mews, Chiswick W4; Jaguar Spares (Swanton Morley), The Beeches, Swanton Morley, nr East Dereham, Norfolk; Norman Motors, 100 Mill Lane, London NW6; Three Point Four, Unit 3, Lands Garage, Langdale Road, Barnsley, South Yorks; Transpeed, 213 Portland Road, Hove, Sussex; CeBeEe Automobile Components, Unit 11, Holy Park Mills, Calverley, Pudsey, West Yorks; Maple Corner Garage, Unit 11, Brinksway Trading Estate, Brinksway, Stockport, Cheshire; Olaf P Lund and Son, 40 Upper Dean Street, Birmingham; Jaguar Focus Ltd, Bellbrook Industrial Estate, Bell Lane, Uckfield, East Sussex; British Sports Car Centre Ltd, 299-309 Goldhawk Road, London W12.

For panels and repair sections: Martin E. Robey Ltd, Poole Road, Camp Hill Industrial Estate, Nuneaton, Warks; RS Panels, Kelsey Close, Attleborough Fields Industrial Estate, Nuneaton, Warks; PSW Panels (Coventry), 76a Albany Road, Earlsdon, Coventry.

For trim: Barry Hankinson, Claypitts House, Much Birch, Hereford; Suffolk and Turley, Unit 7, Attleborough Fields Industrial Estate, Garrett Street, Nuneaton, Warks.

For brake parts and service: J W Bailey, 50 Latimer Gardens, Pinner, Middx.

For steering parts and service: R B Components, Unit 7, Wilden Trading Estate, Wilden Lane, Stourport-on-Severn, Worcs.

For transmission parts and service: Alan R George, Plot 11, Small Firms Compound, Dodwells Bridge Industrial Estate, Hinckley, nr Coventry.

For exhaust systems: JP Exhausts, Old School House, Brook Street, Macclesfield, Cheshire; Exhaustfit, 272 Chipstead Valley Road, Coulsdon, Surrey.

Very few Jaguar books specialise in the compacts, which is surprising in view of the vast number of books written about the Coventry marque. *The Jaguar Saloons: A Collectors Guide* by Chris Harvey is a first class introduction to the models. It is published by MRP at £8.95. *Jaguar Saloon Cars*, by Paul Skilleter, is the definitive book on all Jaguar's non-sports cars and includes a highly comprehensive chapter on the compacts. Published by Haynes it's worthwhile at £24.00. Haynes also publish one of their Super Profiles on the MkII at £4.95. Then there are the inevitable Brooklands Books test reprints. Two are published which cover the MkI/II — *Jaguar Cars 1961-64* and *Jaguar Cars 1964-68*. They are distributed by Brooklands Books and cost £5.95 each.

MkII interior was beautifully finished and well equipped, with plenty of leather and walnut veneer belying low price

There are now two organisations catering for MkI/IIs

MkII Tour de France record saw five Touring class wins

'County' estate was a one-off used as T de F tender car — Duncan Hamilton and Mike Hawthorn wanted to produce it

Rivals when new

Jaguars have always had the reputation for being in a class of their own. The compacts were no exception. When the first 2.4 entered the market place in 1956 it cost £1430. That was the same sort of money as an Armstrong-Siddeley Sapphire 235, an Austin A105, a Borgward Isabella T/S, a Humber Super Snipe, a Lanchester Sprite, a Rover 90, a Riley Pathfinder, a Sunbeam MkIII saloon and a Wolseley Six-Ninety saloon. Not one of those cars, with the possible exceptions of the Borgward and Riley (which were considerably slower than the new Jaguar) could have been called sporting.

When the 3.4-litre was launched the following year it only cost £1739, even with overdrive. You couldn't buy a Mercedes-Benz 180 (the company's cheapest model) for that. Most rivals were from a different era — all that is, except the Citroën ID19, which, together with the later DSs that followed, were the only consistant Jaguar rivals throughout the compact's era. While undoubtedly less sporting, the Citroëns were suitably avant garde to appeal to the Jaguar market but even they — complex foreign cars at a time when 'British was always best' — made little impression on Jaguar sales figures.

With the arrival of the 3.8 MkII in 1959 at £1779, Jaguar established themselves in a totally unthreatened market. Nothing could come close except at twice the price.

The mid-sixties saw more variety but still no direct competition to Jaguar. Italy offered the Alfa Romeo Giulia saloons, Fiat 2300 series and Lancia Flavias — all fine cars but no threat for detail or outright performance. The first real opposition must surely have been the Rover and Triumph 2000s which were decently quick, refined, British and a bit cheaper than the 2.4 Jaguar.

The cost-cutting features of the 240 and 340 made them even more competitively priced especially after the XJ arrived in 1969.

The Daimler V8 250 was always an expensive car but held its own private, unrivalled market place.

Prices

Condition and engine capacity (in that order) are the two most influential factors governing compact prices. A rough example of any model will fetch little more then a few hundred pounds. Sound cars which would scrape through an MoT will fetch around the £1000 mark. At £2500, model type starts to become important. You might be lucky enough to find a good 3.8 for that money while a V8 250 or 2.4 Mk 2 would have to be in significantly better condition.

At around £4000 come excellent Mk Is (only a concours example would sell for more, and then not by much), exceptional V8 250s and 2.4 Mk IIs, excellent unrestored 3.4 and 3.8 Mk IIs which should require no essential outlay, and badly restored 3.4/3.8s.

Prices above £4000 are haywire. Concours cars of any model will command prices between £6000 and £8000 with mint Coombes cars possibly reaching nearly £10,000. It is possible to buy as good a car for £4000 as it is for £6000, and a well-known restorer will ask £8000 for a car for which a less well-known garage will ask £5000.

The economy 240s, although younger than the Mk I/IIs and V8 250s, command less money on account of their thin bumpers and Ambla upholstery, 340s as much as 3.4 Mk IIs due to their rarity and performance. Works sunroofs, wire wheels and manual gearboxes with overdrive all add to values.

BUYERS' GUIDE

JAGUAR
A stylish performer

RESTORED PHOTOS: MARTYN BARNWELL

David Lillywhite investigates purchase and restoration of Mk 2s, 240s and 340s in today's market.

A decade or so ago, when John Thaw seemed content squealing Mk I Granadas around corners in The Sweeney, the Jaguar Mk 2 was thought of as a nice car, but something of a liability in terms of running and restoration costs. Gradually prices increased, pushed up by the investors who thought a shiny thing under a cotton cover would earn them easy money. It got to the point where classic car 'investment' magazines were being published and even a wrecked Mk 2 was beyond the reach of the average enthusiast.

Now John Thaw is an Oxford detective with a cultured accent and a rather nice looking Jaguar Mk 2, and the falling prices of the cars have let the real enthusiasts back into the market.

The car was a development of the much less sought after Mk 1, often critised for its slightly bulky styling and suspect handling (still not a bad car though). The Mk 1, introduced in 1955, was Jaguar's first attempt at a unitary construction and it's said that the development costs really stretched their finances to the limit. It was known as the 'small-Jaguar' and was initially only available with a six-cylinder 2.4 litre engine version of the 3.4 unit first used in the XK 120. Public demand for more power resulted in a 3.4 version being introduced in 1957 though, and the two versions remained in production until the introduction of the Mk 2 in October 1959. The motoring press loved the new car, one of the many glowing headlines being "Subdued or unleashed, the Jaguar 3.8 is strictly fabulous".

Technically speaking, the new model was merely a facelifted version of its predecessor. Glass area was increased by a drastic slimming of the door pillars and a deeper windscreen, and the front wings were modified but, not surprisingly after all the initial bodyshell investment, the underpinnings remained the same.

The Mk 1's suspension was carried over into the new model with nothing more than a few minor (but crucial) geometry changes. The coil-over-damper twin wishbone front suspension was mounted on an insulated subframe, but unlike early Mk 1s, disc brakes were standard fitment. Leaf springs suspended the back axle, but from their rear ends in a cantilever system, rather than having the axle sitting in the middle of each spring in conventional fashion. The idea of the cantilever system was to reduce stress on the new-to-Jaguar monocoque body design, and a Panhard rod and radius arms were fitted to keep the axle in line. Discs at the rear completed the servo-assisted braking system.

Unfortunately, the Mk 2 gained a little weight compared to its predecessor, but both the 2.4 and the 3.4 litre engines had their power increased slightly, to 120bhp and 210bhp respectively. All the same, the 2.4 could no longer quite make it to 100mph, although the 3.4 was capable of almost 120mph. The pick of the bunch, though, performance wise was the new 3.8 with 220bhp and performance previously unheard of from a luxury saloon.

The power of the 3.8 was such that a limited slip differential was fitted as standard (optional on US models), but there was a choice of Borg Warner automatic gearbox

Specifications

	2.4	3.8
Power	120bhp	220bhp
Torque	144ft.lb	240ft.lb
Weight	28.5cwt	30cwt
Length	15ft. 0.75in	15ft. 0.75in
Width	5ft. 6.75in	ft. 6.75in
0-60mph	17.3secs	8.5secs
Top speed	96.3mph	125mph
Average mpg	18	15.7

● *Three superb examples. The blue car is a 340 (note the bumpers) belonging to Lee Graze, the white 3.8 belongs to Mike Thomas and the red 3.8 (note the modified rear arches) belongs to Derek Skinner.*

WHAT TO PAY

For the very brave home restorer, 'basket case' Mk 2s can now be found for well under £1000. Driveable cars needing complete rebuilds will be anything from £1000 for a 2.4 up to £2000 for a 3.8 but of course these may cost as much to restore as a wreck if a concours car is your aim. Sound, useable 2.4 automatics will fetch around £5000 but an equivalent manual overdrive 3.8 will be nearer £10,000 and an immaculate one will fetch approximately £15000 compared to £12000 and £9000 for immaculate 3.4s and 2.4s. Prices of 240s and 340s are similar to equivalent condition 2.4s.

There were actually a few estate versions made by a coachbuilding company going by the name of Jones Bros. but the chances of finding one of these are extremely slim. Modified saloons, such as those converted by Coombs in the 1950s and 1960s would also be very interesting finds and with valid history (which is very difficult to come by) will fetch more than a standard car.

It is possible to find ex-USA cars with wrecked interiors and mechanicals but sound bodies, perfect for restoration. Better still, South African and Australian cars also tend to be relatively rust-free and have the added bonus of being right hand drive.

● *This is how an original specification Jaguar Mk 2 should look. Note the painted wire wheels - a popular optional extra - and the rear spats which are often removed, usually to fit wider wheels.*

● *Comparison of a Mk 2 with a 340. The slim bumpers of the 240s and 340s make quite a difference..*

or four speed manual with optional overdrive across the range. The manual gearbox was one of the few major items to change during the life of the Mk 2, going from a Moss 'box with no synchromesh on first to an all-synchromesh unit in September 1985.

The new gearbox is often said to be the last improvement made to the Mk 2 before cost-cutting resulted in leather upholstery becoming an option and the distinctive fog lamps being replaced by dummy vents. However, in July 1967 variable ratio 'Varamat-ic' steering replaced the original power assisted steering box option on 3.4 and 3.8 models, a very worthwhile modification.

Purists will state that the Mk 2 ceased production in September 1967. While it is true to say that after this date the Mk 2 name was dropped, the 2.4 and 3.4 models did carry on, with very few changes, as the more powerful 240 and the 340. These carried on the late-model Mk 2 tradition of Ambla rather than leather interior (leather was an option though) and dummy vents in place of fog lamps, the only obvious change being new slimline bumpers, giving the car a slightly more modern, but less characterful, appearance. These cars, never as sought after as the 'real' Mk 2s, overlapped the production of the XJ6 and, by April 1969 the 240 and 340 were no more.

Buying

Never let it be said that certain less-than-scrupulous types didn't take great advantage of the investor interest in Mk 2s a few years ago. When prices were at their peak, a bit of glass fibre, filler and a high gloss paint job could virtually be regarded as a licence to print money. Fortunately, most such cars have given themselves away by now (we all know how long filler over rust lasts) but the difference between a fully restored Jaguar and a presentable but well used version is wider than the smile of your bank manager when you ask for another car restoration loan.

The trouble is, a good Mk 2 is just so good. The engine is, or should be, so smooth and the cabin so comfortable, that any rattles, squeaks or whining bearings tend to be very prominent. Bear this in mind if you're looking for a

● *Who can fail to be impressed by such a sight? Veneers that are lifting or cracked are expensive to have restored.*

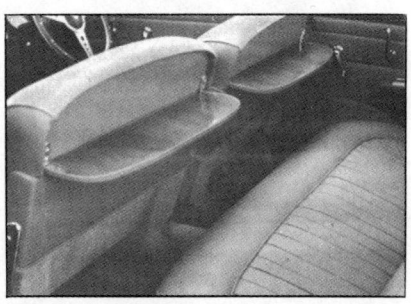

● *Fold down tables are a nice touch in the Mk 2s and 240/340s. The wood trim has been professionally rejuvenated.*

roadworthy but tatty example; the cost of getting it running nicely may approach that of a 'basket case' restoration.

Compared to modern cars, Jaguar Mk 2 bodywork is thick. In fact, even compared to other cars of the 1950s, it's pretty hefty but, typical of early monocoque shells, mud traps abound and rot really is the bugbear of the Mk 2. Over the years many examples will have been bodged, so start by examining the panel fit very carefully and if it's possible to get the car onto a flat piece of ground then squint along the track of the wheels to check the chassis alignment.

Start your piece-by-piece scrutiny of the car underneath, first under the front wheel arches where, at the the front, the triangulated 'crow's foot' crossmember is exposed to everything that the wheels can throw at it. Mud will happily sit on this thoughtfully provided platform, rotting the structural members, the front of the outer wing and the inner wing. To the rear of the front wheel is where the fun really starts though; the outer sill is joined to the outer wing, the front door hinge pillar (A-post), the bulkhead and the inner sill. The joint between outer sill and outer wing is smoothed over (by the factory) with lead, and, in extreme cases, this may be the only metal holding the whole structure together. What happens is the area directly behind the front wheel rots, opening the area between the A-post, inner sill and the bottom of the front wing to the elements. Left alone to rot, the previously mentioned structures will part company, leaving just the floor - probably rusty in the footwells anyway - and the transmission tunnel to take the strain for the front of the car.

The frightening part about it is that there are Mk 2s about that look really good from the outside and even have MoTs, but are structurally dangerously weak across the front. It's an easy fault to hide and a difficult one to rectify... a bad combination.

The remaining sections of sill will hold fewer surprises - good honest rot, if there is such a thing, mainly below the floorline in the case of the inners. Jacking points may well be rotten and, at the rear ends of the sills, rot in the inner wheel arches can extend into the sills, the rear door posts and the rear spring mounts. Some of this may be made more visible by removing the rear wheel spats (presuming they haven't already been removed and the rear arches modified). The spat are removed - with prior permission from the owner of course - by releasing the two Dzus fasteners visible with the rear door open. The spring mounting boxes can collapse in severe cases and the Panhard rod mount on the offside may even fall off if rust has really got a hold.

Still on the rear suspension, corrosion in the damper turrets should be visible by removing the rear seat cushions. In fact, the seat pan itself has been known to crumble to nothing, but this seems to vary widely between cars and can't be taken as a good pointer to the condition of the car as a whole. Parts of the rear valance are double skinned, inevitably leading to rot. Also check the insides of the boot, particularly the spare wheel well, and the lower edge of the boot lid. Examine the bottom edges of the doors and also a spot halfway up each door where the window frame presses on the inside of the skin via a felt pad.

There are a number of cosmetic weak spots at the front. The front valance is most likely to suffer under the front grille and the tops of the massive front wings can rot from the inside, sometimes spreading to the under-windscreen panel and the windscreen pillar. As with most cars, wheelarches suffer and, already touched upon, the area around the lights and indicators corrode thanks to a mud trap under the wing. Despite this, specialists usually advise against replacing the whole front wing unless absolutely necessary. The panel is just so big that, aside from the cost, a new panel is extremely difficult to fit correctly, the shape of the grille accentuating the smallest alignment discrepancies. The 240s and 340s are meant to have an extra 'under-bib' panel behind the front bumper, although nine times out of ten these are missing.

Repair sections are available for all the 'rotspots', not just for the front wings but for the whole car including the chassis members. New doors are available, but to get a good panel fit it's often easier to rebuild the originals. Most of the panels available are generally of a high quality and reasonable price but don't underestimate the skill factor necessary for a full blown restoration of a Mk 2; anyone taking on such a project would need to have had previous experience of total rebuilds because it really is a difficult car to restore.

A large number of the restorations going on in professional workshops across the country were taken on from amateurs who found that they were out of their depth. All the same, an experienced DIY enthusiast, taking care to brace the car properly and prepared to farm some jobs out if necessary, should end up with a good shell. The *Practical Classics* project Mk 2 series (now reproduced in book form) a few years ago comprehensively demonstrates the steps required.

The upholstery is often the last consideration of the enthusiast looking for a new aquisition. Even for more humble modes of transport, ripped seats and headlinings can provide a major headache but the restoration of a Jaguar interior, with its leather (usually) faced seats, wooden dashboard and door cappings and quality wool headlining can really strain resources.

The wood can often be restored very economically on a DIY basis if the veneers haven't started to lift or crack badly. If they have then some home restorers manage to get good results replacing the veneers but with so much of it in a Mk 2, the job is usually left to the professionals which can cost anything between

> '...a bit of glass fibre, filler and a paint job could be regarded as a licence to print money.'

WHAT TO LOOK FOR

● *Rot at the bottom of the rear wing often means that the crucial structural panels underneath are rotten.*

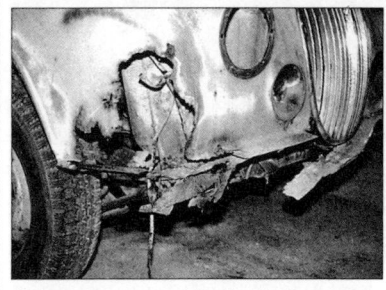

● *An extreme case! Under the front of the wing is a 'crow's foot' structure which supports wing and bumper.*

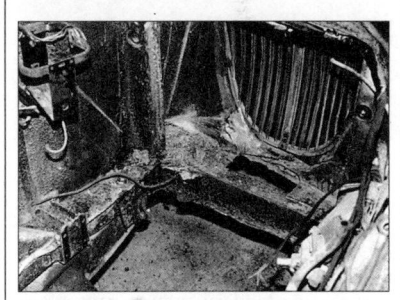

● *With the benefit of an empty engine bay, the rotten front crossmember and grille surround can be seen.*

● *This car has its rear spats removed and is sitting on a jig. Note rot in the forward spring mounts and sill closing panel.*

● *The rear seat pan can rust quite badly, although this otherwise rotten shell is relatively sound under the seat.*

JAGUAR
A stylish performer

● *The massive air cleaner does well to dominate the equally large twin overhead camshaft six-cylinder engine.*

● *The boot looks cavernous. The spare wheel and toolkit are stored under the false floor; check there for rot.*

● *This is a Mk 1 Jaguar. The thick door pillars were never popular and the handling was not as good as the Mk 2's.*

Clubs

Jaguar Enthusiasts Club, Freepost, Patchway, Bristol, BS12 6BR.
Jaguar Drivers Club Ltd, Jaguar House, 18 Stuart Street, Luton, Beds. LU1 2SL (Tel: 0582 419332).

£700 to £1800. Unfortunately Murphy's law invariably dictates that an otherwise good dashboard will have been subjected to a Black and Decker attack from a previous owner who felt the need to supplement the already extensive range of standard fitment gauges and switches with a few tacky extras. It's also worth noting that cost cutting resulted in significantly thinner veneers being fitted to late Mk 2s, and 240s and 340s.

The leather faced seats, particularly from earlier models seem quite robust and even now may only need the application of a renovation kit which feeds and recolours the hide. If the seats are beyond hope then there is no problem obtaining re-covering kits, supplied by a number of specialists, at a price. Covers for the front seats cost in the region of £425 and the rear covers are a similar price; or complete trim kits can be obtained for £1470 from Aldridge trimming, for example. The originality of late models fitted with an Ambla interior is usually sacrificed by restorers in favour of new leather, an acceptable change as leather was offered as an option anyway. It's interesting to note that many people regard the seats in the 240s and 340s as more comfortable than earlier models.

Many Mk 2s were exported to North America and, like many other British classics, these models are gradually making their back to Britain, making left to right hand drive conversions popular business. Ironically, dedicated Mk 2 fans are also having cars converted from right to left hand drive to take on to the Continent. Either way, the conversion is quite feasible, there is still an abundance of secondhand spares on the market, all the relevant holes in the bulkhead are symmetrical and there are even companies who will convert the steering box. To really do the job properly the order of the switches and gauges in the centre of the dashboard needs to be reversed as well.

Engine and transmission

A surprisingly high proportion of cars have been the subject of larger capacity engine swaps, particularly the 2.4, for obvious reasons. The 3.8 might seem the obvious choice and it's probably the most popular, but for all out refinement the 3.4 power plant is the one to go for. The thing to watch for though, is the quality of the conversion which should include the exhaust, radiator and front springs (the larger capacity engines are heavier). It's even possible that a 4.2 from the bigger Jaguars might have been used, but although this fits there can be problems such as trying to squeeze the cooling fan in.

All the six-cylinder, twin overhead camshaft engines are extremely well-built, but the first thing that usually strikes newcomers to the Jaguar marque is the sheer physical size of the power units. When you consider that the gearbox has to come out with the engine, you might feel like taking a muscle-toning course before proceeding with any mechanical work.

The curse of most Jaguar saloons is the cheap price that they all seem to fall to at some time in their lives. At this point they get bought by people who could never otherwise afford a Jaguar and still can't afford to run them, thus regular oil changes and servicing goes out of the window, as does the usual longevity of the engine. To quote extreme cases, Geof Maycock of Autocats knows of two Mk 2s that, several years ago, were used in the Essex police force covering a quarter of a million miles each. One of these cars needed a new bottom timing chain, the other needed nothing replacing at all. He has also seen an engine that had covered almost 80,000 miles with no signs of wear, but he once had to recondition a 26,000-mile unit. The difference? It's all in the oil changes.

With the engine ticking over, don't expect complete silence, after all the engine was designed in the early 1950s, but listen out for a rattle from the front which may be one or more of the three timing chains. The bottom timing chain in particular is tricky to renew without quite extensive stripping down. Top-end noise may be due to oversize valve clearances; they're set with shims so with two camshafts and six cylinders, resetting them is not a job to be taken lightly. Nothing can really be described as a weakness except the cost of a full engine rebuild; around £2500 by a professional. Most spares are available, but pistons for the 2.4 have to be specially made at a cost of approximately £500.

If the engine is losing oil, rather than burning it, the leaks are likely to be from the back of the cam covers, the rear of the sump or, most seriously from the crankshaft rear seal which will necessitate a major strip down. As ever, remove the oil filler cap to check for excessive fumes, watch for blue smoke from the exhaust and listen for rumblings or knocking under load from the bearings. Don't forget that an oil pressure gauge is fitted; it should register at least 40lb with the fully-warmed engine running at around 3000rpm.

Carburettors are usually good old twin 1.75in. SUs, although 2.4s were fitted with

Solex units. Many of the Solex equipped cars will now have SUs anyway and 2in. SU carburettors are often fitted to the 3.4 and 3.8 in particular. This leaves little or no room for the original air filter, so pancake filters are usually used instead. With all the carburettors, the main problem is going to be throttle spindle wear; waggle the linkage about and check that the spindle doesn't move excessively in the carburettor body.

The early Moss gearbox fitted to pre-September 1965 cars is devoid of synchromesh on first gear and the remaining three gears aren't much better. A careful, practised movement is required to produce a crunch-free change on a 'box that has seen a bit of action and there will always be some bearing whine, but not excessive amounts. This shouldn't make it sound as if the Moss is weak; unless it is over-revved in first it is virtually indestructible and if problems do arise parts are available. Once again, cars originally equipped with the Moss unit may well have been fitted with the all-synchromesh gearbox of later models which has a significantly nicer feel to it, although parts availability is not up to the Moss 'box standards.

The torquey six-cylinder engine is ideally suited to automatic transmission; before June 1965 a Borg Warner DG was used, parts for which can be very expensive. A professional rebuild can cost around £600 to £700 and the torque converter £250. The Type 35 unit that replaced the DG is somewhat cheaper. Neither auto 'box is particularly smooth by modern standards, often being a little slow to kickdown, but reliability is rarely a problem. A good pointer with any automatic gearbox is the colour of the transmission fluid; ideally it should be red, brown is acceptable, but black indicates worn bands which only usually occurs after quite a high mileage.

The optional Laycock de Normanville overdrive, well sought after nowadays, is generally very reliable although when problems do arise they can be quite difficult to solve. Ensure that overdrive engages quickly and smoothly every time. If the cars are driven hard the rear axle can suffer, the splines on the half-shafts wearing to produce a 'clonk' when starting to reverse. The Thornton Powr-Lok limited slip differential fitted to the 3.8 can suffer from friction plate wear after high mileages, but parts are available. The axle, in fact, is still used in the modern day XJS.

Suspension and brakes

That massive engine does no favours to the suspension but, well set up, the Mk 2 handles resonably well. The crucial phrase in that sentence though was "well set up"; the weight of the car means that rubber bushes don't last as long as might be expected and when they're worn, especially combined with tired dampers, the car will feel awful. If a car has been standing for a while then it's very likely that all the suspension bushes will need changing, although fortunately the bottom ball joints on the front suspension are extremely hefty and may escape the need for replacement. Reckon on over £250 worth of parts to change all the bushes, mounts and ball joints, though, so listen carefully for any knocking noises, but don't blame sagging rear suspension on the springs as the drop is quite likely to be due to tired rubber spring mounts (or chassis mount

Driving

Who can fail to be impressed with those leather seats and polished veneer? And what about that exhaust note; pure bliss! It comes as a bit of a shock when trying to get this gorgeous creature out of its parking space, struggling with the heavy steering. Once on the open road the steering improves, although it still takes quite a haul on the wheel to get round the twisty bits - understeer's the name of the game - but fast cruising on straight roads is effortless. The torque of the 3.8 is phenomenal but don't underrate the others, they're still quite an experience. The same applies to the 240 and 340 which have significantly more power than their 2.4 and 3.4 equivalents.

corrosion!). New shock absorbers (relatively cheap at around £30 each) will often transform a tired Mk 2's handling, or for further improvement it is worthwhile uprating the front suspension.

Unfortunately, whatever the state of the suspension, the steering is always going to be rather heavy for low-speed manoeuvring and rather low-geared for hustling along twisty A-roads. It's something that you get used to though, and the steering boxes and idlers give few problems even with the added complexity of the optional power steering. The feel of the steering really is the Mk 2's greatest fault though, so it's no surprise that there are firms who will convert cars to either the Varamatic system or rack and pinion. R.S. Coachworks for example will fit an XJ6 rack and pinion system for £1500 which, especially with radial tyres, will improve the handling no end.

The powerful Dunlop brake calipers used on Mk 2s can create problems because the seals are on the pistons rather than in the bores of the caliper. This allows rust to form on the unused portion of the bore, fine until the pads wear and the piston seals move forward onto the rusty sections. The best long-term solution is to have the calipers fitted with stainless steel liners for which Autocats charge £38 per cylinder. New master cylinders are rather expensive at around £75 and servos can be reconditioned or replaced with Lockheed's modern equivalent.

Many people may turn their noses up at modified cars but, having driven standard and modified cars, my personal opinion is that after spending much time and money on a car I would want it to be as pleasurable to drive as possible. As a restoration project, I can only say that you must have considerable experience, be careful and never skimp on a job.

Then you will end up with one of the most desirable cars ever produced.

Many thanks to RS Coachworks, sales and restoration specialists, of Reading (Tel: 0734 832319) for arranging the cars featured and much advice. Also to Jaguar restoration specialists, Autocats of Rayleigh, Essex (Tel: 0268 782306), for even more advice and a guided tour of every rust spot known to Mk 2 Jaguars. Thanks also to the owners of the cars, the New Mill Restaurant of Eversley, Berks. and former *Practical Classics* editor/publisher Paul Skilleter, now of the Jaguar Quarterly magazine. ■

● *With rear doors open the Dzus fasteners securing the spats can be seen. Remove to examine underbody more easily.*

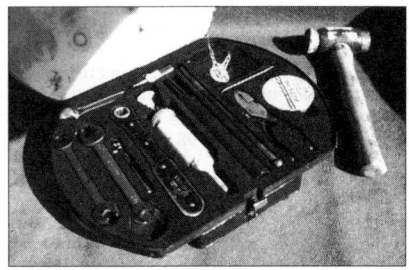

● *An original toolkit is always nice to have, although it's possible to buy new ones now for approximately £120.*

COOMBS REPLICA

In their day, the Guildford based company of John Coombs were well known for tuning Jaguars and actually had the official approval of the factory. The car shown here, belonging to Martin Steeles isn't an original Coombs car, but has been made into a replica by RS Coachworks. Other than the louvred bonnet and modified rear wheelarches, it also has a (non-period) 3/4 race specification engine, a tubular exhaust manifold, five speed all-synchromesh gearbox, rack and pinion steering, uprated front suspension, anti-tramp bars on the rear axle, four-pot calipers, six-inch wide wire wheels and an electric sunroof.

IDENTIFICATION

With so many Mk 2s having been modified, it's worth checking the chassis number carefully. 2.4s start at **100001 (rhd)/ 125001 (lhd)**; 3.4s at **150001 (rhd)/ 175001 (lhd)**; 3.8s at **200001 (rhd)/ 210001 (lhd)**; 240s at **1J1001 (rhd)/ 1J30001 (lhd)**; and 340s at **1J50001 (rhd)/ 1J80001 (lhd)**. Suffix **'DN'** indicates overdrive, **'BW'** automatic transmission and **'P'** power steering.

On the engine, the serial numbers will be prefixed by **'BG'** or **'BJ'** for a 2.4, **'KG'** or **'KJ'** for a 3.4 and **'LA'** or **'LE'** for a 3.8. Don't forget to compare the numbers to those on the documents.

OWNER'S VIEW

Richard Sutton describes the colourful life led by BOD147C, first as a day-to-day car and latterly as a racer

The Mk II Jaguar is unique among classic cars. It's cheap to buy, great fun to drive, is as capable a sports car as it is a limousine, you can race it, show it, use it as a business car or day to day family car. To cap all that, it can be cheap to run (petrol excluded), utterly reliable and is stunningly pretty.

During the last 10 years BOD147C, a red 3.4 manual/overdrive car, has been used by an average family man as an average family car, by an advertising agency as a courtesy car, an Army officer as a staff car, an antique dealer as a hack, a car club as a pace car and, during the last year, as a competitive racer for the Jaguar Drivers Inter-Area Challenge and relay races.

Previous owners disclose that during the car's last 40,000 miles, and until it was prepared for racing, it had received no mechanical or bodywork attention, and had never failed to start, nor ever broken down.

The third from last owner, Alan Hope, now a freelance graphic designer, explains: "I bought BOD in March 1982 from a long time owner in Salisbury for £1625. It was in sound and attractive condition with no looming expenditure — a good buy actually. I bought it purely because my neglected Volvo had come off the road for some time. It was consuming attention and I needed another car quickly — my work relied on it. I had to find an inexpensive, fast, comfortable, and above all reliable car, preferably with a bit of style, before the end of the week. BOD happened to come on the market at just the right time.

"Over the next year and substantial mileage it never let me down and required no expenditure at all. It was always good fun to drive, was amazingly comfortable on long trips and went down well with my clients.

"I sold it only when it was obvious that owning two cars was too much of an extravagance, and the Volvo was far more economical. In fact, that was the only thing I didn't like about the Jag — it was just too thirsty. It averaged about 17mpg, maybe 22mpg on a long trip, which is far too little when you do my kind of mileage and have to pay for the petrol yourself."

Major change

BOD was then bought for £1500 by John Sutton, a Major in the Army who, like Alan, relied on the car both for his work and as a family car. "It was a lovely car. When my son persuaded me to buy it I was sceptical to say the least, but the year that followed proved that running the Mk II as an everyday car was very practical and enjoyable. It was inexpensive, never needed any time or money spent on it over the 15,000 miles that followed, and passed two MoTs without any problem. Actually there was one persistent irritant — a leaking windscreen, which was finally cured by injecting liquid rubber behind the screen moulding.

"As a long distance cruiser the car was superb — the excellent overdrive facility paid off here — and the car was ideal for taking senior members of the corps to functions. My only quibbles were a small boot, useless jack and poor wet weather traction. A new set of Continental radials solved the last problem and greatly improved the car's roadholding and handling.

"I was unfortunately obliged to lend the car to my son for his use in the antique business, and that was the last I saw of it."

For the next six months, BOD's life was hard and varied but it continued to give reliable and exciting service in the hands of John Sutton's son, Richard. That's me.

I drove thousands of miles in it, carted furniture about in it, took far too many friends at once to race meetings in it, holiday toured in it, slept in it, sprinted in it, tried to seduce in it, was stopped in it (by the police as well). Indeed, the only thing I couldn't, and didn't, do in it was keep a low profile.

Current owners Bob Perry and Bob Miller have turned their MkII into a competitive racer that can be used every day

It was sold for £1100 in October 1983 to Bob Miller and Bob Perry, partners in business and Jaguar racing. By this time, although structurally and mechanically sound, BOD was beginning to look a little tired round the edges and the paintwork was poor. Thus, it was an ideal racing car proposition.

Once at the Bobs' base in High Wycombe, local Jaguar specialist Dave Symonds was commissioned to remove the engine and gearbox, fit a new sill and strengthen the rear spring hangers and boxes. In the meantime the two Bobs rebuilt a separate 3.8-litre engine they'd recently acquired and fitted BOD's old 3.4 engine into Bob Perry's father's faithful Mk II 2.4 which shares the same garage.

BOD was then returned to the two Bobs who fitted the 3.8 engine together with BOD's original Moss gearbox as well as a new Jaguar limited slip differential. The 3.8 engine had received a new top end, was gas flowed, fully balanced and fitted with a lightened flywheel. Straight-through exhausts, twin 2ins SU carburettors and a competition clutch assembly were added.

The car's entire suspension assembly was rebuilt to critical standards — the specification is, of course, highly confidential, but the obscure set up 'probably'

John Sutton: "BOD was very practical and enjoyable"

Bob Perry: "Between races BOD never needed attention"

includes Daimler limousine components! The braking system was entirely rebuilt but, except for linings remains very standard, and the steering overhauled. A roll cage and new steering wheel were fitted.

With XJ 205 wheels and tyres the Bob team went to Silverstone for their first race at last year's JDC meeting. They finished sixth out of a field of 18 Highly commendable.

Important modifications were made throughout the season. Although the standard brakes were up to the job they were very hard worked and the system was substantially altered mid-season. A big servo, large master cylinder, four-pot front calipers and cooling ducts were added. Stopping ability is now 'phenomenal'. "I doubt that any modern production car is better," says Perry.

Further modifications included the fitting of a Kenlowe fan and cutting off one inch from each of the viscous fan blades. This is reckoned to give 7bhp.

Little maintenance

The car raced competitively thoughout the season and was probably the most reliable car in the paddock. Bob Perry explains: "Everyone always seemed to be fiddling with their cars before and after races. We never had to touch BOD. We would drive it to the circuit, put in a full day's racing and drive it home again without really having to lift the bonnet. In fact between races it needed no attention at all, but we always bled the brakes and changed the oil." A case in point was last October's JDC 6-hour relay race in which BOD completed nearly half of the five car team's allotted 223 laps as team-mates failed with mechanical problems.

As a road car BOD is nothing short of mind blowing. A top speed estimated at 135mph is modest anything, and acceleration is comparable with V12 E-types. Bob Perry also has a far from standard 3.8 E-type which he reckons is slower.

In its current state of tune, which includes standard compression ratio and gearbox, the car is also a highly useable everyday car. BOD is often used as 'the family car' and was taken on holiday to Somerset last year and returned 15mpg. It could be worse.

So what does such a versatile and desirable car cost? The Bob team have spent £5000 which includes BOD's initial purchase price. Just what other modern, or even classic car, can do so much at the price?

RUNNING REPORT
HOW OUR OWN CLASSICS PERFORM DAY-TO-DAY

Blundell's MkII looks the part, with its optional wire wheels

1961 Jaguar MkII 3.4

Do you know the song that goes 'Your kneebone's connected to your thighbone,... your thighbone's connected to your hipbone... your hipbone's connected to... etc? Well, refurbishing part of my Jaguar's interior has been a bit like that song. Only for parts of the body read parts of the car's interior – seat covers... armrest... the centre console... cushion foams...

The headlining, door panels and rear seats were all in good condition for a 31-year-old car, but the front seats were cracked and split and the carpets were faded and worn away in the front. So I bought a carpet set (£260 approx) and front seat covers (£450 approx) in suede green from G W Bartlett and asked Thompsons of Wandsworth, London, to do the fitting. Thompsons quoted £340 for this and so for just over £1000 I expected to get a pretty good MkII interior. But the front seats needed cushion foams at £60, and the rear seats had to be lacquered to match the new brighter-coloured front seats (£130) and a split arm rest now looked too tatty and had to be repaired (£35) and the moquette on the centre console had to be renewed to match the carpets as did the moquette on the reclining parts of the seats (£135) and so the final total came to over £1400. But it was worth it: Thompsons did a superb job – much better than if I had done it myself – and the interior is now very pleasant indeed. Maybe a year or two of use is needed for the front seats to mellow.

When I took the car to Spain on holiday last August (*Running Report,* May) I had been worried about the clutch, which engaged on the end of its travel with no more adjustment left. So in October I bought a new clutch plate, cover and release bearing for about £250 from FB Components in Oxford and took the car back to friend Charles Teall in Oxfordshire for him to fit. The power unit was removed by first removing the front suspension crossmember, then releasing the engine and lowering it onto a low flat trolley, and then lifting the front of the car up as high as it would go in his garage until the engine, gearbox and overdrive could be rolled out on the trolley.

While Charles changed the clutch I took the opportunity to paint the engine compartment with black Hammerite, replace the battery leads, and thoroughly clean the engine and re-paint the head with pale green cylinder head paint, available from the Jaguar Enthusiasts' Club.

The car was ready for Christmas, and on Friday December 27, after excesses that defy remembering, I drove up to Scotland to walk off a few turkey lunches in the mountains. On the way I visited Glasgow Museum of Transport, which has a superb collection of ship models and trams; admission is free. There is a very good display of post-war British cars arranged as if in a showroom – including a white MkII.

The next three days I spent driving and walking in the mountains. Fortunately the weather was mild for December, as try as I might to unkink the heater hoses, I could not get more than a trickle of lukewarm air into the cabin, even with the fan on and the vent open. Perhaps MkII heaters never worked that well? I have read that Sir William Lyons always drove with an overcoat on in winter. The Jaguar is marvellous on open Scottish roads like the A82 and A85, going effortlessly at between 60 and 70mph, just flicking it into and out of overdrive as necessary for corners, the difficult gear change and heavy steering associated with London driving forgotten. **John Blundell**
Audiophile Magazine

Re-covering front seats was more of an involved task than at first it was thought, but the operation has proved worthwhile

Jaguar Factory Manuals

Workshop Manuals

Model	Part no.
Jaguar 1.5, 2.5, 3.5 1946-48	
XK 120/140/150 & Mk. 7/8/9 (SC)	
XK 120/140/150 & Mk. 7/8/9 (HC)	
Mk.2 2.4,3.4,3.8, 240,340 (SC)	E121/7
Mk.2 2.4,3.4,3.8, 240,340 (HC)	E121/7
Mk.10 3.8/4.2 & 420G (SC)	E136/2
Mk.10 3.8/4.2 & 420G (HC)	E136/2
'S' type 3.4 & 3.8	E133/1
Jaguar Automatic Transmission	E113/6
E-type 3.8/4.2 Ser. 1 & 2 (SC)	E123/8/B
E-type 3.8/4.2 Ser.1 & 2 (HC)	& E156/1
E-type V12 Ser.3 (ring binder)	E165/3
E-type V12 Ser.3 (SC)	E165/3
Jaguar 420	E143/2
XJ6 2.8 & 4.2 Ser.1	E155/3/H
XJ6 Ser.2	E188/4
XJ12 Ser.1	E172/1
XJ12 Ser.2 & DD6	E190/4
XJ6 & XJ12 Ser.3	AKM9006
Jag/Daimler 6 Ser. 3 Intro. Supp.	AKM8061
XJS 3.6 & XJSC 3.6 (ring binder)	AKM9063
XJS V12 (+HE supp.) (SC)	AKM3455
XJS V12 (+HE supp.)HC	AKM3455

Parts Catalogue

Model	Part no.
Mk.2 3.4, 3.8 & 340 (SC)	J34
Mk.2 3.4, 3.8 & 340 (HC)	J34
'S' Type 3.4 & 3.8	J35
E-type 3.4 Ser. 1	J30
E-type 4.2 Ser. 1	J37
E-type Ser. 2 GT	IPL5/2
E-type V12 Ser. 3	RTC9014
Jaguar 420 (ring binder)	J39
XJ6 Ser. 1	RTC9016
XJ6 Ser. 2	RTC9883

Owners' Handbooks

Model	Part no.
Jaguar XK120	
Jaguar XK140	E101/2
Jaguar XK150	E111/2
Mk.2 3.4	E116/1
Mk.2 3.8	E115/1
Mk.10 3.8	E124/4
E-type tuning & prep. for competition	
E-type 3.8 Ser. 1	E122/7
E-type 4.2, 2+2 Ser. 1	E131/6
E-type 4.2 Ser. 2	E154/5
E-type V12 Ser. 3	E160/2
E-type V12 Ser. 3 (US)	A181/2
XJ12 Ser. 1	E171/2
XJ3.4 & 4.2 Ser. 2	E200/8
XJC SER. 2	E184/1
XJ5.3 Ser 2 Sal. & Coupe	E196/3
XJ5.3 Ser. 2 (fuel injection)	E196/4
XJ12 Ser. 3	AKM4181
XJ12 Ser. 3 (US)	AKM4179
XJS (Eng/Ger/Ital) pub. '79	AKM3454
XJS (Eng./Fren/Dutch) pub. '75	AKM3454/2
XJS (US) pub. '77	AKM3980

Maintenance Handbooks

Model	Part no.
Jag/Daimler 6 Ser. 3	AKM8045
Jag 6 Ser. 3 (US)	AKM8104/80
Jag/Daimler 12 Ser.3 (US & GB)	AKM8046

Note: SC - soft cover HC - hard cover

From specialist booksellers or, in case of difficulty, direct from the distributors:

Brooklands Books Ltd., PO Box 146, Cobham, Surrey KT11 1LG, England
Phone: 0932 865051   Fax 0932 868803
Brooklands Books Ltd., 1/81 Darley St., PO Box 199, Mona Vale, NSW 2103, Australia
Phone: 2 997 8428 Fax: 2 979 5799
Car Tech, 11481 Kost Dam Road, North Branch, MN 55056, USA
Phone: 800 551 4754 & 612 583 3471   Fax: 612 583 2023